T0259856

Lecture Notes in Computer Science 14774

Founding Editors

Gerhard Goos
Juris Hartmanis

The series Lecture Notes in Computer Science (LNCS), including its subseries Lecture Notes in Artificial Intelligence (LNAI) and Lecture Notes in Bioinformatics (LNBI), has established itself as a medium for the publication of new developments in computer science and information technology research, teaching, and education.

LNCS enjoys close cooperation with the computer science R & D community, the series counts many renowned academics among its volume editors and paper authors, and collaborates with prestigious societies. Its mission is to serve this international community by providing an invaluable service, mainly focused on the publication of conference and workshop proceedings and postproceedings. LNCS commenced publication in 1973.

Russ Harmer · Jens Kosiol
Editors

Graph Transformation

17th International Conference, ICGT 2024
Held as Part of STAF 2024
Enschede, The Netherlands, July 10–11, 2024
Proceedings

 Springer

Editors
Russ Harmer ⓘ
National Centre for Scientific Research
and Laboratoire de l'Informatique du
Parallélisme
Lyon, France

Jens Kosiol ⓘ
Philipps-Universität Marburg and Universität
Kassel
Kassel, Germany

ISSN 0302-9743   ISSN 1611-3349 (electronic)
Lecture Notes in Computer Science
ISBN 978-3-031-64284-5   ISBN 978-3-031-64285-2 (eBook)
https://doi.org/10.1007/978-3-031-64285-2

This Springer imprint is published by the registered company Springer Nature Switzerland AG
The registered company address is: Gewerbestrasse 11, 6330 Cham, Switzerland

If disposing of this product, please recycle the paper.

Preface

This volume contains the proceedings of ICGT 2024, the 17th International Conference on Graph Transformation, held during 10–11 July, 2024, at the University of Twente in Enschede, NL. ICGT 2024 was affiliated with STAF (Software Technologies: Applications and Foundations), a federation of leading conferences on software technologies. ICGT 2024 took place under the auspices of the European Association of Theoretical Computer Science (EATCS), the European Association of Software Science and Technology (EASST), and the IFIP Working Group 1.3, Foundations of Systems Specification.

The ICGT series aims at fostering exchange and the collaboration of researchers from different backgrounds working with graphs and graph transformation, either by contributing to their theoretical foundations or by applying established formalisms to classic or novel areas. The series not only serves as a well-established scientific publication outlet but also as a platform to boost inter- and intra-disciplinary research and to stimulate new ideas. The use of graphs and graph-like structures as a formalism for specification and modelling is widespread in all areas of computer science as well as in many fields of computational research and engineering. Relevant examples include software architectures, pointer structures, state-space and control/data flow graphs, UML and other domain-specific models, network layouts, topologies of cyber-physical environments, quantum computing, and molecular structures. Often, these graphs undergo dynamic change, ranging from reconfiguration and evolution to various kinds of behaviour, all of which may be captured by rule-based graph manipulation. Thus, graphs and graph transformation form a fundamental universal modelling paradigm that serves as a means for formal reasoning and analysis, ranging from the verification of certain properties of interest to the discovery of fundamentally new insights.

ICGT 2024 continued the series of conferences previously held in Barcelona (Spain) in 2002, Rome (Italy) in 2004, Natal (Brazil) in 2006, Leicester (UK) in 2008, Enschede (The Netherlands) in 2010, Bremen (Germany) in 2012, York (UK) in 2014, L'Aquila (Italy) in 2015, Vienna (Austria) in 2016, Marburg (Germany) in 2017, Toulouse (France) in 2018, Eindhoven (The Netherlands) in 2019, online in 2020 and 2021, in Nantes (France) in 2022 and Leicester (UK) in 2023, following a series of six International Workshops on Graph Grammars and Their Application to Computer Science from 1978 to 1998 in Europe and in the USA.

This year, the conference solicited papers across three categories: research papers, describing new and unpublished contributions to the theory and applications of graph transformation; tool presentation papers, demonstrating the main features and functionalities of graph-based tools; and blue skies papers, reporting on new research directions or ideas that are at an early or emerging stage.

From an initial 22 abstract announcements, 21 submissions were received, and were each reviewed by three Programme Committee members and/or additional reviewers. Following an extensive discussion phase, the Programme Committee selected 10 research papers, one tool presentation paper and two blue skies papers for publication in these proceedings.

The topics of the accepted papers cover a wide spectrum, including papers advancing the classical theory of graph transformation, investigating more recent approaches to transformation like global transformations in the context of (port) graph transformation, addressing graph models like bigraphs and bond graphs, as well as the application of graph transformation in new areas such as taint analysis and the rewriting of constraint models.

In addition, we solicited proposals for 'journal-first' talks, allowing authors to present relevant (previously published) work to the community. Finally, we were delighted to host invited talks by Mariëlle Stoelinga and Tiago Prince Sales, both from the University of Twente.

We would like to thank everyone who contributed to the success of ICGT 2024, including the members of our Programme Committee, our additional reviewers and our invited speakers. We are grateful to Reiko Heckel, the Chair of the ICGT Steering Committee, for his valuable suggestions; the organising committee of STAF 2024, for hosting and supporting ICGT 2024; conf.researchr.org, for hosting our website; and EasyChair, for supporting the review process and the preparation of the proceedings.

May 2024   Russ Harmer
                                                                            Jens Kosiol

Organization

Program Committee

Nicolas Behr	Université Paris Cité, CNRS, IRIF, France
Paolo Bottoni	Sapienza University of Rome, Italy
Andrea Corradini	Università di Pisa, Italy
Juan De Lara	Universidad Autonoma de Madrid, Spain
Juergen Dingel	Queen's University, Canada
Rachid Echahed	CNRS and University of Grenoble, France
Joerg Endrullis	Vrije Universiteit Amsterdam, Netherlands
James Fairbanks	University of Florida, USA
Maribel Fernandez	King's College London, UK
Fabio Gadducci	Università di Pisa, Italy
Russ Harmer	CNRS, Lyon, France
Reiko Heckel	University of Leicester, UK
Jens Kosiol	Philipps-Universität Marburg, Germany
Barbara König	University of Duisburg-Essen, Germany
Leen Lambers	Brandenburgische Technische Universität Cottbus-Senftenberg, Germany
Yngve Lamo	Western Norway University of Applied Sciences, Norway
Mark Minas	Universität der Bundeswehr München, Germany
Fernando Orejas	Universitat Politècnica de Catalunya, Spain
Detlef Plump	University of York, UK
Christopher M. Poskitt	Singapore Management University, Singapore
Arend Rensink	University of Twente, Netherlands
Andy Schürr	TU Darmstadt, Germany
Gabriele Taentzer	Philipps-Universität Marburg, Germany
Kazunori Ueda	Waseda University, Japan
Steffen Zschaler	King's College London, UK

Additional Reviewers

Castelnovo, Davide	Machowczyk, Adam
Fritsche, Lars	Miculan, Marino
Kratz, Maximilian	Sakizloglou, Lucas
Lauer, Alexander	Stoltenow, Lara
Lieb, Alexander	Söldner, Robert

Graphs, Logics and Transformations for Effective Risk Analysis (Invited Paper)

Mariëlle Stoelinga[1,2] (ID)

[1]University of Twente, Enschede, the Netherlands
m.i.a.stoelinga@utwente.nl
[2]Department of Software Science, Radboud University, Nijmegen, the Netherlands

Risk management is a fundamental process to ensure the reliable operation of systems, services, processes, and missions in our society [1]. Examples range from self-driving cars, power grids, credit card payments, and military missions. Proper risk management techniques enable organizations to achieve their goals in an effective way and take effective mitigating measures.

Risk models support the risk management process in the identification, prioritization, and quantification of risks via effective preventive and corrective actions. Numerous industrial risk models exist. In this talk, I will focus on fault trees [6] and attack trees [3], which are both top-down models that break high-level system risks into their causes, until the root causes are found. While fault trees focus on safety risks, i.e., unintended failures, attack trees take into account security risks, i.e., disruptions due to malicious attacks.

In this presentation, I will take a graph-theoretic perspective on fault trees, attack trees, and their combination.

- First, I will present a formal semantics, which is surprisingly intricate given the fact that there are only a handful of logical gates to propagate failures and attacks.
- Next, I will propose several algorithms to analyse quantitative attack trees, based on BDDs and stochastic model checking, highlighting the role of graph transformations to make this process more efficient [2].
- Finally, I will present risk query logics [4, 5], which allows engineers to query large attack and fault tree models.

Together these ingredients allow organizations to make better decisions on mitigating measures, making decisions more systematic, transparent and evidence-based—the increased constraints imposed by international standards, together with the ever-growing penetration of AI components in high-tech systems make rigorous and powerful risk management more important than ever.

References

1. Hopkin, P.: Fundamentals of Risk Management: Understanding, Evaluating and Implementing Effective Risk Management, 5th edn. Kogan Page (2018)
2. Junges, S., Guck, D., Katoen, J.P., et al.: Fault trees on a diet: automated reduction by graph rewriting. Formal Aspects Comput. **29**, 651–703 (2017).
3. Mauw, S., Oostdijk, M.: Foundations of attack trees. In: Won, D.H., Kim, S. (eds.) ICISC 2005. LNCS, vol. 3935, pp. 186–198. Springer, Heidelberg (2006). https://doi.org/10.1007/11734727_17
4. Nicoletti, S., Hahn, E., Stoelinga, M.: BFL: a logic to reason about fault trees. In: Dependable Systems and Networks (DSN), pp. 441–452. IEEE (2022)
5. Nicoletti, S.M., Lopuhaä-Zwakenberg, M., Hahn, E.M., Stoelinga, M.: ATM: a logic for quantitative security properties on attack trees. In: Ferreira, C., Willemse, T.A.C. (eds.) SEFM 2023. LNCS, vol. 14323, pp. 205–225. Springer, Cham (2023). https://doi.org/10.1007/978-3-031-47115-5_12
6. Ruijters, E., Stoelinga, M.: Fault tree analysis: a survey of the state-of-the-art in modeling, analysis and tools. Comput. Sci. Rev. **15–16**, 29–62 (2015)

Contents

Tool and Blue Skies Presentations

Theoretical Advances

Linear-Time Graph Programs
for Unbounded-Degree Graphs

Ziad Ismaili Alaoui[(✉)] and Detlef Plump

Department of Computer Science, University of York, York, UK
{z.ismaili-alaoui,detlef.plump}@york.ac.uk

Abstract. Achieving the complexity of graph algorithms in conventional languages with programs based on graph transformation rules is challenging because of the cost of graph matching. Previous work demonstrated that with so-called *rooted* rules, certain algorithms can be executed in linear time using the graph programming language GP 2. However, for non-destructive algorithms which retain the structure of input graphs, achieving a linear runtime required that input graphs have a bounded node degree. In this paper, we show how to overcome this restriction by enhancing the graph data structure generated by the GP 2 compiler and exploiting the new structure in programs. As a case study, we present a 2-colouring program that runs in linear time on connected input graphs with arbitrary node degrees. We prove the linear time complexity and also provide empirical evidence in the form of timings for various classes of input graphs.

Keywords: Rooted graph programs · Efficient graph matching · GP 2 · Linear-time algorithms · Depth-first search · 2-colouring

1 Introduction

Designing and implementing languages for rule-based graph transformation, such as GReAT [1], GROOVE [9], GrGen.Net [10], Henshin [12] or PORGY [8], is challenging in terms of performance. Typically, there is a gap between the runtime that can be achieved with programs in conventional imperative languages and rule-based graph programs. The bottleneck for graph transformation is the cost of graph matching. In general, matching the left-hand graph L of a rule within a host graph G requires time $|G|^{|L|}$, where $|X|$ is the size of a graph X. (This is a polynomial since L is fixed.) As a consequence, linear-time imperative graph algorithms may have a polynomial runtime when they are recast as rule-based graph programs.

To mitigate this problem, the graph programming language GP 2 supports *rooted* graph transformation rules which were first proposed by Dörr [7]. The idea is to distinguish certain nodes as *roots* and to match roots in rules with roots in host graphs. Then only the neighbourhood of host graph roots needs to be searched for matches, allowing, under certain conditions, to match rules

© The Author(s), under exclusive license to Springer Nature Switzerland AG 2024
R. Harmer and J. Kosiol (Eds.): ICGT 2024, LNCS 14774, pp. 3–20, 2024.
https://doi.org/10.1007/978-3-031-64285-2_1

in constant time. The GP 2 compiler [2] maintains a list of pointers to roots in the host graph, hence allowing to access roots in constant time if the number of roots throughout a program's execution is bounded. In [3], *fast* rules were identified as a class of rooted rules that can be applied in constant time if host graphs contain a bounded number of roots and have a bounded node degree.

The first linear-time graph problem implemented by a GP 2 program with fast rules was 2-colouring. In [3,4], it is shown that this program colours connected graphs of bounded degree in linear time. Since then, the GP 2 compiler has received some major improvements, in particular relating to the runtime graph data structure used by the compiled programs [6]. These improvements made a linear time worst-case performance possible for a wider class of programs, in some cases even on input graph classes of unbounded degree. See [5] for an overview.

Despite this progress, programs that retain the structure of input graphs, such as the said 2-colouring program, up to now required non-linear runtimes on graphs of unbounded degree. The problem is that during a depth-first search, the number of failed attempts to match an edge incident to a node may increase over repeated visits to this node. As a consequence, in graph classes of unbounded degree, this number may grow quadratically in the graph size, leading to a quadratic program runtime. In graph classes of bounded degree, this undesirable behaviour is ruled out because the maximal number of failed matching attempts per node is constant.

In this paper, we present an update to the GP 2 compiler which mitigates this performance bottleneck. In short, the solution is to improve the graph data structure generated by the compiler in that the edges incident with a node are stored in separate linked lists if they have different *marks* (red, green, blue, dashed or unmarked). This allows the matching algorithm to find an incident edge in constant time. For example, if an unmarked edge is required, a single access to the list of unmarked incident edges will either find such an edge or determine that none exists.

In addition to the new graph data structure, a technique is needed to exploit the improved storage in programs. We demonstrate in a case study of the 2-colouring problem how the new graph representation allows to achieve a linear runtime on graphs with arbitrary node degrees. Our program expects connected input graphs and either detects that a graph is not 2-colourable or colours the nodes blue and red such that all non-loop edges link nodes of different colour.

We provide a detailed proof that this program runs in linear time on arbitrary connected input graphs[1]. To the best of our knowledge, such a demonstration of a rule-based linear-time 2-colouring algorithm does not exist in the literature. We also present the results of timing experiments with the colouring program on six different graph classes containing graphs with up to one million nodes and edges.

---

[1] Full proofs available at: https://uoycs-plasma.github.io/GP2/documents/ICGT_20 24_GP_2_Extended.pdf.

2   The Problem with Unbounded-Degree Graphs

We refer to [5] for a description of the GP 2 programming language. Previous versions of non-destructive GP 2 programs based on depth-first-search showed a linear time complexity on graph classes of bounded degree but a non-linear runtime on graph classes of unbounded degree [5].

For example, the program is-connected in Fig. 1 checks whether a graph is connected. Input graphs have arbitrary node and edge labels of type list, unmarked edges and grey-marked nodes. The program fails on a graph if and only if the graph is disconnected.

Rule init picks an arbitrary grey node as a root (if the input graph is non-empty) and then the loop DFS! performs a depth-first search of the connected component of the node chosen by init. The rule forward marks each newly visited node blue, and back unmarks it once it is processed. Procedure DFS ends when back fails to match, indicating that the search is complete. Rule match checks whether a grey-marked node still exists in the graph following the execution of DFS!. This is the case if and only if the input graph contains more than one connected component. In this situation the program invokes the command fail, otherwise it terminates by returning the graph resulting from the depth-first search.

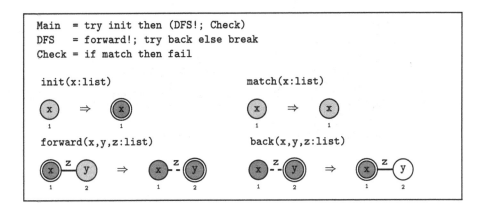

Fig. 1. The old program is-connected. (Color figure online)

It can be shown that the program is-connected runs in linear time on classes of graphs with bounded node degree [5]. However, as the following example shows, the program may require non-linear time on unbounded-degree graph classes. Figure 2 shows an execution of is-connected on a star graph with 8 edges (see also Fig. 11). The numbers below the graphs show the ranges of attempts that the matching algorithm may perform. For instance, in the second graph of the top row, either a match is found immediately among the edges that connect the central node with the grey nodes, or the dashed edge is unsuccessfully tried first. In order to find a match for the rule forward, the matching algorithm considers, in the worst case, every edge incident with the root. When the node central to the graph is rooted and the rule forward is called, the matching algorithm may first attempt a match with the dashed back edge and all edges incident with an unmarked node. Therefore, the maximum number of matching attempts for forward grows as the root moves back to the central node. As can be seen from this example, the worst-case complexity of matching forward throughout the program's execution is $2|E| + \sum_{i=1}^{|E|} i = \mathcal{O}(|E|^2)$ where E is the set of edges.

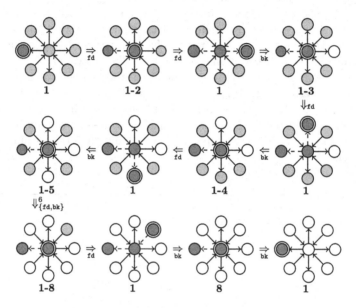

Fig. 2. Matching attempts with the forward rule. fd and bk denote forward and back, respectively. (Color figure online)

3   The Updated GP 2 Compiler

To address the problem described in Sect. 2, we changed the GP 2 compiler described in [6], which we refer to as the *2020 compiler*. We call the version introduced in this paper the *new compiler*[2].

3.1   New Graph Data Structure

The 2020 compiler stored the host graph's structure as one linked list containing every node in the graph, with each node storing two additional linked lists: one

---

[2] Available at: https://github.com/UoYCS-plasma/GP2.

for incoming edges and one for outgoing edges. When iterating through edge lists to find a particular match for a rule edge, the 2020 compiler had to traverse through edges with marks incompatible with that of the rule edge. This resulted in performance issues, especially if nodes could be incident to an unbounded number of edges with marks incompatible with the edge to be matched. For example, consider the rule `blue_red` from Fig. 6. Initially, the matching algorithm matches node 1 from the interface with a root node in the host graph. Subsequently, it iterates through the node's edge lists to locate a match for the red edge. In the 2020 compiler, all edges incident to this node were stored within two lists, one for each orientation, irrespective of their marks. However, if the node were incident to a growing number of unmatchable edges (because of mark changes), the matching algorithm would face, in the worst case, a growing number of iterations through the edge lists to find a single red edge.

	in	out	loop
unmarked	...	...	...
dashed	...	...	...
red	...	...	...
green	...	...	...
blue	...	...	...

Fig. 3. Two-dimensional array of linked lists of edges. (Color figure online)

When considering a match for a rule edge, host edges with incorrect orientation and incompatible marks do not match; thus, the matching algorithm need not iterate through them. By organising edges into homogeneous linked lists as array entries based on their marks and orientations, the matching algorithm can selectively consider linked lists of edges of correct orientation and mark. More precisely, in the new compiler, we update the graph structure of the 2020 compiler by replacing the two linked lists with a two-dimensional array. Each element of the array stores a linked list containing edges of a particular mark and orientation. We also consider loops to be a distinct type of orientation, separate from non-loop outgoing and incoming edges. The 2D array, therefore, consists of 5 rows (unmarked, dashed, red, blue, green) and 3 columns (incoming, outgoing, loop), totalling 15 cells, each one storing a single linked list.

Consider again the rule `blue_red` of `2-colouring` from Fig. 3 and Fig. 6. In the new compiler, the matching algorithm can access the linked list of non-loop incoming red edges and that of non-loop outgoing red edges in constant time and only consider these edges. Other edges, such as blue edges incident to the matched node, are stored in separate linked lists and thus are not considered. In this specific instance, it can be shown that there is at most one red edge in the host graph throughout the execution of `2-colouring` (Proposition 3). Therefore, there can be at most one edge in either list, and a matching attempt will either find such an edge or determine that none exists, both in constant time. However, under the 2020 compiler, the presence of non-red edges in the list could result in longer, non-constant search times, as, in the worst-case scenario, all edges except the red one would need to be iterated over.

Procedure	Description	Complexity
alreadyMatched	Test if the given item has been matched in the host graph.	$O(1)$
clearMatched	Clear the is matched flag for a given item.	$O(1)$
setMatched	Set the is matched flag for a given item.	$O(1)$
firstHostNode	Fetch the first node in the host graph.	$O(1)$
nextHostNode	Given a node, fetch the next node in the host graph.	$O(1)$
firstHostRootNode	Fetch the first root node in the host graph.	$O(1)$
nextHostRootNode	Given a root node, fetch the next root node in the host graph.	$O(1)$
firstInEdge(m)	**Given a node, fetch the first incoming edge of mark m.**	$O(1)$
nextInEdge(m)	**Given a node and an edge of mark m, fetch the next incoming edge of mark m.**	$O(1)$
firstOutEdge(m)	**Given a node, fetch the first outgoing edge of mark m.**	$O(1)$
nextOutEdge(m)	**Given a node and an edge of mark m, fetch the next outgoing edge of mark m.**	$O(1)$
firstLoop(m)	**Given a node, fetch the first loop edge of mark m.**	$O(1)$
nextLoop(m)	**Given a node and an edge of mark m, fetch the next loop edge of mark m.**	$O(1)$
getInDegree	Given a node, fetch its incoming degree.	$O(1)$
getOutDegree	Given a node, fetch its outgoing degree.	$O(1)$
getMark	Given a node or edge, fetch its mark.	$O(1)$
isRooted	Given a node, determine if it is rooted.	$O(1)$
getSource	Given an edge, fetch the source node.	$O(1)$
getTarget	Given an edge, fetch the target node.	$O(1)$
parseInputGraph	Parse and load the input graph into memory: the host graph.	$O(n)$
printHostGraph	Write the current host graph state as output.	$O(n)$

Fig. 4. Updated runtime complexity assumptions. Modified procedures are highlighted in grey, and added ones, in blue. n is the size of the input. (Color figure online)

3.2 New Programs

The is-connected program of Fig. 1 does not yet run in linear time under the new compiler. Figure 5 shows the runtimes of the program on star graphs and, for comparison, linked lists (see Figs. 10 and 14). As a consequence, we need a new technique to exploit the improved graph data structure in programs. Indeed, a rule that matches a rooted node adjacent to an unvisited node via an unprocessed edge would require the matching algorithm to iterate through all unprocessed edges incident to the root, which includes cross edges. To mitigate this problem, we implement the *forward* operation by marking an unprocessed edge incident to the root in a colour such that it is the only edge incident to the root marked of that colour. Then, we check the mark of the node adjacent to the root via the uniquely-coloured edge: if it is marked such that it is visited, we ignore it by marking the uniquely-coloured edge in the mark denoting that it is processed, otherwise, we move the root to the unvisited node and mark the edge as being processed.

In order to reason about programs, it is primordial to lay down assumptions on the complexity of certain elementary operations. We define the *search plan* of a rule as the procedure generated to compute a match satisfying the application condition, should one exist. Figure 4 showcases the complexity assumptions of the basic procedures of the search plan, adapted from [5]. The grey rows indicate

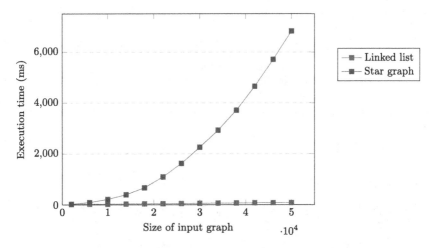

Fig. 5. The program `is-connected` running on the new compiler. (Color figure online)

existing procedures updated by the changes introduced in this paper and the blue rows, new procedures. The proof of Theorem 2 of the new `2-colouring` program relies on these complexity assumptions, and the empirical evidence showcased in Fig. 15 corroborates them.

4 Case Study: Two-Colouring

Vertex colouring is a well-known graph problem which has applications in domains such as scheduling, compiler optimisation and register allocation [11]. In 2012, Bak and Plump came up with a GP 2 program for the 2-colouring problem [3]. The program achieved linear time complexity on bounded-degree graph classes but exhibited quadratic time complexity on graph classes of unbounded degree [4].

In this section, we discuss the `2-colouring` program of Fig. 6, which achieves linear runtime on both bounded- and unbounded-degree graph classes using the same input and output conditions as Bak and Plump's program. This became possible by the improvements to the compiler described in Sect. 3, namely, the separation of edge lists with respect to marks and orientation.

The program `2-colouring`[3] expects a host graph satisfying the input conditions of Definition 1. It attempts to 2-colour the graph by performing a depth-first search (DFS) from an initial node, colouring each newly visited node in the colour contrasting that of the node it is visited from (either red or blue). The program fails if an edge is found to be incident with two non-grey nodes of the same colour. Figure 7 provides a sample execution of `2-colouring`.

---

[3] The concrete syntax of the program available at: https://gist.github.com/ismaili-ziad/51cc29fa3ea49a49d1922acd560ce3ee.

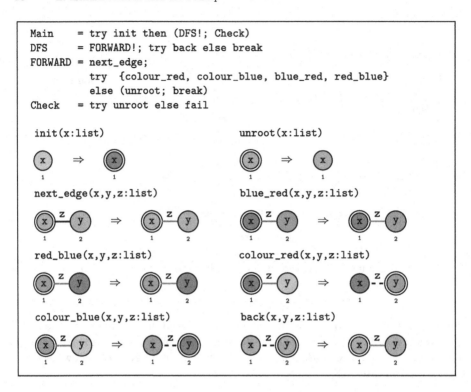

```
Main      = try init then (DFS!; Check)
DFS       = FORWARD!; try back else break
FORWARD = next_edge;
            try {colour_red, colour_blue, blue_red, red_blue}
            else (unroot; break)
Check     = try unroot else fail
```

Fig. 6. The program 2-colouring (magenta represents the mark *any*). (Color figure online)

In contrast to previous implementations of the 2-colouring, the linearity of this program's runtime is primarily attributed to an invariant ensuring that a single edge incident to the root is marked red, allowing rules implementing a depth-first search to process both forward and cross edges, instead of forward edges only. Invariant 3 shows that there can be, at most, one red edge at a time in the host graph throughout the execution of the program, allowing for instantaneous access with the compiler's recent optimisations. Furthermore, a blue mark on an edge indicates that its processing has ended, eliminating the need for it to be matched again as a forward or cross edge in the DFS traversal.

We first demonstrate that 2-colouring is totally correct. Then, we show that the program is linear with respect to the size of the input graph on any class of input graphs. Finally, we provide empirical evidence on various graph classes of bounded and unbounded degrees to corroborate our claim.

For the purposes of this section, the set of nodes of a graph G is denoted as V_G. The cardinality of a set X is represented by $|X|$. We use the notation $A \Rightarrow_r B$ to indicate that B results from applying r on A. For convenience, we define COLOUR to be the set {colour_red, colour_blue} and IGNORE, the set {blue_red, red_blue}. When we refer to the application of COLOUR, we mean

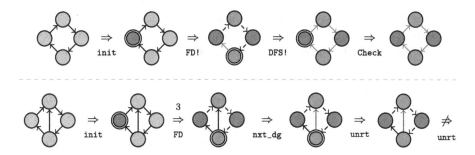

Fig. 7. Sample executions of **2-colouring** on a 2-colourable (top) input graph and a non-2-colourable (bottom) input graph. **FD**, **nxt_dg** and **unrt** denote **FORWARD**, **next_edge** and **unroot**, respectively. (Color figure online)

that either **colour_red** or **colour_blue** is applied. Similarly, applying **IGNORE** means applying either **blue_red** or **red_blue**. By a *rule call* or a *rule invocation*, we mean a completed or failed rule or procedure application, the **break** operation or the **fail** operation.

We first lay down the definition of an input graph.

Definition 1 (Input graph). *An* input graph, *within the context of the two-colourability problem, is an arbitrarily-labelled* connected *GP 2 host graph such that:*

1. *every node is marked grey,*
2. *every node is non-rooted, and*
3. *every edge is unmarked.*

Let us first examine the correctness of **2-colouring**.

Invariant 1. *Throughout the execution of* **2-colouring** *on a given input graph, the following invariant holds: edges marked as blue and nodes marked as either blue or red retain their respective marks unchanged.*

Invariant 2. *Throughout the execution of* **2-colouring**, *the following invariant holds: there is at most one root in the host graph and is incident to at most one dashed edge.*

Proposition 1. *Upon the execution of* **2-colouring** *on an input graph G, the rule* **unroot** *in DFS! is applied if and only if G is not 2-colourable.*

Proof. The lemma is trivially true if G is empty. Assume G contains at least one node. Let us split the proof into two distinct cases.
Case 1. G is 2-colourable. Let X and Y be disjoint subsets of V_G with respect to the bipartition of G such that $X \cup Y = V_G$ and every edge in G shares endpoints in both X and Y. Thus, no pair of distinct nodes within the same subset are adjacent in G. Clearly, given that the program is structure-preserving, X and

Y remain the same throughout the execution of 2-colouring. Let G' be the initialised graph such that $G \Rightarrow_{\text{init}} G'$. Suppose that init roots and marks a node blue in X. Consider the graph H such that $G' \Rightarrow_{\text{FORWARD}}^n H$ for some $n \geq 0$. By induction on n, we show that two non-grey nodes sharing the same mark are never adjacent.

When $n = 0$ (base case), the body DFS! has not been invoked once. Hence, H consists of (possibly zero) grey nodes and a single blue root node in X created by init, which does not violate the property.

Now, assume the property holds in H for some n; we show it still holds for $n + 1$. At the invocation of FORWARD, the rule next_edge is called. If there is no unmarked edge incident to the root node, the rule fails to match, and the loop exits, satisfying the condition. Otherwise, next_edge matches node 1 to the root node and node 2 to some arbitrarily marked node adjacent to the root via an unmarked edge. Since the host graph is 2-colourable, next_edge preserves marks and it is assumed, by the induction hypothesis, that no pair of adjacent nodes are both marked blue or red, node 2 must be part of the subset opposite to the root's and marked differently. That is, if node 1 is in X, node 2 is in Y, and vice versa. Otherwise, it would imply two nodes within the same subset share an edge, violating the assumption on G. At the invocation of try {colour_red, colour_blue, blue_red, red_blue}, following next_edge, node 2 must either have a mark opposite to the root's (i.e. blue if the root is red and vice versa) or be grey. Hence, one rule within that body must apply and move the root to the opposite subset while alternating its mark, preserving the property.

Therefore, upon the execution of FORWARD, either next_edge fails to apply and exits the loop, or it does apply and a rule in the try condition consequently matches and applies. Given that the rule back does not affect marks, unroot in DFS! is never invoked.

Case 2. G is not 2-colourable. By definition, there exists no assignment of marks from the set $\{blue, red\}$ to every node in G such that every edge has endpoints of different marks. Thus, a program that colours nodes in G either blue or red has to violate the two-colourability condition in that at least one pair of adjacent nodes share the same colour. Indeed, the node-marking rules of DFS! are colour_red and colour_blue, and each rule colours a node in the contrasting colour to that of the adjacent root. As such, an eventual application of these rules will result in two non-grey nodes sharing the same mark. Let w be node 2 following such an application of either colour_red or colour_blue on the host graph. w is either blue or red, rooted and adjacent to some node x that shares its mark. Any edge that connects w and x at this instance must be unmarked since a mark would imply that the edge was previously matched by next_edge, and subsequently by a rule in try {colour_red, colour_blue, blue_red, red_blue}. One of two situations occurs at the next invocation of FORWARD: either next_edge matches w and x (1), or it matches w and a different neighbour distinct from x (2).

1. Each rule in COLOUR and IGNORE is invoked, but none matches. Consequently, the unroot rule, applicable to w, is invoked and applied, leading to the call of break and the termination of the DFS! loop.

2. Let y be the neighbour of w distinct from x matched by the rule next_edge. Three subcases emerge: y admits of the same mark as w (a); y is marked in the colour opposite to w (b); y is grey (c).

(a) The same reasoning as Situation (1) applies.

(b) As both w and y are of contrasting colours and connected by a red edge, IGNORE applies. This ends the current instance of FORWARD, although not necessarily terminating the loop, and brings us back to the beginnings of Situations (1) and (2).

(c) After COLOUR is applied, w is unrooted, and y becomes the new root, marked with the colour opposite to w. Additionally, the edge matched by the rule is dashed. Invariant 2 establishes that the root node is incident to at most one dashed edge. Consequently, any subsequent application of back successfully backtracks the root to its ancestral node in the depth-first search tree.

The loop DFS! terminates upon the failure of the back rule. There are two possible scenarios: either the root eventually backtracks to node w before DFS! terminates, returning us to Situations (1) and (2), or DFS! breaks prior to w being rooted again. The latter can only happen if back is no longer applicable.

However, since COLOUR creates a path of dashed edges with a single endpoint rooted, the inapplicability of back can only occur due to the removal of the root, i.e. the application of unroot. In either case, unroot is invoked, causing FORWARD! to break, back to fail, and DFS! to terminate.

Therefore, executing 2-colouring on a non-2-colourable input graph results in the eventual application of unroot in DFS!. □

The next lemma demonstrates that termination of 2-colouring is ensured by showing that the body of each loop reduces a measure that assigns a non-negative integer to each host graph, thereby showing that the loop body eventually fails.

Lemma 1 (Termination of 2-colouring). *On any host graph, the program 2-colouring terminates.*

Proof. The program 2-colouring contains two looping procedures: DFS! and FORWARD!. To show termination, consider a measure $\#(X)$ consisting of the number of unmarked edges in the host graph X. The rule next_edge is invoked at the beginning of FORWARD, and if it fails to match, the loop breaks. Clearly, an application of next_edge reduces the measure $\#$. Since the number of edges is finite and no rule in 2-colouring creates or unmarks an edge, FORWARD! terminates. We now show that the upper body DFS! terminates. This time, consider $\#(X)$ to consist of the number of non-blue edges in the host graph X. The loop DFS! breaks if and only if back is called and fails to apply. Let H be the resulting graph of an application of back on G, that is, $G \Rightarrow_{back} H$. Clearly, $\#(H) < \#(G)$ since the rule marks a dashed edge blue and preserves the size (i.e. $|H| = |G|$). Similarly, given that blue edges retain their marks throughout the execution of 2-colouring (Invariant 1), the number of edges is finitely fixed and FORWARD! is known to terminate, dashed edges are eventually exhausted, the break command is called, and DFS! terminates.   □

Lemma 2 shows the program 2-colours the graph, should it be 2-colourable. It demonstrates that the existence of an unvisited (grey) node following the termination of DFS! on a 2-colourable graph leads to a contradiction.

Lemma 2. *Consider a* 2-colourable *input graph G. Upon termination of DFS! on G, the host graph contains no grey nodes.*

Proof. The lemma is trivially true if G is empty since the rule init fails to apply and the body (DFS!; Check) is not invoked. Suppose G consists of at least one node, thereby making init applicable, and consider G' and H such that $G \Rightarrow_{\text{init}} G' \Rightarrow_{\text{DFS!}} H$. For the sake of contradiction, assume that a grey node exists in H. The rule init creates a blue node in G'. Note, upon inspection of the rules, that the program is structure-preserving. Therefore, both G' and H are 2-colourable and connected. As per Invariant 1, once a node is turned blue, its mark is no longer modified. This implies at least one grey node in H (assumption) is adjacent to some non-grey node by the connectedness of H. Let u and v be the non-grey (blue or red) and grey nodes, respectively. We show that u and v are matched by a rule in COLOUR prior to the termination of DFS!, thereby contradicting the assumption.

Given that u is non-grey in G, and non-grey nodes preserve their mark, it must have been matched by either init or a rule in COLOUR. Either way, u must have been a non-grey root node. Recall that DFS! terminates if and only if back fails to apply, which can only occur following the termination of the loop FORWARD!. The latter breaks if either condition holds: next_edge fails to apply; no rule in COLOUR ∪ IGNORE applies. The second condition implies that the body (unroot; break) is then called. However, as shown in Proposition 1, unroot is never invoked if G is 2-colourable. Therefore, the second condition can never hold; that is, a rule in COLOUR ∪ IGNORE is always applicable upon its invocation. Regarding the first condition, let us examine the implicit data structure DFS! generates. The dashed edges form a path of non-grey nodes, wherein an endpoint is rooted. This models a stack of nodes where the root represents the top element. Specifically, init initialises the stack, COLOUR executes the *push* operation, and back performs the *pop* operation. Given that a root node can be incident to, at most, one dashed edge, it is guaranteed to backtrack to its ancestral node in the depth-first search tree as back is applied. The inapplicability of back can be seen as an exhaustion of the stack, which can only occur if there are no dashed edges in the host graph (as previously stated, the application of unroot is not a possibility). It is easy to observe that the remaining root at the end of DFS! is the initial node of the stack (i.e. the first pushed node). If u is rooted, then it is the initial node (i.e. the node init was applied to). The invocation of back follows the termination of FORWARD!, which only occurs if next_edge is not applicable, provided the graph is 2-colourable. However, next_edge is applicable on u and

v, hence a contradiction. If u is non-rooted, it must have been a root at some point during the execution of DFS! prior to it being popped from the so-called stack. Again, an analogous argument shows that this leads to a contradiction.

Since u and v ought to have been matched by next_edge, one of two outcomes must have occurred: either COLOUR successfully applied, or the statement (unroot; break) was invoked. However, since G is 2-colourable, the unroot rule cannot be applied. Therefore, COLOUR must have been applicable, resulting in v being marked non-grey. As established in Invariant 1, non-grey nodes retain their colour assignment. Hence, v cannot be grey in H.

Given the following properties:

- nodes in the host graph can only be grey, blue or red;
- a node adjacent to a non-grey node cannot be grey;
- G is 2-colourable and connected;
- blue and red nodes retain their mark; and
- the rule init, prior to DFS!, creates one non-grey node in the host graph;

it follows that every node in H (i.e. the resulting graph following the execution of DFS!) is non-grey.   □

Building upon the previous lemmata and propositions, we now show the correctness of 2-colouring.

Theorem 1 (Correctness of 2-colouring). *The program* 2-colouring *is totally correct with respect to the following specifications:*

Input: *An input graph.*

Output: *The program fails if and only if the input is not 2-colourable. Otherwise, it outputs a 2-colouring of the input such that every node is either blue or red, and no pair of adjacent nodes share the same mark.*

Proof. Termination follows from Lemma 1. Loop edges do not affect the 2-colouring and are omitted in the program. Let G be the input graph. If G is empty, init fails to match, and the program terminates, outputting the empty graph and satisfying the specifications. Suppose G consists of at least one node. We then split the remainder of this proof into two cases.

Case 1. G is 2-colourable. Since it is nonempty, init applies and DFS! is invoked. Consider G' and H such that $G \Rightarrow_{\text{init}} G' \Rightarrow_{\text{DFS!}} H$. It follows from Lemma 2 that H only contains red and blue nodes. Furthermore, as per Proposition 1, the rule unroot in DFS is never invoked, indicating that no violation of the 2-colouring has been encountered and the existence of a single root in H. Upon termination of DFS!, the procedure Check is called, and the rule unroot unroots the unique root node. The program then terminates and returns the 2-coloured host graph.

Case 1. G is not 2-colourable. Since it is nonempty, init applies and DFS! is invoked. Analogously to the previous argument, consider G' and H such that $G \Rightarrow_{\text{init}} G' \Rightarrow_{\text{DFS!}} H$. It follows from Proposition 1 that unroot is applied at the last execution of FORWARD!, breaking the DFS! loop. Therefore, there is no root node in H. The procedure Check is called, and since H contains no root, the rule unroot fails to match, and the fail command is invoked, failing the entire program. □

We now examine the complexity of 2-colouring. Prior to doing so, we establish another invariant of the program in Invariant 3 so as to argue for the constant-time matching of rules in COLOUR ∪ IGNORE.

Invariant 3. *Throughout the execution of 2-colouring, there is at most one red edge in the host graph.*

Theorem 2 (Complexity of 2-colouring). *On any class of input graphs, the program 2-colouring terminates in time $\mathcal{O}(|V| + |E|)$, where $|V|$ is the number of nodes and $|E|$, the number of edges.*

Proof. We first show that, for every rule, there is at most one matching attempt with respect to the complexity assumptions of the updated compiler (Fig. 4).

The rule init matches a single node. If the graph is nonempty, init immediately applies on the first match as every node is grey. And is therefore constant. Otherwise, no node is considered, and the matching fails immediately. Therefore, there is at most one matching attempt for init. The rule unroot matches a single node. Given that there is a bounded number of roots in the graph (Invariant 2), the number of matching attempts is also bounded.

Since there is at most one root in the host graph, node 1 of next_edge, blue_red, red_blue, colour_red, colour_blue and node 2 of back match in constant time. The rule next_edge matches any unmarked edge incident to the root incident to an *any*-marked node. Since any node adjacent to the root is marked, and non-loop edges are stored in distinct lists with respect to their marks, there is at most one matching attempt for next_edge. An analogous argument can be made for all rules beside init and unroot, as it is known from Invariant 3 that there is at most one red edge in the host graph, and Invariant 2 establishes that a root node is incident to at most one dashed edge, hence limiting the number of possible matches to a single one. Therefore, there is at most one matching attempt for every rule in 2-colouring. Now, let us look at the number of calls during the execution of the program for each rule. We define a call to be the invocation of a rule. For the purpose of this proof, let n and m be the number of nodes and edges, respectively.

The rule init is only called once at the beginning. It succeeds if the input graph is nonempty; otherwise, it fails. Let us show that back is called at most $m + 1$ times. The number of calls is the sum of successful applications and unsuccessful ones. Since the loop DFS! terminates at the inapplicability of back, there can be, at most, one unsuccessful application. Observe that the rule back marks an edge blue. It has been established in Invariant 1 that blue edges retain their mark. Hence, since there are m edges, there can be at most m successful applications.

Similarly, we demonstrate that the rules blue_red, red_blue, colour_red and colour_blue are called at most $n + m$ times each. Since the reasoning applies analogously to each of these rules, let r represent one of them. It can be observed that an unsuccessful application of r does not necessarily trigger the invocation of (unroot; break), as the latter is called if and only if every single rule in COLOUR∪IGNORE fails. Let us then look at the number of successful applications first. The rules red_blue and blue_red mark an edge blue. As previously stated, blue edges retain their mark. Therefore, there can be at most m successful applications of COLOUR. The rules colour_red and colour_blue mark a grey node blue or red. Non-grey nodes retain their mark, thus implying that there can be at most $n - 1$ successful applications of IGNORE (there are n nodes and init has already turned one node non-grey, hence $n - 1$). Given that the failure of every rule in COLOUR∪IGNORE triggers the invocation of (unroot; break) and terminates both FORWARD! and DFS!, each rule can fail to apply at most as many times as some rule in COLOUR ∪ IGNORE succeeds and one more time, marking the terminations of the loops FORWARD! and DFS!. Hence, each rule application can be unsuccessful at most $(n - 1) + (m) + 1 = n + m$ times.

The rule next_edge marks an unmarked edge, implying that there can be at most m successful applications. An unsuccessful application of next_edge terminates the loop FORWARD! and invokes the rule back, which either succeeds or terminates the DFS! loop. Thus, there can be at most as many unsuccessful applications of next_edge as there are successful applications of back, that is, m. The number of calls of next_edge is bounded to $m + m = 2m$. Finally, it is easy to see that unroot is called at most twice. Once in DFS! (its call terminates the loop) and once in Check.

Figure 8 offers an overview of the maximum number of calls for each rule of the program. Therefore, taking into account that all rules are constant, the overall time complexity of the program 2-colouring is

$$2 \cdot 1 + 2 \cdot 1 + 2 \cdot m + 4 \cdot (n + m) + 2 \cdot (n - 1) + 2 \cdot m + (m + 1)$$
$$= 6n + 9m + 3 = \mathcal{O}(|V| + |E|).$$

□

Rules	Unsuccessful	Successful
init	1	1
unroot	1	1
next_edge	m	m
blue–red	$n + m$	m
red–blue	$n + m$	m
colour_red	$n + m$	$n - 1$
colour_blue	$n + m$	$n - 1$
back	1	m

Fig. 8. Bounds on the number of rule calls for each rule throughout an entire execution. n is the number of nodes and m, edges.

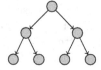

Fig. 9. Grid graph.

Fig. 10. Binary tree.

Fig. 11. Star graph.

Fig. 12. Cycle graph.

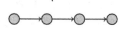

Fig. 13. Complete graph.

Fig. 14. Linked list.

Figure 15 showcases the empirical benchmarks of 2-colouring on the graph classes of Figs. 9, 10, 11, 12, 13 and 14. The measured runtimes do not account for graph parsing, building and printing, as these operations have a linear time complexity with respect to the input size (Fig. 4). Compilation time is also not included.

As evidenced, both unbounded-degree graph classes, complete graphs and binary trees, exhibit linear runtime performance in these tests. It is interesting to

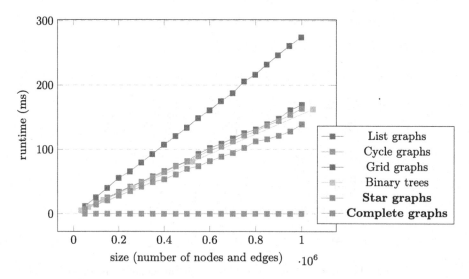

Fig. 15. Measured performance of the program 2-colouring. Unbounded-degree graph classes are boldened in the legend. (Color figure online)

note that complete graphs exhibit almost constant runtime. Since any complete graph K_n with $n \geq 3$ is not 2-colourable, the GP 2 program can detect a violation early during execution. This consistent behaviour is primarily attributed to the deterministic nature of the compiler implementation. In theory, matches in GP 2 are nondeterministic, and it is conceivable that a GP 2 compiler strictly adhering to this nondeterminism would visit every node in an arbitrary complete graph before encountering such a violation.

5   Conclusion

We have presented an approach to implement a 2-colouring algorithm with a rule-based graph program that has a linear runtime on input graphs with arbitrary node degrees. Removing the condition of bounded-degree input graphs has been an open problem since the publication of the first paper on rooted graph transformation [3]. So far, only certain reduction programs that destroy their input graphs could be designed to run in linear time on unbounded-degree graphs [5].

Our solution consists in both improving the graph data structure generated by the GP 2 compiler and devising a technique to exploit the new representation in programs. Previously, the graph data structure of the C program generated by the compiler stored lists of incoming and outgoing edges with each node. These lists were searched by the matching algorithm to quickly find an edge corresponding to an incoming or outgoing edge in the left-hand graph of a processed rule. However, in the presence of incident edges with different marks, these searches became linear-time operations that prevented constant-time rule matching. With the new data structure, finding an incident edge with a particular mark requires only constant time.

We exploit the new graph representation by designing a rule assigning a unique mark to an edge (next_edge in 2-colouring resp. is-connected) and constructing other rules that perform actions on the uniquely marked edge and its linked nodes depending on the marks of the nodes.

In future work, we plan to overcome the remaining restriction that programs such as 2-colouring require connected input graphs. By creating a separate node list for each node mark, it should be possible to find an unprocessed connected component of the host graph in constant time. For example, after 2-colouring a connected component, an uncoloured connected component could be found by searching for an arbitrary grey node.

We speculate that it will ultimately be possible to implement all DFS-based linear-time graph algorithms by linear-time GP 2 programs. Such algorithms include, for example, the non-destructive recognition of acyclic graphs, the topological sorting of acyclic graphs, the construction of Eulerian cycles, and the generation of strongly connected components.

References

1. Agrawal, A., Karsai, G., Neema, S., Shi, F., Vizhanyo, A.: The design of a language for model transformations. Softw. Syst. Model. **5**(3), 261–288 (2006). https://doi.org/10.1007/s10270-006-0027-7
2. Bak, C.: GP 2: efficient implementation of a graph programming language. Ph.D. thesis, Department of Computer Science, University of York, UK (2015). https://etheses.whiterose.ac.uk/12586/
3. Bak, C., Plump, D.: Rooted graph programs. In: Proceedings of 7th International Workshop on Graph Based Tools (GraBaTs 2012). Electronic Communications of the EASST, vol. 54 (2012). https://doi.org/10.14279/tuj.eceasst.54.780
4. Bak, C., Plump, D.: Compiling graph programs to C. In: Echahed, R., Minas, M. (eds.) ICGT 2016. LNCS, vol. 9761, pp. 102–117. Springer, Cham (2016). https://doi.org/10.1007/978-3-319-40530-8_7
5. Campbell, G., Courtehoute, B., Plump, D.: Fast rule-based graph programs. Sci. Comput. Program. **214**, 102727 (2022). https://doi.org/10.1016/j.scico.2021.102727
6. Campbell, G., Romö, J., Plump, D.: The improved GP 2 compiler. Technical report, Department of Computer Science, University of York, UK (2020). https://arxiv.org/abs/2010.03993
7. Dörr, H. (ed.): Efficient Graph Rewriting and Its Implementation. LNCS, vol. 922. Springer, Heidelberg (1995). https://doi.org/10.1007/BFb0031909
8. Fernández, M., Kirchner, H., Pinaud, B.: Strategic port graph rewriting: an interactive modelling framework. Math. Struct. Comput. Sci. **29**(5), 615–662 (2019). https://doi.org/10.1017/S0960129518000270
9. Ghamarian, A., de Mol, M., Rensink, A., Zambon, E., Zimakova, M.: Modelling and analysis using GROOVE. Int. J. Softw. Tools Technol. Transfer **14**(1), 15–40 (2012). https://doi.org/10.1007/s10009-011-0186-x
10. Jakumeit, E., Buchwald, S., Kroll, M.: GrGen.NET – the expressive, convenient and fast graph rewrite system. Int. J. Softw. Tools Technol. Transf. **12**(3–4), 263–271 (2010). https://doi.org/10.1007/s10009-010-0148-8
11. Skiena, S.S.: The Algorithm Design Manual, 3rd edn. Springer, Cham (2020). https://doi.org/10.1007/978-3-030-54256-6
12. Strüber, D., et al.: Henshin: a usability-focused framework for EMF model transformation development. In: de Lara, J., Plump, D. (eds.) ICGT 2017. LNCS, vol. 10373, pp. 196–208. Springer, Cham (2017). https://doi.org/10.1007/978-3-319-61470-0_12

A Bigraphs Paper of Sorts

Blair Archibald[(✉)] and Michele Sevegnani

School of Computing Science, University of Glasgow, Glasgow, UK
{blair.archibald,michele.sevegnani}@glasgow.ac.uk

Abstract. Bigraphs are an expressive graphical modelling formalism to represent systems with a mix of both spatial and non-local connectivity. Currently it is possible to write nonsensical models, *e.g.* with a Room nested inside a Person rather than Person nested inside a Room, or to create a hyperedge from what should be a binary link. A sorting scheme can be used to filter badly-formed bigraphs from those that are well formed. While the theory of bigraph sorts is well developed, none of the existing methods leads to a practical implementation. Instead they are based on tables of descriptions or semi-mathematical notations. We look at sorting bigraphs through a practical lens: developing a new sorting language, and show how an extension to the existing theory of bigraphs, in the form of well-sorted interfaces, paves the way for an implementation of well-sorted bigraphs. We discuss the trade-offs of this approach, and show how it allows sorts to be specified for existing bigraph models found in the literature.

Keywords: Bigraphs · Sortings · Type systems

1 Introduction

Bigraphs [19] are an expressive graphical modelling formalism designed to represent systems that have both spatial aspects, *e.g.* a Person in a Room, and non-local aspects, *e.g.* communication between Person entities in (possibly) different Rooms. In fact, they are sometimes too expressive: allowing nonsense models to be created, *e.g.* where a Room is nested *inside* a Person! Sorting schemes [19] have been proposed as a way to filter badly formed models—in a similar way to how a type system excludes badly formed programs. In a sorting scheme, entities are assigned both a *type* (called a control), *e.g.* Person, and a *sort*, *e.g.* moveable, and a set of constraints determines how entities of different sorts may interact. For example we might constrain that stationary sorts are never nested below moveable to exclude our Room within a Person issue.

While there have been many theoretical discussions of sorting schemes [3, 5,6,9,15,18,19], and they have been used to describe well-formed models of systems such as CCS, Petri-nets, and π-calculus, there is currently no practical tooling available for computationally specifying or working with sorts. The lack

Supported by an Amazon Research Award on Automated Reasoning.

of tooling is obvious from the current descriptions of sorting schemes which are often simply plain text descriptions of their expected operation. For example Table 1, partially recreated from [23], shows some of the types of constraints we would like to specify: child relationships (*e.g.* children of x has sort y), and cardinality constraints (*e.g.* x has one z child). While these textual descriptions are useful, they are not immediately amenable to automatic analysis, and it is up to the modeller to ensure these constraints are met.

Table 1. Example of sorting conditions for a bigraph encoding of the Actor Model recreated (partially) from [23]. Notation $\widehat{be}$ indicates sum sorts, *i.e.* b or e, and θ_{Act} is a set of sorts.

Constraint	Description
ϕ_1	all children of a θ-regions have sort θ, where $\theta \in \theta_{Act}$
ϕ_2	all children of an a-node have sort $\widehat{be}$

In this paper, we unlock the potential for automated implementation of sort checking/inference by:

- Defining a simple, yet powerful, language for describing entities and their sorts.
- Showing how the standard operators for building bigraphs, *e.g.* written in terms of tensor products, compositions, and substitutions, can be extended to only allow building well-sorted bigraphs.
- Using models found in the literature, show how our language captures existing modelling domains.

This approach paves the way for an implementation of bigraph sorts, *e.g.* in BigraphER [22], but at present no implementation exists. Throughout we utilise a running example of Petri-nets [20] modelled in bigraphs.

The paper is organised as follows: Sect. 2 gives necessarily background on bigraphs and sorting schemes. In Sect. 3 we show a new language for expressing sorts, and in Sect. 4 we show how this language enables sortings for elementary bigraphs (and therefore any bigraph). Section 5 applies the language to a set of applications. We discuss the limitations, future work, and conclusion in Sect. 6.

2   Background

2.1   Bigraphs

A bigraph consists of two orthogonal structures defined on the same set of entities: a *place graph* (a forest) that describes the *nesting* of entities, and a *link graph* that provides non-local hyperlinks between entities.

An example bigraph that models a Petri net is in Fig. 1. This differs from existing Petri net models [16], by allowing an unbounded number of tokens in

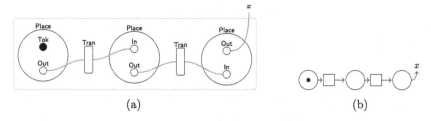

Fig. 1. (a) Bigraph modelling an (open[1]) Petri net; (b) Petri net representation. ([1] Open meaning the output of this system can connect, via x, to another Petri net [4].)

each place and using extra entities for linking rather than defining a family of places/transitions (one for each linking structure, *e.g.* 1 link in, 2 links out). We draw entities as different (coloured) shapes. Containment illustrates the spatial nesting relationship, *e.g.* Tok is contained by Place, while green hyperedges (1-to-n links) represent non-spatial connections, in this case giving the wiring of places and transitions. Entities have a fixed *arity* (number of links/ports), *e.g.* Out has arity 1, but links may be disconnected/closed, *i.e.* a 1-to-0 link. We use *port* to refer to the link point of an entity, and link to mean a collection of ports (and names).

Each place graph has m *regions*, shown as the dashed rectangles, and n *sites*, usually shown as filled dashed rectangles. Regions represent parallel parts of the system, and sites represent abstraction, *i.e.* an unspecified bigraph (including the empty bigraph) exists there. Similarly, link graphs have a (finite) set of inner names and outer names, *e.g.* $\{x\}$.

Bigraphs are compositional structures, and we build larger bigraphs from smaller bigraphs. A bigraph is described by an interface $B : \langle n, X \rangle \rightarrow \langle m, Y \rangle$ specifying bigraph B maps n sites, to m regions, and inner names X to outer names Y. Composition of bigraphs, denoted ∘, consists of placing regions in sites (when $n = m$), and connecting inner and outer-faces on like-names. Composition combines bigraphs *vertically*, but we can also combine bigraphs *horizontally* through the tensor product ⊗. This tensor product extends both the sites/regions and name sets for the interfaces. ⊗ is only defined when the sets of interface names are disjoint. We introduce additional algebraic representations, *e.g.* symmetries, for bigraphs in the following sections.

2.2 Sorting

Given the term-rewriting/algebraic nature of bigraphs, initial sorting schemes were designed in a similar manner to many-sorted algebras but with more freedom due to the need to classify both places and links, and the flexibility of changing the argument order for constructors (to reflect the graph-like nature).

Milner describes a sorting scheme [19] based on an assignment of sorts to entities (controls) and a *formation rule* as in Table 1 that defines constraints on the sorts. One type of constraint is a *stratified (place) sorting* that determines,

$$\langle sort \rangle ::= \textsf{sort} \ \langle sname \rangle \ | $$
$$\textsf{sort} \ \langle sname \rangle = \langle constructors \rangle_|^*$$
$$\langle constructors \rangle ::= \langle cname \rangle [\{\langle lpat \rangle_|^*\}] \ [\langle pat \rangle]$$
$$\langle pat \rangle ::= \langle pat \rangle \times \langle pat \rangle \ | \ \langle pat \rangle + \langle pat \rangle \ |$$
$$\langle pat \rangle^* \ | \ (\langle pat \rangle) \ | \ \langle baseS \rangle$$
$$\langle baseS \rangle ::= \langle sname \rangle \ | \ \mathbb{1}$$
$$\langle lpat \rangle ::= \langle sname \rangle \rightarrow \langle pat \rangle$$
$$\langle sname \rangle ::= \textsf{[a-z][A-z0-9]+}$$
$$\langle cname \rangle ::= \textsf{[A-Z][A-z0-9]+}$$

Fig. 2. Grammar for sort and entity specification. Expressions contained in [ ] are optional. Expressions $e_|^*$ means any number of e expressions separated by |.

for a given sort, the particular sorts of the children. An extension to binding bigraphs [6] is possible. This approach uses properties of port-sortings to form link-sortings, and details how to construct a sub-sorting relation (which we do not explore here).

Birkedal *et al.* [5] give a categorical approach to specifying sorts based on selecting an appropriate functor that determines correctly shaped bigraphs within the category of all bigraphs. While useful, it is not clear how to practically specify the shape category. This approach is reminiscent of type-graphs [24] where the well sortedness of a graph can be checked by verifying an appropriate mapping from instance to type graph exists. This type of approach does not work well for bigraphs as bigraphs are categorical arrows, not objects, and so all structural checks need to be through defining valid decompositions[1].

3   A Pattern Language of Sorts

To allow sorts to be useful for practical bigraph modelling scenarios, we must move sort definitions from textual descriptions and into a formal language amenable to parsing and analysis.

For usability, we introduce a new language similar to algebraic data type definitions seen in programming languages but with some major differences:

1. There is support for specifying not just record (product) like structures, but also explicit support for typing of links.
2. Most languages order constructor parameters, *e.g.* $s(x, y) \neq s(y, x)$. We treat $s(x, y)$ and $s(y, x)$ as equivalent as bigraphs do not have fixed child orders.

The grammar of our sort definition language is given in Fig. 2. We use strings starting with lowercase to identify sorts, and strings starting with an uppercase to identify entity constructors. We use fonts, *e.g.* A, to distinguish entities from

---

[1] Similar issues are seen in [1].

sorts that we denote with font **s**. $\mathbb{1}$ is the unit sort. We also have a sort $\mathbb{0}$, the empty sort, that is used to mark entities as atomic. A user never specifies $\mathbb{0}$ directly so it is not included in the grammar.

Sorts may be defined either constructorless or with entity constructors. Constructorless sorts, *e.g.* **sort s** is useful for defining link sortings since we may wish to give a link port a different sort than the entity the port is on. Constructor-based sorts have the form **sort t = A s | B w** that specifies entities A and B have sort **t**, and the patterns for their children: in this case a single **s** sorted child for A and **w** sorted child for B. Each entity belongs to a single sort, and it is an open question if this could be weakened, *e.g.* to allow sub-typing relationships.

Sort patterns can be combined in two ways: **s × t** creates a product sort, *i.e.* specifying the need for (exactly) two child bigraphs of sort **s and t**; while **s + t** is a sum sort, *i.e.* specifying a child must have sort **s or t**. For product sorts, there is no need for **s** and **t** to differ, *e.g.* **B n × n** specifies that a B entity must have **exactly two n** sorted children, and in this way allows cardinality constraints to be encoded. Finally, **s*** is the (infinite) product sort of **s**, *e.g.* **A s*** allows A to have any number of **s** sorted children—including none. Sorts may be defined recursively, *e.g.* **sort s = A s*** is well defined.

Atomic entities, that never have children are specified as the constructor name and no pattern. When analysing sorts, we treat, for example, A as A $\mathbb{0}$.

Link patterns use the notation, *e.g.* **sort s = A{t → w + y}**, that specifies A has a[2] port with port-sort **t** that is part of a link that additionally has either a **w** or **y** sorted port on it. In any link pattern **t → w**, we call **t** the *domain sort* and **w** the *range sort*. Domain sorts must be a base sort. For entities with arity greater than one, we simply list all port sort patterns.

Formally, the grammar defines sorting signatures in the form $\Sigma = (\Theta, \mathcal{K}, \mathcal{L})$, with Θ mapping constructors to sorts, $\mathcal{K}$ constructors to sort patterns, and $\mathcal{L}$ constructors to link patterns.

Example 1. (Sorting Petri nets). Our model has sorts:

$$\textsf{sort m = Tok} \qquad \textsf{sort l = In}\{1 \to \textsf{o} \times 1^*\} \mid \textsf{Out}\{1 \to \textsf{i} \times 1^*\}$$
$$\textsf{sort i} \qquad\qquad \textsf{sort t = Tran}\{\textsf{i} \to 1^*, \textsf{o} \to 1^*\}$$
$$\textsf{sort o} \qquad\qquad \textsf{sort p = Place m}^* \times 1^*$$

Which encodes properties including: 1. Places may contain any number of tokens, including none, 2. All controls except Place are atomic, 3. Places only connect to transitions (via explicit links In and Out that encode a direction), and transitions only to places, *i.e.* we cannot connect two transitions. 4. Transitions may connect to/from nowhere (sources/sinks) as 1^* allows closed links. 5. Places may have no links, *e.g.* we allow disconnected places, but could be strengthened using $1 \times 1^*$ to force at least a single incoming or outgoing link.

---

[2] Ports are not ordered in bigraphs.

$$\llbracket 0 \rrbracket = \emptyset \tag{1}$$

$$\llbracket 1 \rrbracket = \{ (\!|\!) \} \tag{2}$$

$$\llbracket s \rrbracket = \{ (\!| s |\!) \} \text{ where } s \text{ is a base sort} \tag{3}$$

$$\llbracket s + t \rrbracket = \llbracket s \rrbracket \cup \llbracket t \rrbracket \tag{4}$$

$$\llbracket s \times t \rrbracket = \{ A \uplus B \mid A \in \llbracket s \rrbracket, B \in \llbracket t \rrbracket \} \tag{5}$$

$$\llbracket s^* \rrbracket = \llbracket 1 \rrbracket \cup \llbracket s \rrbracket \cup \llbracket s \times s \rrbracket \cup \llbracket s \times s \times s \rrbracket \cup \ldots \tag{6}$$

Fig. 3. Mapping between sort patterns and set-of-multisets representation.

3.1   Compatibility of Patterns

To ensure a bigraph is well sorted we need to determine when two sorts are *compatible*. For example, we want s compatible with s and also s + v *etc.* We now develop a theory that determines when two sorts are compatible. We have chosen this notion of compatibility as it captures a wide range of examples (Sect. 5), but it may be that a different notion of compatibility is useful in future, *e.g.* one that allows subsorts to be specified.

Multiple patterns can represent the same sorting constraints. For example A w × y and A y × w both specify that some entity A has one w sorted and one y sorted child, *i.e.* there is some notion of commutativity in the patterns.

Inspired by Fowler *et al.* [11,12], who need to reason over a similar pattern language—in this case not to specify structure constraints, but instead to specify possible elements within an asynchronous mailbox—we map our pattern language into a sets-of-multisets structure that gives us the required axioms/structural equivalences for free. This mapping is given in Fig. 3. We use $(\!| s, a, a |\!)$ to denote a multiset containing elements s, a and a, with the usual multiset functions defined, *e.g.* $(\!| a |\!) \uplus (\!| a, b |\!) = (\!| a, a, b |\!)$.

Using this mapping, the pattern s × (t + w) is encoded as:

$$
\begin{aligned}
\llbracket s \times (t + w) \rrbracket &= \{ A \uplus B \mid A \in \llbracket s \rrbracket, B \in \llbracket t + w \rrbracket \} \\
&= \{ A \uplus B \mid A \in \{ (\!| s |\!) \}, B \in (\llbracket t \rrbracket \cup \llbracket w \rrbracket) \} \\
&= \{ A \uplus B \mid A \in \{ (\!| s |\!) \}, B \in \{ (\!| t |\!), (\!| w |\!) \} \} \\
&= \{ (\!| s |\!) \uplus (\!| t |\!), (\!| s |\!) \uplus (\!| w |\!) \} \\
&= \{ (\!| s, t |\!), (\!| s, w |\!) \}
\end{aligned}
$$

Sort 1 is a unit for × but not for +. This allows expressing when entities *may* have no children (or links might be closed), *e.g.* s + 1 is either empty or of sort s. Sort 0 is a unit for + and an absorbing element for ×. + is idempotent, + and × are commutative, and × distributes over +.

Given the set-of-multisets encoding of patterns, checking pattern compatibility, which we denote ⋈, consists mainly of checking for a non-empty intersection of the encoded patterns. We have:

$$s \bowtie t \ \textbf{iif} \ \llbracket s \rrbracket = \llbracket t \rrbracket \ \textbf{or} \ \llbracket s \rrbracket \cap \llbracket t \rrbracket \neq \{\}$$

The use of non-empty intersection handles the special case of 0 as sort 0 is never compatible with anything other than itself (handled by the equality case), and we use this to denote unsortable elements. $\bowtie$ is reflexive and symmetric, but not transitive[3], so is not an equivalence relation. For example, we can check $(s \times t) \bowtie (s\,(s + t))$:

$$\llbracket s \times t \rrbracket = \{(\!|s, t|\!)\} \qquad\qquad \llbracket s \times (s + t) \rrbracket = \{(\!|s, s|\!), (\!|s, t|\!)\}$$

Which are compatible given the intersection $(\!|s, t|\!)$.

Useful compatibilities include $\mathbb{1} \bowtie \alpha^*$ for any sorting pattern α (by definition above), *e.g.* it is explicitly zero or more entities. A sort $\alpha \times \alpha^*$ forces at least one α.

4   Sorting Abstract Bigraphs

We work with abstract bigraphs, where vertices of the graph structure do not have specific identifiers only entity types. Working with the set of primitive bigraphs—from which all other bigraphs can be built—, we show how the sorts change through composition, tensor and closure/renaming of links. As every bigraph is a combination of these primitives, we can sort any bigraph. These bigraphs are well-sorted-by-design, *i.e.* it is impossible to build a bigraph that does not correspond to the (user-defined) sorting scheme.

Bigraph tensor product $\otimes$ does not commute, and any region movement is through an explicit symmetry operator γ. As we want pattern *under* a single entity to commute—*e.g.* $A\,s \times t$ and $A\,t \times s$ should specify the same pattern—we need an additional product type, $\Diamond$, that specifies disjoint patterns that cannot be swapped (unless under an explicit symmetry operator). These only ever appear in the decomposition of a bigraph and a user never writes these directly.

We extend sort pattern compatibility to support $\Diamond$ as follows:

$$(s_0 \Diamond s_1 \Diamond \ldots) \bowtie (t_0 \Diamond t_1 \Diamond \ldots) \quad \text{iff} \quad s_0 \bowtie t_0, s_1 \bowtie t_1, \ldots$$

That is, we apply pattern compatibility component-wise. 0 is a unit for $\Diamond$ and for ease of presentation we assume 0 elements are always cancelled out.

Given a set of user-defined sorts Σ (via the pattern language introduced in Sect. 3) we can now determine the sort of any bigraph. To achieve this, we extend the bigraph interface definition to also track the sort pattern. For a bigraph B mapping n sites, to m regions, and inner names X to outer names Y we have a (pattern) sorted signature:

$$B : \langle n : s_0 \Diamond \ldots \Diamond s_{n-1}, X : \Gamma \rangle \to \langle m : t_0 \Diamond \ldots \Diamond t_{m-1}, Y : \Gamma' \rangle$$

which has a parameter of n sites with sorts compatible with pattern $s_0 \Diamond \ldots \Diamond s_{n-1}$, and requires a context of m regions compatible with $t_0 \Diamond \ldots \Diamond t_{m-1}$.

---

[3] For example, $s \bowtie (s + t), (s + t) \bowtie (t + w)$, but s is not compatible with $(t + w)$..

For links, we create a *sorting context* Γ that stores the link patterns for names in X, *e.g.* $\Gamma = [x : \mathsf{s} \to \mathsf{t}]$. The ordering of names in Γ does not matter and this can be treated like a set[4]. A bigraph can produce a different context Γ' for the set of names Y, which may be smaller, *i.e.* if B closes/substitutes a name. We say two sorting contexts Γ and Δ are sort compatible $\Gamma \bowtie \Delta$ when:

$$X = \{x_0, \ldots, x_{n-1}\} = \mathrm{dom}\,\Gamma = \mathrm{dom}\,\Delta \qquad \text{and} \qquad \Gamma(x_i) \bowtie \Delta(x_i)$$

with $0 \le i < n$ and treating link patterns as $\Diamond$ products:

$$(\mathsf{s} \to \mathsf{t}) \bowtie (\mathsf{s}' \to \mathsf{t}') \quad \text{iif} \quad (\mathsf{s}\,\Diamond\,\mathsf{t}) \bowtie (\mathsf{s}'\,\Diamond\,\mathsf{t}')$$

That is, two contexts are compatible if they have the same names, and the names agree on the sorts (up to sort compatibility).

To make it easier to work with sorted interfaces, we introduce the following notation:

Definition 1 (Sorted Interface). $\langle\!\langle s, \Gamma \rangle\!\rangle$ *denotes an interface with place sorting* s *and link sorting* Γ. *As sorts determine the* shape *of a bigraph, it is possible to recover a bigraph interface from a sorted interface. For example,* $\langle\!\langle v \Diamond w, [a : \mathsf{s} \to \mathsf{t}] \rangle\!\rangle$ *must have bigraph interface* $\langle 2, \{a\} \rangle$.

Notation. When only describing place or link sorts specifically we sometimes use the reduced notation, *e.g.* $\langle\!\langle \mathsf{s} \Diamond \mathsf{t} \rangle\!\rangle$, or $\langle\!\langle [\mathsf{s} \to \mathsf{t}] \rangle\!\rangle$ when the other components are trivial (single sorted place $\not\vdash$, or $\Gamma = [\,]$). Similarly, $\langle\!\langle \, \rangle\!\rangle$ is a shorthand for $\langle\!\langle \not\vdash, [\,] \rangle\!\rangle$. We write $A \,\sharp\, B$ to indicate sets $A \cap B = \emptyset$. We define iterated operators as follows:

$$\square_{i<n}\mathsf{s}_i = \mathsf{s}_0 \,\square\, \cdots \,\square\, \mathsf{s}_{n-1}$$

with $\square \in \{\times, \Diamond, +\}$. Bound n is dropped when clear from the context. A sorting context in the form $[x_0 : \mathsf{s}_0 \to \mathsf{t}_0, \ldots, x_{n-1} : \mathsf{s}_{n-1} \to \mathsf{t}_{n-1}]$ is indicated by $[x_i : \mathsf{s}_i \to \mathsf{t}_i]_{i<n}$. Similarly, we write $[\mathsf{s}_i \to \mathsf{t}_i]_{i<n}$ to denote n link patterns. Throughout this section, we use $\Gamma, \Delta, \Theta, \ldots$ and $\alpha, \tau, \mu, \ldots$ to range over sorting contexts and families of sort patterns, *e.g.* variables that can later be unified with a particular sort pattern, respectively.

4.1   Elementary Place Graphs and Constructors

We introduce primitive (sorted) building blocks for bigraphs and then show how any bigraphs can be created from the combination of these.

Bigraph $1 : \langle 0, \emptyset \rangle \to \langle 1, \emptyset \rangle$ represents the *empty* bigraph consisting of a single region. It has sorted equivalent

$$\frac{}{1 : \langle\!\langle \, \rangle\!\rangle \to \langle\!\langle \mathbb{1} \rangle\!\rangle}\ \text{[One]}$$

---

[4] We use the $[\,]$ notation to distinguish between name sets and sorting contexts.

where $\mathbb{1}$ is our unit sort.

For any interface, we can form the identity bigraph $\mathsf{id}_{\langle n,X\rangle} : \langle n, X\rangle \to \langle n, X\rangle$ that consists only of regions/sites and maps from names to themselves. Identity bigraphs have the property (by definition) that they do not change the sorts of their inputs. We sort them with:

$$\frac{\alpha = \lozenge_{i<n}\alpha_i \qquad \mathrm{dom}\,\Gamma = X \qquad \Gamma(x_i \in X) = \tau_i \to \mu_i}{\mathsf{id}_{\langle n,X\rangle} : \langle\!\langle \alpha, \Gamma \rangle\!\rangle \to \langle\!\langle \alpha, \Gamma \rangle\!\rangle} \;\; [\textsc{Id}]$$

Bigraph $merge_n : \langle n, \emptyset\rangle \to \langle 1, \emptyset\rangle$ places n sites into a single region and is often used before a composition, *e.g.* to allow all children to fall into one site. For sorts, $merge_n$ has the ability to convert the product-of-patterns $\lozenge$ into the sort product $\times$:

$$\frac{}{merge_n : \langle\!\langle \alpha_0 \lozenge \cdots \lozenge \alpha_{n-1} \rangle\!\rangle \to \langle\!\langle \alpha_0 \times \cdots \times \alpha_{n-1} \rangle\!\rangle} \;\; [\textsc{Merge}]$$

As with identity bigraphs, this is a family of merge operators (with α_i any sort pattern) based on the specific sorts/size of the merge.

Symmetries are sorted as follows:

$$\frac{}{\gamma_{m,n} : \langle\!\langle (\lozenge_{i<m}\alpha_i)\,\lozenge(\lozenge_{j<n}\alpha_j) \rangle\!\rangle \to \langle\!\langle (\lozenge_{j<n}\alpha_j)\,\lozenge(\lozenge_{i<m}\alpha_i) \rangle\!\rangle} \;\; [\textsc{Sym}]$$

Finally, for each user defined entity and a sorting signature $\Sigma = (\Theta, \mathcal{K}, \mathcal{L})$, we sort ions—which are bigraphs consisting of a single entity, *i.e.* an entity constructor—$\mathsf{K}_X : \langle 1, \emptyset\rangle \to \langle 1, X = \{x_0, \ldots, x_{n-1}\}\rangle$ with the following rule:

$$\frac{\Theta(\mathsf{K}) = \mathtt{w} \qquad \mathcal{K}(\mathsf{K}) = \mathtt{v} \qquad \mathcal{L}(\mathsf{K}) = [\mathtt{s}_i \to \mathtt{t}_i]_{i<n}}{\mathsf{K}_X : \langle\!\langle \mathtt{v} \rangle\!\rangle \to \langle\!\langle \mathtt{w}, [x_i : \mathtt{s}_i \to \mathtt{t}_i]_{i<n} \rangle\!\rangle} \;\; [\textsc{Ion}]$$

For example, given signature $\mathtt{sort\ s} = \mathsf{A}\{\mathtt{t} \to \mathtt{v}\}\,\mathtt{n} \mid \mathsf{B}\,\mathtt{m} + \mathtt{n}$ we get ions:

$$\mathsf{A}_x : \langle\!\langle \mathtt{n} \rangle\!\rangle \to \langle\!\langle \mathtt{s}, [x : \mathtt{t} \to \mathtt{v}] \rangle\!\rangle \quad \text{and} \quad \mathsf{B} : \langle\!\langle \mathtt{m} + \mathtt{n} \rangle\!\rangle \to \langle\!\langle \mathtt{s} \rangle\!\rangle$$

4.2 Combining Place Graphs

With the elementary primitives in place, we show how to combine bigraphs into larger bigraphs. The bigraph theory is based on a symmetric monoidal category and so the main operators are tensor $\otimes$—that places two bigraphs side-by-side—and composition $\circ$—that puts regions into sites, and joins common names.

The sorted tensor introduces products-of-patterns (these are separate regions so should not commute):

$$\frac{F : \langle\!\langle \alpha, \Gamma \rangle\!\rangle \to \langle\!\langle \tau, \Gamma' \rangle\!\rangle \qquad G : \langle\!\langle \varphi, \Delta \rangle\!\rangle \to \langle\!\langle \mu, \Delta' \rangle\!\rangle \qquad \begin{array}{c} \mathrm{dom}\,\Gamma \,\natural\, \mathrm{dom}\,\Delta \\ \mathrm{dom}\,\Gamma' \,\natural\, \mathrm{dom}\,\Delta' \end{array}}{G \otimes F : \langle\!\langle \alpha \lozenge \varphi, \Gamma \cup \Delta \rangle\!\rangle \to \langle\!\langle \tau \lozenge \mu, \Gamma' \cup \Delta' \rangle\!\rangle} \;\; [\textsc{Tens}]$$

Here names must be disjoint. The tensor rule only builds larger sort types but does not do any sort checking.

Composition, which places a bigraph into the parameter of another bigraph, performs sort compatibility checking during composition as follows:

$$\frac{F : \langle\!\langle \alpha, \Gamma \rangle\!\rangle \to \langle\!\langle \varphi, \Gamma' \rangle\!\rangle \qquad G : \langle\!\langle \tau, \Delta \rangle\!\rangle \to \langle\!\langle \mu, \Delta' \rangle\!\rangle \qquad \begin{matrix} \varphi \bowtie \tau \\ \Gamma' \bowtie \Delta \end{matrix}}{G \circ F : \langle\!\langle \alpha, \Gamma \rangle\!\rangle \to \langle\!\langle \mu, \Delta' \rangle\!\rangle} \text{[Comp]}$$

The $\bowtie$ constraints enforce that composition only occurs when the sort patterns of outer and inner face (both link and place sorts) are compatible.

4.3 Elementary Link Graphs

For links, we have two elementary bigraphs: closure, which stops a name from moving to the context, and substitution which combines (a set of) inner names into a single outer name. Practically, this is how we join links, *e.g.* in Fig. 4, a substitution joins two links (from A and B) to a single link with outer name y.

To sort the outer names, we need to determine if we can produce a valid *extension* of a sort pattern, *i.e.* the minimum we would need to include in the context to make a sort compatibility $\bowtie$ hold. We make use of the sets-of-multisets encoding to check when this is the case. We use $t \preceq s$ when there exists *any* element in $[\![t]\!]$ that is a sub-multiset of an element in $[\![s]\!]$. This gives, for example, $s \preceq s \times t$ as $[\![s \times t]\!] = \{(\!(s, t)\!)\}, [\![s]\!] = \{(\!(s)\!)\}$ and $(\!(s)\!) \subseteq (\!(s, t)\!)$. $\preceq$ does not imply pattern compatibility, although this may hold in some cases.

Fig. 4. Example bigraph $B = (merge_2 \otimes y/\{x_0, x_1\}) \circ (A_{x_0} \otimes B_{x_1})$.

To make it clear how to use $\preceq$ in practice, we show how to sort a link by example, and then generalise to a sorting rule for substitutions.

Example 2. Consider bigraph B in Fig. 4, and the following (s, t and w are constructorless sorts):

$$\text{sort a} = \text{A}\{\text{s} \to \text{s} \times \text{t}\} \mid \text{B}\{\text{s} \to \text{s} \times (\text{t} + \text{w})\}$$

We need to check there is a possible sorting of ports of A, B, and y that makes the link well-sorted. The link has known domain sorts $\text{p} = \text{s} \times \text{s}$ from taking the domain sort from each link pattern. We then check each link constraint is met by the domain sorts in turn. For $\text{A}\{\text{s} \to \text{s} \times \text{t}\}$ we form the pattern $\text{m}_0 = \text{s} \times \text{s} \times \text{t}$, *i.e.* we introduce the domain sort into the range sort pattern. We then check this constraint is a valid context extension of the domain sorts, *i.e.* checking $p \preceq m_0$. This holds as $(\!(s, s)\!) \subseteq (\!(s, s, t)\!)$ as required. For $B\{s \to s \times (t + w)\}$ we have $m_1 = s \times s \times (t + w)$ and $p \preceq m_1$ because $(\!(s, s)\!) \subseteq (\!(s, s, t)\!)$ as required. Note that $(\!(s, s, w)\!) \in [\![s \times s \times (t + w)]\!]$, but we only need one sub-multiset to hold. At this point we know that there are valid sorts on the links and the final piece is to determine a suitable sort for y (the extended context). We find this through:

$$[\![\bigcap_{i<2} [\![m_i - p]\!]]\!]^{-1} \quad \text{with} \quad [\![s - t]\!] = \{A \setminus B \mid A \in [\![s]\!], B \in [\![t]\!]\} \tag{7}$$

where $[\![\;]\!]^{-1}$ is the inverse of the encoding defined in Fig. 3. As $s - t$ always returns a single set of multisets, the intersection of n sets of multisets is a single (or empty) set of multisets, and there is a image for any single set of multisets in Fig. 3, the inverse is always defined. While $[\![\;]\!]$ is not injective in general, *e.g.* $[\![[\![1 \times 1]\!]]\!]^{-1} = 1$ it captures the intended semantics (sort compatibility), *i.e.* we want to treat $1 \bowtie (1 \times 1)$.

In our case:

$$[\![m_0]\!] = [\![s \times s \times t]\!] = \{(\!(s, s, t)\!)\}$$
$$[\![m_0 - p]\!] = \{(\!(s, s, t)\!) \setminus (\!(s, s)\!)\} = \{(\!(t)\!)\}$$

$$[\![m_1]\!] = [\![s \times s \times (t + w)]\!] = \{(\!(s, s, t)\!), (\!(s, s, w)\!)\}$$
$$[\![m_1 - p]\!] = \{(\!(s, s, t)\!) \setminus (\!(s, s)\!), (\!(s, s, w)\!) \setminus (\!(s, s)\!)\} = \{(\!(t)\!), (\!(w)\!)\}$$

$$\{(\!(t)\!)\} \cap \{(\!(t)\!), (\!(w)\!)\} = \{(\!(t)\!)\}$$
$$[\![\{(\!(t)\!)\}]\!]^{-1} = t$$

We now know that the port sorts can be extended to make a valid link, and that the extension needs to be $y : s \times s \to t$ *i.e.* the link taking our port patterns to t. The substitution $y/\{x_0, x_1\}$ can be safely made and we only need to consider the name y in future.

Formally, for a set of names $X = \{x_0, \ldots, x_{n-1}\}$, with $n > 0$, and a name y, substitutions $y/X : \langle 0, X \rangle \to \langle 0, \{y\} \rangle$ are sorted by the following rule:

$$\frac{\alpha = \times_{i<n} \alpha_i \qquad \forall_{i<n} (\alpha \preceq \alpha_i \times \tau_i) \qquad \mu = \mathcal{S}(\alpha, [x_i : \alpha_i \to \tau_i]_{i<n})}{y/X : \langle\!\langle [x_i : \alpha_i \to \tau_i]_{i<n} \rangle\!\rangle \to \langle\!\langle [y : \alpha \to \mu] \rangle\!\rangle} \text{ [RNAME]}$$

which states all existing ports on the links need to be valid on all link constraints, and computes the remainder by generalising the construction given in Eq. (7) to define $\mathcal{S}$

$$\mathcal{S}(\alpha, [x_i : \alpha_i \to \tau_i]_{i<n}) = [\![\bigcap_{i<n} ((\alpha_i \times \tau_i) - \alpha)]\!]^{-1}$$

In practice we defer applying $\mathcal{S}(\;)$ until we have concrete sorts. When $X = \emptyset$, *i.e.* the empty substitution $y : \langle 0, \emptyset \rangle \to \langle 0, \{y\} \rangle$, we apply the rule below:

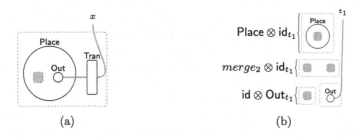

(a)   (b)

Fig. 5. (a) Worked example: simple Petri net model; (b) Partial algebraic decomposition (terms compose vertically).

$$\frac{\mu \text{ fresh sort}}{y : \langle\!\langle\rangle\!\rangle \rightarrow \langle\!\langle [y : \mathbb{1} \rightarrow \mu] \rangle\!\rangle} \text{ [NEW]}$$

Intuitively, as there are no ports in the link, no constraints need to be propagated up to the context, as represented by the fresh sort in the premise.

Finally, for closures $/x$ we ensure the range sort of x maps to a sort compatible with $\not\Vdash$ at which point nothing else needs to be added to the link.

$$\frac{\tau \bowtie \mathbb{1}}{/x : \langle\!\langle [x : \alpha \rightarrow \tau] \rangle\!\rangle \rightarrow \langle\!\langle\rangle\!\rangle} \text{ [CLOSE]}$$

4.4   Worked Example

We show how to sort the simple Petri net model shown in Fig. 5 with respect to the sorts in Example 1. An algebraic form for the example (partially shown in Fig. 5b) is:

$$P = \overbrace{(\text{Place} \otimes id_{t_1})}^{P'} \circ \underbrace{\overbrace{(merge_2 \otimes id_{t_1})}^{M} \circ \overbrace{(id \otimes \text{Out}_{t_1})}^{O}}_{P''}$$

$$B = \overbrace{(/y \otimes id_{1,x})}^{B'} \circ \underbrace{\overbrace{(y/\{t_1, t_2\} \otimes id_x \otimes merge_2)}^{M'} \circ \overbrace{(P \otimes \text{Tran}_{t_2,x})}^{T}}_{B''}$$

We check the sortings beginning with bigraph P. Simple applications of rules TENS, ION, MERGE, and ID give us the sorts for the elementary subterms

$$P' : \langle\!\langle \mathbf{m}^* \times \mathbb{1}^*, \Theta \rangle\!\rangle \rightarrow \langle\!\langle \mathbf{p}, \Theta \rangle\!\rangle \qquad M : \langle\!\langle \tau \Diamond \lambda, \Gamma \rangle\!\rangle \rightarrow \langle\!\langle \tau \times \lambda, \Gamma \rangle\!\rangle$$

$$O : \langle\!\langle \alpha \rangle\!\rangle \rightarrow \langle\!\langle \alpha \Diamond \mathbb{1}, \Delta \rangle\!\rangle$$

where $\Theta = [t_1 : \gamma]$, $\Gamma = [t_1 : \beta]$, $\Delta = [t_1 : \mathbf{d}]$, and $\mathbf{d} = \mathbb{1} \rightarrow \mathbf{i} \times \mathbb{1}^*$. Term P'' is sorted by applying rule COMP

$$\frac{O : \langle\!\langle \alpha \rangle\!\rangle \rightarrow \langle\!\langle \alpha \Diamond \mathbb{1}, \Delta \rangle\!\rangle \qquad M : \langle\!\langle \tau \Diamond \lambda, \Gamma \rangle\!\rangle \rightarrow \langle\!\langle \tau \times \lambda, \Gamma \rangle\!\rangle \qquad \begin{array}{c} \tau \Diamond \lambda \bowtie \alpha \Diamond \mathbb{1} \\ \tau \bowtie \alpha \quad \lambda \bowtie \mathbb{1} \\ \Gamma \bowtie \Delta \\ \Longrightarrow \lambda = \mathbb{1} \quad \beta = \mathbf{d} \end{array}}{P'' : \langle\!\langle \alpha \rangle\!\rangle \rightarrow \langle\!\langle \alpha \times \mathbb{1}, \Delta \rangle\!\rangle}$$

where we highlight the unifiers required for sort inference. Families of sorts are always compatible so no substitution is needed to satisfy $\tau \bowtie \alpha$. By further applying COMP we can sort P

$$\frac{P'' : \langle\!\langle \alpha \rangle\!\rangle \to \langle\!\langle \alpha \times 1, \Delta \rangle\!\rangle \qquad P' : \langle\!\langle \mathtt{m}^* \times 1^*, \Theta \rangle\!\rangle \to \langle\!\langle \mathtt{p}, \Theta \rangle\!\rangle}{P : \langle\!\langle \mathtt{m}^* \times 1^* \rangle\!\rangle \to \langle\!\langle \mathtt{p}, \Delta \rangle\!\rangle} \qquad \begin{array}{c} \mathtt{m}^* \times 1^* \bowtie \alpha \times 1 \\ \Theta \bowtie \Delta \\ \Longrightarrow \alpha = \mathtt{m}^* \times 1^* \\ \gamma = \mathtt{d} \end{array}$$

Here we have removed many fresh sorting variables since the ion context, $e.g.$ the sort of Place, forces a sort for α (inferred using a similar process to Eq. (7) but for place graphs, $i.e.$ intuitively since we have 1^*, removing an 1 still leaves an 1^*) and likewise for the identity links on t_1.

Extending to add the transition entity through an application of TENS, we obtain $T : \langle\!\langle \mathtt{m}^* \times 1^* \rangle\!\rangle \to \langle\!\langle \mathtt{p} \Diamond \mathtt{t}, \Delta' \rangle\!\rangle$ where $\Delta' = [t_1 : \mathtt{d}, t_2 : \mathtt{e}, x : \mathtt{f}]$, $\mathtt{e} = \mathtt{i} \to 1^*$, and $\mathtt{f} = \mathtt{o} \to 1^*$.

At B'' we are composing into a substitution and so we follow the link sorting procedure as before:

$$\frac{\dfrac{\dfrac{\gamma' \preceq \alpha_1 \times \tau_1}{\dfrac{\gamma' \preceq \alpha_2 \times \tau_2 \qquad \mu = \mathcal{S}(\gamma', \Gamma'')}{y/\{t_1, t_2\} : \langle\!\langle \Gamma'' \rangle\!\rangle \to \langle\!\langle y : \gamma' \to \mu \rangle\!\rangle}}}{M' : \langle\!\langle \phi \Diamond \psi, \Gamma' \rangle\!\rangle \to \langle\!\langle \phi \times \psi, \Theta' \rangle\!\rangle} \qquad T : \langle\!\langle \mathtt{m}^* \times 1^* \rangle\!\rangle \to \langle\!\langle \mathtt{p} \Diamond \mathtt{t}, \Delta' \rangle\!\rangle}{B'' : \langle\!\langle \mathtt{m}^* \times 1^* \rangle\!\rangle \to \langle\!\langle \mathtt{p} \times \mathtt{t}, \Theta'' \rangle\!\rangle} \; : \; \begin{array}{c} \phi \Diamond \psi \bowtie \mathtt{p} \Diamond \mathtt{t} \\ \phi \bowtie \mathtt{p} \quad \psi \bowtie \mathtt{t} \\ \Gamma' \bowtie \Delta' \\ \Longrightarrow \phi = \mathtt{p} \quad \psi = \mathtt{t} \\ \alpha_1 = 1 \quad \alpha_2 = \mathtt{i} \quad \alpha_3 = \mathtt{f} \\ \tau_1 = \mathtt{i} \times 1^* \quad \tau_2 = 1^* \\ \mu = 1^* \end{array}$$

where $\gamma' = \alpha_1 \times \alpha_2$, $\Gamma'' = [t_1 : \alpha_1 \to \tau_1, t_2 : \alpha_2 \to \tau_2]$, $\Gamma' = \Gamma'' \cup [x : \alpha_3]$, $\Theta' = [y : \gamma' \to \mu, x : \alpha_3]$, and $\Theta'' = [y : 1 \times \mathtt{i} \to 1^*, x : \mathtt{f}]$.

Finally, we apply rule CLOSE to check y may be closed—which is possible as $1^* \bowtie \nvDash$—and so the example is well sorted with final sort for B:

$$\frac{\dfrac{\dfrac{\tau \bowtie \mathbb{1}}{/y : \langle\!\langle [y : \alpha \to \tau] \rangle\!\rangle \to \langle\!\langle \rangle\!\rangle} \; : \; B' : \langle\!\langle \gamma, \Delta \rangle\!\rangle \to \langle\!\langle \gamma, \Delta' \rangle\!\rangle}{B'' : \langle\!\langle \mathtt{m}^* \times 1^* \rangle\!\rangle \to \langle\!\langle \mathtt{p} \times \mathtt{t}, \Theta'' \rangle\!\rangle}}{B : \langle\!\langle \mathtt{m}^* \times 1^* \rangle\!\rangle \to \langle\!\langle \mathtt{p} \times \mathtt{t}, [x : \mathtt{o} \to 1^*] \rangle\!\rangle} \qquad \begin{array}{c} \gamma \bowtie \mathtt{p} \times \mathtt{t} \\ \Delta \bowtie \Theta'' \\ \Longrightarrow \gamma = \mathtt{p} \times \mathtt{t} \\ \alpha = 1 \times \mathtt{i} \quad \tau = 1^* \quad \beta = \mathtt{f} \end{array}$$

where $\Delta' = [x : \beta]$ and $\Delta = \Delta' \cup [y : \alpha \to \tau]$. Notice this *must* be a top-level bigraph, $i.e.$ cannot compose elsewhere, as we have no constructors that accept the sort $\mathtt{p} \times \mathtt{t}$.

5   Different Sorts of Application

Using our new syntax, we show how to encode sorting schemes for a range of existing models in the literature.

Fig. 6. Arithmetic Nets example: $\bot = 0 + x$

Arithmetic Nets. A model of Arithmetic Nets is given in [19, Chapter 6], and an example is in Fig. 6. These nets wire components to denote arithmetic result flow. We show a reduced model with only Zero and Plus components[5]. While similar to Petri-nets, differ by the lack of component nesting.

Milner assigns these nets sorts s (source) and t (target) and creates a rule set (in plain text, not amenable to implementation) that: 1. every link has only one source, 2. a *link* has sort s if it has a source on it, 3. and closed links always have sort s. As we only sort *ports* (and constrain through links) we cannot express the last two constraints directly, nor can we describe a closed link specifically. The essence of these sorts can be maintained by ensuring we never have s on both the left and right of a link pattern and only ever as a single sort (*i.e.* never as s^*):

> sort s
>
> sort t
>
> sort e $=$ Zero$\{s \rightarrow t^*\}$ | Plus$\{t \rightarrow s, t \rightarrow s, s \rightarrow t^*\}$

Zero entities only have one source (connected to any number of targets), while Plus entities have two targets/inputs (from one source node each), and creates a source/output (connected to any number of future targets).

Small changes in the sorting can specify significantly different nets. For example:

> sort e $=$ Zero$\{s \rightarrow t^*\}$ | Plus$\{t \rightarrow s, t \rightarrow s, s \rightarrow t \times t^*\}$

Ensures a Plus node always connects somewhere (but allows the source of Zero to be closed).

> sort e $=$ Zero$\{s \rightarrow t\}$ | Plus$\{t \rightarrow s, t \rightarrow s, s \rightarrow t\}$

Makes all links binary (and unclosed), *e.g.* the source of Zero always connects to *exactly* one target port of a Plus.

> sort e $=$ Zero$\{s \rightarrow t\}$ | Plus$\{t \rightarrow (\not{k} + s), t \rightarrow (\not{k} + s), s \rightarrow t^*\}$

Allows some targets/inputs to Plus to be unspecified (closed); perhaps treating them implicitly as 0 internally.

---

[5] Milner also gives the successor function, and a node that forwards an input to an output.

CCS. Milner gives a model and sorting scheme for CCS in [19]. The main sorting constraint is that *alternations* always contain *processes* and *processes* always contain *alternations*. We express this in our sortings using:

$$\texttt{sort a} = \mathsf{Send}\{\mathsf{a} \rightarrow \mathsf{a}^*\}\, \mathsf{p}^* \mid \mathsf{Get}\{\mathsf{a} \rightarrow \mathsf{a}^*\}\, \mathsf{p}^*$$
$$\texttt{sort p} = \mathsf{Alt}\, \mathsf{a}^*$$

That is, send and get *can* (but do not need to) connect other sends and gets via hyperedges, and they always contain processes. Alternations can contain many alternative processes including the *nil* process (modelled implicitly by the empty region 1).

λ-Calculus. Currently our approach does not support binding bigraphs (where names can have locality constraints), however we can still specify the sorts for a λ-calculus models such as that in [17]. A possible sorting is:

$$\texttt{sort exp} = \mathsf{Var}\{\mathsf{exp} \rightarrow \mathsf{exp}^*\} \mid \mathsf{Lam}\{\mathsf{exp} \rightarrow \mathsf{exp}^*\}\, \mathsf{exp} \mid \mathsf{App}\, \mathsf{l} \times \mathsf{r}$$
$$\texttt{sort l} = \mathsf{Left}\, \mathsf{exp}$$
$$\texttt{sort r} = \mathsf{Right}\, \mathsf{exp}$$

This encodes the three main components of λ calculus: variables, which connect to all like-named[6] variables/binders, abstractions that bind a (new) name, and function application that includes the important constraint that App must contain exactly one left and one right component: something that is currently difficult to enforce without a sorting scheme.

Virus Spread. Our sortings are not only useful for computational models. A model for the spread of a virus through a network is given in [14] and adapted to bigraphs in [2]. Network nodes have a specific status—safe, attacked, or infected—and connect to other nodes through a nesting of links (similar to the Petri net model) which allows a virus to spread. A possible sorting is:

$$\texttt{sort n} = \mathsf{Safe}\, \mathsf{l}^* \mid \mathsf{Attacked}\, \mathsf{l}^* \mid \mathsf{Infected}\, \mathsf{l}^*$$
$$\texttt{sort l} = \mathsf{Link}\{\mathsf{l} \rightarrow \mathsf{l}\}$$

Here the status of a node is encoded in the entity type. We allow flexible network configurations through Link entities, but enforce in the sorting that these links are always binary—something that has been difficult to express without sorts.

Formal Results. Full formal analysis is out of scope, but one result of interest is showing how the category of sorted bigraphs relates to the category of unsorted bigraphs. Bigraphical categories have interfaces as objects and bigraphs as morphisms. The goal of sorting is to essentially filter badly formed morphisms.

For sorted bigraphs we have objects $\langle\!\langle\,\rangle\!\rangle$ and morphsims defined by the sorting rules and (user defined) sorts. We can define a functor mapping sorted interfaces

---

[6] We do not need an explicit notion of names as links perform the role of binders.

to unsorted equivalents (there is always a way to recover this as described in Definition 1). Identities are preserved by sorting rule ID (which sorts any interface). Composition is preserved by rule COMP, *i.e.* whenever two sorted bigraphs compose, because they have the correct interface in the non-sorted version (and this is the only requirement), they must also compose there.

6   Conclusions

For practical modelling it is important to be able to restrict the range of bigraphs that can be created, *e.g.* to avoid nesting physical entities within virtual spaces, and utilising sorting schemes is a promising approach. In contrast to existing approaches that lack computational descriptions, we have defined a language for specifying sorts, and shown how, by extending the usual bigraph interfaces to sorted variants, a pathway to practical sorting is possible.

Expressivity and Limitations. While the new sorting approach is flexible, and captures a wide range of practical models, there are some constraints that are difficult to express.

We only define sort patterns for *entities*, and while we can use this to infer the sorts of names/sites/regions we cannot *specify* the sort for a region, *e.g.* we cannot say all regions are s sorted. This can lead to cases, such as the Petri net of Sect. 4.4, where we have a well-sorted bigraph that cannot compose anywhere.

We only constrain the sorts of *direct* children, and we cannot encode constraints such as "there must be a s sorted grandchild". This type of constraint sometimes appears in the literature [7]. More generally, we cannot express global constraints, *e.g.* that only one instance of a particular sort exists in an entire model (singleton types). Implementing binding bigraphs as a sorting would require similar expressiveness for both placement and links. One enabler for this in future might be to make use of spatial logics [8,10].

For links, we can express when a link is *allowed* to be closed but cannot force a link to be closed. We also only have notions of port-sorts and it remains open if this is enough to express existing link-sort constraints (*e.g.* the models in Sect. 5), and how useful link-sorts are in practice.

Future Work. We plan to implement this approach within the BigraphER tool [22]. We will explore the interplay of sorts with reaction rules (that specify dynamics), in particular showing how to manage instantiation maps (that can manipulate sites during rewriting) so that sorts are preserved. We believe restricting to solid bigraphs [13], which BigraphER already does to handle probabilistic/stochastic bigraphs, is also beneficial for sorting since it ensures the sorts of the left-hand of a rule is unambiguous.

We will also consider how sorts allow generation of random bigraphs for model testing; explore the theoretical connections to existing sorting schemes and bigraph concepts *e.g.* RPOs; and extend our rules to handle variants of bigraphs including bigraphs with sharing [21] and conditional bigraphs [1].

References

1. Archibald, B., Calder, M., Sevegnani, M.: Conditional bigraphs. In: Gadducci, F., Kehrer, T. (eds.) ICGT 2020. LNCS, vol. 12150, pp. 3–19. Springer, Cham (2020). https://doi.org/10.1007/978-3-030-51372-6_1

2. Archibald, B., Calder, M., Sevegnani, M.: Probabilistic bigraphs. Formal Aspects Comput. **34**(2), 1–27 (2022). https://doi.org/10.1145/3545180

3. Bacci, G., Grohmann, D.: On decidability of bigraphical sorting. In: International Workshop on Graph Computation Models (2010)

4. Baez, J.C., Master, J.: Open petri nets. Math. Struct. Comput. Sci. **30**(3), 314–341 (2020). https://doi.org/10.1017/S0960129520000043

5. Birkedal, L., Debois, S., Hildebrandt, T.: On the construction of sorted reactive systems. In: van Breugel, F., Chechik, M. (eds.) CONCUR 2008. LNCS, vol. 5201, pp. 218–232. Springer, Heidelberg (2008). https://doi.org/10.1007/978-3-540-85361-9_20

6. Bundgaard, M., Sassone, V.: Typed polyadic pi-calculus in bigraphs. In: Bossi, A., Maher, M.J. (eds.) Proceedings of the 8th International ACM SIGPLAN Conference on Principles and Practice of Declarative Programming, July 10-12, 2006, Venice, Italy, pp. 1–12. ACM (2006). https://doi.org/10.1145/1140335.1140336

7. Calder, M., Sevegnani, M.: Modelling IEEE 802.11 CSMA/CA RTS/CTS with stochastic bigraphs with sharing. Formal Aspects Comput. **26**(3), 537–561 (2014). https://doi.org/10.1007/S00165-012-0270-3

8. Ciancia, V., Latella, D., Loreti, M., Massink, M.: Specifying and verifying properties of space. In: Theoretical Computer Science - 8th IFIP TC 1/WG 2.2 International Conference, TCS 2014, Rome, Italy, September 1-3, 2014. Proceedings, pp. 222–235 (2014). https://doi.org/10.1007/978-3-662-44602-7_18

9. Conchúir, S.T.O.: Explicit Substitution and Sorted Bigraphs. Ph.D. thesis, Trinity College, Dublin, Ireland (2009). http://www.tara.tcd.ie/handle/2262/83173

10. Conforti, G., Macedonio, D., Sassone, V.: Static BiLog: a unifying language for spatial structures. Fundam. Informaticae **80**(1-3), 91–110 (2007). http://content.iospress.com/articles/fundamenta-informaticae/fi80-1-3-06

11. de'Liguoro, U., Padovani, L.: Mailbox types for unordered interactions. In: Millstein, T.D. (ed.) 32nd European Conference on Object-Oriented Programming, ECOOP 2018, July 16-21, 2018, Amsterdam, The Netherlands. LIPIcs, vol. 109, pp. 15:1–15:28. Schloss Dagstuhl - Leibniz-Zentrum für Informatik (2018).https://doi.org/10.4230/LIPICS.ECOOP.2018.15

12. Fowler, S., Attard, D.P., Sowul, F., Gay, S.J., Trinder, P.: Special delivery: programming with mailbox types. Proc. ACM Program. Lang. **7**(ICFP), 78–107 (2023). https://doi.org/10.1145/3607832

13. Krivine, J., Milner, R., Troina, A.: Stochastic bigraphs. In: Bauer, A., Mislove, M.W. (eds.) Proceedings of the 24th Conference on the Mathematical Foundations of Programming Semantics, MFPS 2008, Philadelphia, PA, USA, May 22-25, 2008. Electronic Notes in Theoretical Computer Science, vol. 218, pp. 73–96. Elsevier (2008). https://doi.org/10.1016/j.entcs.2008.10.006

14. Kwiatkowska, M.Z., Norman, G., Parker, D., Vigliotti, M.G.: Probabilistic mobile ambients. Theor. Comput. Sci. **410**(12–13), 1272–1303 (2009). https://doi.org/10.1016/J.TCS.2008.12.058

15. Leifer, J.J., Milner, R.: Transition systems, link graphs and petri nets. Math. Struct. Comput. Sci. **16**(6), 989–1047 (2006). https://doi.org/10.1017/S0960129506005664

16. Milner, R.: Bigraphs for petri nets. In: Desel, J., Reisig, W., Rozenberg, G. (eds.) ACPN 2003. LNCS, vol. 3098, pp. 686–701. Springer, Heidelberg (2004). https://doi.org/10.1007/978-3-540-27755-2_19

17. Milner, R.: Local bigraphs and confluence: Two conjectures: (extended abstract). In: Amadio, R.M., Phillips, I. (eds.) Proceedings of the 13th International Workshop on Expressiveness in Concurrency, EXPRESS 2006, Bonn, Germany, August 26, 2006. Electronic Notes in Theoretical Computer Science, vol. 175, pp. 65–73. Elsevier (2006). https://doi.org/10.1016/J.ENTCS.2006.07.035

18. Milner, R.: Pure bigraphs: structure and dynamics. Inf. Comput. **204**(1), 60–122 (2006). https://doi.org/10.1016/j.ic.2005.07.003

19. Milner, R.: The Space and Motion of Communicating Agents. Cambridge University Press (2009)

20. Peterson, J.L.: Petri nets. ACM Comput. Surv. **9**(3), 223–252 (1977). https://doi.org/10.1145/356698.356702

21. Sevegnani, M., Calder, M.: Bigraphs with sharing. Theor. Comput. Sci. **577**, 43–73 (2015). https://doi.org/10.1016/j.tcs.2015.02.011

22. Sevegnani, M., Calder, M.: BigraphER: rewriting and analysis engine for bigraphs. In: Chaudhuri, S., Farzan, A. (eds.) CAV 2016. LNCS, vol. 9780, pp. 494–501. Springer, Cham (2016). https://doi.org/10.1007/978-3-319-41540-6_27

23. Sevegnani, M., Pereira, E.: Towards a bigraphical encoding of actors (June 2014). http://eprints.gla.ac.uk/94772/

24. Taentzer, G., Rensink, A.: Ensuring structural constraints in graph-based models with type inheritance. In: Cerioli, M. (ed.) FASE 2005. LNCS, vol. 3442, pp. 64–79. Springer, Heidelberg (2005). https://doi.org/10.1007/978-3-540-31984-9_6

Generalized Weighted Type Graphs for Termination of Graph Transformation Systems

Jörg Endrullis[(⊠)] and Roy Overbeek

Vrije Universiteit Amsterdam, Amsterdam, The Netherlands
{j.endrullis,r.overbeek}@vu.nl

Abstract. We refine the weighted type graph technique for proving termination of double pushout (DPO) graph transformation systems. We increase the power of the approach for graphs, we generalize the technique to other categories, and we allow for variations of DPO that occur in the literature.

Keywords: Graph rewriting · Termination · Weighted Type Graphs · DPO

1 Introduction

Termination is a central aspect of program correctness. There has been extensive research on termination [2,5,11,12,23,29] of imperative, functional, and logic programs. Moreover, there exists a wealth of powerful termination techniques for term rewriting systems [41]. Term rewriting has proven useful as an intermediary formalism to reason about C, Java, Haskell and Prolog programs [22,24].

There are limitations to the adequacy of terms, however. Terms are usually crude representations of program states, and termination might be lost in translation. This holds especially for programs that operate with graph structures, and for programs that make use of non-trivial pointer or reference structures.

Nonetheless, such programs can be elegantly represented by graphs. This motivates the use of graph transformation systems as a unifying formalism for reasoning about programs. Correspondingly, there is a need for automatable techniques for reasoning about graph transformation systems, in particular for (dis)proving properties such as termination and confluence.

Unfortunately, there are not many techniques for proving termination of graph transformation systems. A major obstacle is the large variety of graph-like structures. Indeed, the termination techniques that do exist are usually defined for rather specific notions of graphs. This is in stark contrast to the general philosophy of algebraic graph transformation [18], the predominant approach to graph transformation, which uses the language of category theory to define and study graph transformations in a graph-agnostic manner.

© The Author(s), under exclusive license to Springer Nature Switzerland AG 2024
R. Harmer and J. Kosiol (Eds.): ICGT 2024, LNCS 14774, pp. 39–58, 2024.
https://doi.org/10.1007/978-3-031-64285-2_3

In this paper, we propose a powerful graph-agnostic technique for proving termination of graph transformation systems based on weighted type graphs. We base ourselves on work by Bruggink et al. [8]. Their method was developed specifically for edge-labelled multigraphs, using double pushout (DPO) graph transformation, and the graph rewrite rules were required to have a discrete interface, i.e., an interface without edges. We generalize and improve their approach in several ways:

(a) We strengthen the weighted type graphs technique for graph rewriting with monic matching. In the DPO literature, matches are usually required to be monic since this increases expressivity [26]. The method of Bruggink et al. [8] is not sensitive to such constraints. It always proves the stronger property of termination with respect to unrestricted matching. Consequently, it is bound to fail whenever termination depends on monic matching. Our method, by contrast, is sensitive to such constraints.

(b) We generalize weighted type graphs to arbitrary categories, and we propose the notion of 'traceable (sub)objects' as a generalization of nodes and edges to arbitrary categories.

(c) We formulate our technique on an abstract level, by describing minimal assumptions on the behaviour of pushouts involved in the rewrite steps. This makes our technique applicable to the many variations of DPO (see Remark 2.2), for instance, restrictions to specific pushout complements.

Outline

The general idea behind our technique, and outline of the paper, is as follows. DPO rewrite systems induce rewrite steps $G \Rightarrow H$ on objects if there exists a diagram as shown on the right (the top span of the diagram is a rule of the

$$
\begin{array}{ccccc}
L & \xleftarrow{l} & K & \xrightarrow{r} & R \\
\downarrow{\scriptstyle m} & \text{PO} & \downarrow & \text{PO} & \downarrow \\
G & \longleftarrow & C & \longrightarrow & H
\end{array}
$$

rewrite system). We define a decreasing measure $\mathbf{w}$ on objects, and show that $\mathbf{w}(G) > \mathbf{w}(H)$ for any such step generated by the system. We want to do this in a general categorical setting. This means that we cannot make assumptions about what the objects and morphisms represent. For simplicity, assume for now that the system contains a single rule $\rho : L \leftarrow K \rightarrow R$.

A first approximation of our measure $\mathbf{w}$ is $\mathbf{w} = \#\mathrm{Hom}(\cdot, T)$ for some distinguished object T, where a termination proof consists of an analysis that shows that for any step $G \Rightarrow H$ generated by ρ, $\#\mathrm{Hom}(G, T) > \#\mathrm{Hom}(H, T)$. The technical challenge in this case consists in establishing a lower bound on $\#\mathrm{Hom}(G, T)$ and an upper bound for $\#\mathrm{Hom}(H, T)$, using only ρ and generic facts (and possibly assumptions) about pushout squares. The fact that the two pushout squares of a DPO step share the vertical morphism $u : K \rightarrow C$ is useful, because naively, $G = L \cup_K C$ and $H = R \cup_K C$, meaning that K and C are factors common to G and H.

Whatever the precise details of the approximation, it would not be very powerful, because it simply counts morphisms into T, effectively assigning each a weight of 1. Instead, we would like to be able to differentiate between morphisms into T. We therefore add more structure to T by equipping it with a set of weighted T-valued elements $\mathbb{E}$, which are morphisms of the form $e : dom(e) \to T$, together with some weight $\mathbf{w}(e)$ associated to them. Then morphisms $\phi : G \to T$ are assigned a weight depending on how they relate to these weighted elements.

A priori there are different ways of doing this, but we will count for each $e \in \mathbb{E}$ the number $n(e)$ of commuting triangles of the form shown on the right and compute $\mathbf{w}(\phi) = \prod_{e \in \mathbb{E}} \mathbf{w}(e)^{n(e)}$. Next, the weight function $\mathbf{w}$ on objects X sums the weights of the individual morphisms in $\mathrm{Hom}(G, T)$. These combinations are done formally and more generally using a semiring structure. We define this notion of a weighted type graph in Sect. 3.

$$\begin{array}{ccc} dom(e) & \!\!\!-\alpha\rightarrow\!\!\! & G \\ & \!\!\!{=}\!\!\! & \\ {}_e\searrow & & \swarrow_{\phi'} \\ & T & \end{array} \qquad (1)$$

The technical challenge described for the simple measure $\mathbf{w} = \#\mathrm{Hom}(\,\cdot\,, T)$ above returns, but takes a different form. To measure $\mathbf{w}(G)$, we need to measure every $\phi : G \to T$, for which in turn, we need to count, for every $e \in \mathbb{E}$, the number of commuting triangles of the form shown in Diagram (1). In Sect. 4 we will identify some assumptions that allow us to do precisely this. Intuitively, we rely on pushouts not creating certain structure out of thin air. For general pushouts in **Graph**, for instance, any node or edge in the pushout object G traces back to a node or edge in L or C – the same is not true for loops in G, which can be created by the pushout. For pushouts along monomorphisms in **Graph**, however, these "traceable elements" do include loops. What we require, then, is that $\mathbb{E}$ is traceable in this sense w.r.t. the class of pushouts under consideration. We show that this requirement, combined with some additional assumptions, is sufficient to precisely determine $\mathbf{w}(G)$ and to determine an upper bound on $\mathbf{w}(H)$ for any induced rewrite step.

In Sect. 5, we show how to analyze systems for termination, using weighted type graphs that satisfy the right traceability conditions. In Sect. 6 we demonstrate our technique on a number of examples. We conclude by comparing to related work in Sect. 7, followed by a discussion of some directions for future research in Sect. 8. All proofs may be found in the extended version of this article [21, Appendix C]. In [21, Appendix E] we provide some benchmarks from a Scala implementation.

2 Preliminaries

We assume an understanding of basic category theory, in particular of pushouts, pullbacks and monomorphisms $\rightarrowtail$. Throughout this paper, we work in a fixed category $\mathbf{C}$. For an introduction to DPO graph transformation, see [16, 20, 27].

We write **Graph** for the category of *finite* multigraphs, and **SGraph** for the category of *finite* simple graphs (i.e., without parallel edges). The graphs can be edge-labelled and/or node-labelled over a fixed label set.

Definition 2.1 (DPO Rewriting [20]). *A* **DPO** *rewrite rule* ρ *is a span* $L \xleftarrow{l} K \xrightarrow{r} R$ *. A diagram, as shown on the right, defines a* **DPO rewrite step** $G \Rightarrow_{\mathrm{DPO}}^{\rho,m} H$, *i.e., a step from* G *to* H *using rule* ρ *and match morphism* $m : L \to G$.

$$
\begin{array}{ccccc}
L & \xleftarrow{l} & K & \xrightarrow{r} & R \\
\downarrow m & & \downarrow & & \downarrow \\
G & \longleftarrow & C & \longrightarrow & H
\end{array}
$$

with PO labels on both squares.

Remark 2.2. Usually Definition 2.1 is accompanied with constraints. Most commonly, l and m are required to be (regular) monic. Sometimes also the left pushout square is restricted, e.g., it may be required that the pushout complement is **minimal** or **initial** [3,4,7]. Because such restrictions affect whether a DPO rewrite system is terminating, we will define our approach in a way that is parametric in the variation of DPO under consideration (see Definition 5.2).

Definition 2.3 (Termination). *Let* R, S *be binary relations on* $Ob(\mathbf{C})$. *Relation* R *is* **terminating relative to** S, *denoted* $\mathrm{SN}(R/S)$, *if every (finite or infinite) sequence* $o_1 \ (R \cup S) \ o_2 \ (R \cup S) \ o_3 \ (R \cup S) \ o_4 \ldots$, *where* $o_1, o_2, \ldots \in Ob(\mathbf{C})$, *contains only a finite number of* R *steps. The relation* R *is* **terminating**, *denoted* $\mathrm{SN}(R)$, *if* R *terminates relative to the empty relation* $\varnothing$.

These concepts carry over to sets of rules via the induced rewrite relations.

We will interpret weights into a well-founded semiring as defined below.

Definition 2.4. *A* **monoid** $\langle S, \cdot, e \rangle$ *is a set* S *with an operation* $\cdot : S \times S \to S$ *and an* **identity element** e *such that for all* $a, b, c \in S$: $(a \cdot b) \cdot c = a \cdot (b \cdot c)$ *and* $a \cdot e = a = e \cdot a$. *The monoid is* **commutative** *if for all* $a, b \in S$: $a \cdot b = b \cdot a$.

Definition 2.5. *A* **semiring** *is a tuple* $\mathcal{S} = \langle S, \oplus, \odot, \mathbf{0}, \mathbf{1} \rangle$ *where* S *is a set,*

(i) $\langle S, \oplus, \mathbf{0} \rangle$ *is a commutative monoid,*
(ii) $\langle S, \odot, \mathbf{1} \rangle$ *is a monoid,*
(iii) $\mathbf{0}$ *is an annihilator for* $\odot$: *for all* $a \in S$ *we have* $\mathbf{0} \odot a = \mathbf{0} = a \odot \mathbf{0}$, *and*
(iv) $\odot$ *distributes over* $\oplus$: *for all* $a, b, x \in S$ *we have* $(a \oplus b) \odot x = (a \odot x) \oplus (b \odot x)$ *and* $x \odot (a \oplus b) = (x \odot a) \oplus (x \odot b)$.

The semiring $\mathcal{S}$ *is* **commutative** *if* $\langle S, \odot, \mathbf{1} \rangle$ *is commutative. We will often denote a semiring by its carrier set, writing* S *for* $\mathcal{S}$.

Definition 2.6. *A* **well-founded semiring** $\langle S, \oplus, \odot, \mathbf{0}, \mathbf{1}, \prec, \leq \rangle$ *consists of*

(i) *a semiring* $\langle S, \oplus, \odot, \mathbf{0}, \mathbf{1} \rangle$; *and*
(ii) *non-empty orders* $\prec, \leq \subseteq S \times S$ *for which* $\prec \subseteq \leq$ *and* $\leq$ *is reflexive;*

such that $\mathrm{SN}(\succ/\geq)$, $\mathbf{0} \neq \mathbf{1}$, *and for all* $x, x', y, y' \in S$ *we have*

$$
\begin{array}{rclr}
x \leq x' \wedge y \leq y' & \implies & x \oplus y \leq x' \oplus y' & \text{(S1)} \\
x \prec x' \wedge y \prec y' & \implies & x \oplus y \prec x' \oplus y' & \text{(S2)} \\
x \leq x' \wedge \mathbf{1} \leq y & \implies & x \odot y \leq x' \odot y & \text{(S3)} \\
x \prec x' \wedge \mathbf{1} \leq y \neq \mathbf{0} & \implies & x \odot y \prec x' \odot y & \text{(S4)}
\end{array}
$$

The semiring is **strictly monotonic** *if additionally*

$$x \prec x' \wedge y \leq y' \quad \Longrightarrow \quad x \oplus y \prec x' \oplus y' \tag{S5}$$

Example 2.7. The **arithmetic semiring** $\langle \mathbb{N}, +, \cdot, 0, 1, <, \leq \rangle$ is a strictly monotonic, well-founded semiring. The **tropical semiring** $\langle \mathbb{N} \cup \{\infty\}, \min, +, \infty, 0, < , \leq \rangle$ and **arctic semiring** $\langle \mathbb{N} \cup \{-\infty\}, \max, +, -\infty, 0, <, \leq \rangle$ are well-founded semirings, but not strictly monotone. (For details see [21, Examples B.1–B.3].)

3 Weighted Type Graphs

In this section, we introduce a notion of weighted type graphs for arbitrary categories, and we employ them to weigh morphisms (into the type graph) and objects in the category.

Definition 3.1. *A* **weighted type graph** $\mathcal{T} = (T, \mathbb{E}, S, \mathbf{w})$ *consists of:*

(i)   an object $T \in Ob(\mathbf{C})$, called a **type graph**,
(ii) a set $\mathbb{E}$ of T-valued elements $e : dom(e) \to T$,
(iii) a commutative semiring $\langle S, \oplus, \odot, \mathbf{0}, \mathbf{1} \rangle$, and
(iv) a **weight function** $\mathbf{w} : \mathbb{E} \to S$ *such that, for all $e \in \mathbb{E}$, $\mathbf{0} \neq \mathbf{w}(e) \geq \mathbf{1}$.*

The weighted type graph $\mathcal{T}$ is **finitary** *if, for every $e \in \mathbb{E}$ and every $G \in Ob(\mathbf{C})$,* $\mathrm{Hom}(dom(e), G)$ *and* $\mathrm{Hom}(G, T)$ *are finite sets.*

Remark 3.2. Bruggink et al. [8] introduce weighted type graphs for the category **Graph** of edge-labelled multigraphs. We generalize this concept to arbitrary categories. Important technical issues aside (discussed below), their notion is obtained by instantiating our notion of weighted type graphs for **Graph** with $\mathbb{E} = \bigcup \{ \mathrm{Hom}(\bullet \xrightarrow{x} \bullet, T) \mid x \text{ an edge label} \}$.

The technical differences are as follows. We require $\mathbf{w}(e) \geq \mathbf{1}$ and $\mathbf{w}(e) \neq \mathbf{0}$ for every $e \in \mathbb{E}$. The work [8] requires

– $\mathbf{w}(e) > \mathbf{0}$ only for loops $e \in \mathbb{E}$ on a distinguished 'flower node'.

It turns out that this condition is insufficient, causing [8, Theorem 2] to fail as illustrated by [21, Example A.1]. The authors of [8] have confirmed and solved this problem in their latest arXiv version [9].

Moreover, observe that $\mathbf{w}(e) \geq \mathbf{1}$ is often a weaker condition than $\mathbf{w}(e) > \mathbf{0}$. For instance, in the tropical semiring (see [21, Example B.2]), $\mathbf{0} < \mathbf{1}$ does not hold since there $\mathbf{0} = +\infty$ and $\mathbf{1} = 0$. The tropical semiring is one of the prime examples in [8], but, in fact, the condition $\mathbf{w}(e) > \mathbf{0}$ rules out the use of the tropical semiring.

Notation 3.3. *We use the following notation:*

(a) For $\alpha : A \to B$, $x : A \to C$, let $\{ - \circ \alpha = x \} = \{ \phi : B \to C \mid \phi \circ \alpha = x \}$;

(b) For $\alpha : B \to C$, $x : A \to C$, let $\{\alpha \circ - = x\} = \{\phi : A \to B \mid \alpha \circ \phi = x\}$;
(c) For $\alpha : B \to C$ and a set S of morphisms $\phi : A \to B$, we define $\alpha \circ S = \{\alpha \circ \phi \mid \phi \in S\}$.

Definition 3.4 (Weighing morphisms and objects). *Let $\mathcal{T} = (T, \mathbb{E}, S, \mathbf{w})$ be a finitary weighted type graph and let $G, A \in Ob(\mathbf{C})$.*

*(i) The **weight of a morphism** $\phi : G \to T$ is $\mathbf{w}_{\mathcal{T}}(\phi) = \bigodot_{e \in \mathbb{E}} \mathbf{w}(e)^{\#\{\phi \circ - = e\}}$.*

This is equivalent to $\mathbf{w}_{\mathcal{T}}(\phi) = \bigodot\limits_{\substack{e \in \mathbb{E} \\ \alpha \in \{\phi \circ - = e\}}} \mathbf{w}(e)$; visually
$$dom(e) \overset{\alpha}{-\!\!-\!\!\to} G \\ {}_{e}\searrow \overset{=}{} \swarrow {}_{\phi} \\ T$$

*(ii) The **weight of finite** $\Psi \subseteq \{\phi \mid \phi : G \to T\}$ is $\mathbf{w}_{\mathcal{T}}(\Psi) = \bigoplus_{\phi \in \Psi} \mathbf{w}_{\mathcal{T}}(\phi)$.*
*(iii) The **weight of an object** $G \in Ob(\mathbf{C})$ is $\mathbf{w}_{\mathcal{T}}(G) = \mathbf{w}_{\mathcal{T}}(\mathrm{Hom}(G, T))$.*
*(iv) The **weight of** $\phi : G \to T$ **excluding** $(\alpha \circ -)$, for $\alpha : A \to G$, is*

$$\mathbf{w}_{\mathcal{T}}(\phi - (\alpha \circ -)) = \bigodot_{e \in \mathbb{E}} \mathbf{w}(e)^{\#\{\tau \in \{\phi \circ - = e\} \mid \nexists \zeta. \ \tau = \alpha \circ \zeta\}}$$

In other words, we count all τ such that for all $\zeta : dom(e) \to A$, we have the following:

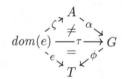

Observe the analogy to weighted automata where the weight of a word is the sum of the weights of all runs, and the weight of a run is the product of the edge weights.

4   Weighing Pushouts

In this section, we introduce techniques for determining exact and upper bounds on the weights of pushout objects in terms of the corresponding pushout span. For these techniques to work, we need certain assumptions on the pushout squares. This concerns, in particular, traceability of elements (Sect. 4.1) and relative monicity conditions (Sect. 4.2), which are used to define the notion of weighable pushout squares (Sect. 4.3). Using the notion of a weighable pushout square, we can bound weights of morphisms out of the pushout object (Sect. 4.4), and consequently bound the weight of the pushout object itself (Sect. 4.5).

4.1   Traceability

We introduce the concept of traceability of elements $X \in Ob(\mathbf{C})$ along pushout squares. Roughly speaking, traceability guarantees that the pushout cannot create occurrences of X out of thin air.

Definition 4.1 (Traceability). *Let Δ be a class of pushout squares. An object $X \in Ob(\mathbf{C})$ is called* **traceable along** Δ *if: whenever we have a configuration as displayed on the right, such that the displayed square is from Δ, then*

$$
\begin{array}{ccc}
A & \xrightarrow{\ \alpha\ } & B \\
\scriptstyle\beta\downarrow & \Delta & \downarrow\scriptstyle\beta' \\
C & \xrightarrow{\ \alpha'\ } & D \\
& & \uparrow\scriptstyle f \\
& & X
\end{array}
$$

(a) there exists a $g : X \to B$ such that $f = \beta'g$, or
(b) there exists an $h : X \to C$ such that $f = \alpha'h$.

If additionally, whenever (a) and (b) hold together, then also

(c) there exists an $i : X \to A$ such that $f = \beta'\alpha i$,

then we say that X is **strongly traceable along** Δ.

In $\mathcal{M}$-adhesive categories (for $\mathcal{M}$ a stable class of monos) [19, Definition 2.4], pushouts along $\mathcal{M}$-morphisms are also pullbacks [28, Lemma 13]. This is a common setting for graph transformation. In this case, the notions of traceability and strong traceability coincide for pushouts along $\mathcal{M}$-morphisms, as the proposition below shows.

Proposition 4.2. *If Δ is a class of pushouts that are also pullbacks, then traceability along Δ implies strong traceability along Δ (for $X \in Ob(\mathbf{C})$).*

Remark 4.3. (Traceability in **Graph***).* In **Graph**, the objects $\bullet$ and $\bullet \xrightarrow{x} \bullet$ (for edge labels x) are the only (non-initial) objects that are (strongly) traceable along all pushout squares. Other objects, such as loops $\bullet\circlearrowright$, are strongly traceable if all morphisms in the square are monomorphisms.

Remark 4.4. (Traceability in **SGraph***).* In the category of simple graphs **SGraph**, object $\bullet$ is strongly traceable along all pushout squares. An object $X = \bullet \xrightarrow{x} \bullet$ (for an edge label x) is traceable along all pushout squares, but not necessarily strongly traceable.

The pushout on the right is a counterexample. However, using Proposition 4.2, we have that X is strongly traceable along pushout squares in which one of the span morphisms α or β is a regular monomorphism, because such pushouts are pullbacks in **SGraph** (and more generally in quasitoposes).

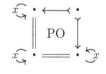

Remark 4.5. As is well known, the category of unlabeled graphs is equivalent to the functor category $[E \rightrightarrows V , \mathbf{Set}]$, where index category $E \rightrightarrows V$ is the category consisting of two objects E and V and two non-identity arrows that are parallel. This makes the category of unlabeled graphs a presheaf category, and the two representable functors for this category are precisely the one node graph $\cdot$ and the one edge graph $\cdot \to \cdot$ (see [40, Sect. 3] for more detail). We identified these two graphs as non-initial objects that are strongly traceable along any pushout (Remark 4.3). Our conjecture is that for any presheaf category for which the index category is small and acyclic, the non-initial strongly traceable objects

for any pushout are precisely the representable functors. For many examples of such presheaf categories, see the overview provided by Löwe [31, Sect. 3.1] (there called graph structures).

4.2 Relative Monicity

Traceability suffices to derive an upper bound on the weight of a pushout. However, we need additional monicity assumptions for lower or exact bounds.

Definition 4.6 (Relative monicity). *A morphism $f : A \to B$ is called*

(i) **monic for** $S \subseteq \mathrm{Hom}(-, A)$ *if* $fg = fh \implies g = h$ *for all* $g, h \in S$,
(ii) X**-monic,** *where* $X \in Ob(\mathbf{C})$, *if* f *is monic for* $\mathrm{Hom}(X, A)$, *and*
(iii) X**-monic outside of** u, *where* $X \in Ob(\mathbf{C})$ *and* $u : C \to A$, *if* f *is monic for* $\mathrm{Hom}(X, A) - u \circ \mathrm{Hom}(X, C)$.

For $\Gamma \subseteq Ob(\mathbf{C})$, f *is called* Γ**-monic** *if* f *is* X-monic for every $X \in \Gamma$. When Γ is clear from the context, $A \longleftrightarrow B$ denotes a Γ-monic morphism from A to B.

Example 4.7 (Edge-Monicity). In **Graph**, let $\Gamma = \{ \cdot \xrightarrow{x} \cdot \mid x \text{ is an edge label} \}$. A morphism $f : G \to H$ that does not identify edges (but may identify nodes) is not necessarily monic, but it is Γ-monic. In this particular case for **Graph**, we will say that f is **edge-monic**.

4.3 Weighable Pushout Squares

The following definition introduces an orientation for pushout squares, distinguishing between horizontal and vertical morphisms. A double pushout rewrite step with respect to a rule consists of two oriented pushout squares. For both squares, β is the shared morphism, and α will be l or r of the rule span, respectively. For the left square, β' is the match morphism. The distinction between horizontal and vertical morphisms will be used in Definition 4.9 below to impose different conditions on these morphisms.

Definition 4.8 (Oriented pushout squares). *An oriented pushout square*

τ *is a pushout square (as shown on the right) with a fixed orientation. We distinguish the morphisms into horizontal and vertical. Morphisms β and β' are* **vertical** *and α and α' are* **horizontal.**

$$\begin{array}{ccc} A & \xrightarrow{\alpha} & B \\ \downarrow{\scriptstyle\beta} & \mathrm{PO} & \downarrow{\scriptstyle\beta'} \\ C & \xrightarrow{\alpha'} & D \end{array}$$

We denote the pushout square by $\langle C \xleftarrow{\beta} A \xrightarrow{\alpha} B; C \xrightarrow{\alpha'} D \xleftarrow{\beta'} B \rangle$ where the order of the morphisms in the span is vertical before horizontal, and the order of the morphisms in the cospan is horizontal before vertical.

The following definition summarizes the conditions required to derive an upper or exact bound on the weight of a pushout object.

Definition 4.9 (Weighable pushout square). *Let* $\mathcal{T} = (T, \mathbb{E}, S, \mathbf{w})$ *be a finitary weighted type graph. Let* δ *be the oriented pushout square from Definition 4.8. We say that* δ *is*

(a) **weighable** *with* $\mathcal{T}$ *if:*
    (i) $dom(\mathbb{E})$ *is strongly traceable along* δ,
    (ii) β' *is* $dom(\mathbb{E})$-*monic, and*
    (iii) α' *is* $dom(\mathbb{E})$-*monic outside of* β.
(b) **bounded above** *by* $\mathcal{T}$ *if* $dom(\mathbb{E})$ *is traceable along* δ.

4.4 Weighing Morphisms Rooted in Pushout Objects

Consider the pushout diagram in Lemma 4.10, below. The lemma decomposes the weight $\mathbf{w}(\phi)$ of the morphism ϕ, rooted in the pushout object, into the product of the weights $\mathbf{w}(\phi \circ \beta')$ and $\mathbf{w}(\phi \circ \alpha' - (\beta \circ -))$. Intuitively, the reasoning is as follows. Let $e \in \mathbb{E}$ be a weighted element. Assume that $e : dom(e) \to T$ occurs in ϕ, that is, there exists $d : dom(e) \to D$ with $e = \phi \circ d$. Then traceability guarantees that this occurrence d can be traced back to $\phi \circ \beta'$ or $\phi \circ \alpha'$. In case the occurrence d traces back all the way to A (to $\phi \circ \beta' \circ \alpha = \phi \circ \alpha' \circ \beta$), then it occurs in both $\phi \circ \beta'$ and $\phi \circ \alpha'$. To avoid unnecessary double counting, we exclude those elements from $\phi \circ \alpha'$, giving rise to $\mathbf{w}(\phi \circ \alpha' - (\beta \circ -))$. In this way, we obtain the following upper bound on the weight of ϕ:

$$\mathbf{w}(\phi) \leq \mathbf{w}(\phi \circ \beta') \odot \mathbf{w}(\phi \circ \alpha' - (\beta \circ -))$$

To obtain the precise weight we need to avoid double counting. First, we need strong traceability to exclude all double counting between $\phi \circ \beta'$ and $\phi \circ \alpha' - (\beta \circ -)$. Second, we need $dom(\mathbb{E})$-monicity of α' outside of β and $dom(\mathbb{E})$-monicity of β' to avoid double counting along these morphisms.

Lemma 4.10 (Weighing morphisms rooted in pushout objects). *Let a finitary weighted type graph* $\mathcal{T} = (T, \mathbb{E}, S, \mathbf{w})$ *be given. Consider an oriented pushout square* δ:

$$\begin{array}{ccc} A & \!-\alpha\!\to\! & B \\ \beta\downarrow & \delta & \downarrow\beta' \\ C & \!-\alpha'\!\to\! D & \!-\phi\!\to\! T \end{array}$$

We define $k = \mathbf{w}_{\mathcal{T}}(\phi \circ \beta') \odot \mathbf{w}_{\mathcal{T}}(\phi \circ \alpha' - (\beta \circ -))$. *Then the following holds:*

(A) We have $\mathbf{w}_{\mathcal{T}}(\phi) = k$ *if* δ *is weighable with* $\mathcal{T}$.
(B) We have $\mathbf{w}_{\mathcal{T}}(\phi) \leq k$ *if* δ *is bounded above by* $\mathcal{T}$.

Remark 4.11 (Related work). Lemma 4.10 is closely related to Lemma 2 of [8] which establishes $\mathbf{w}(\phi) = \mathbf{w}(\phi \circ \beta') \odot \mathbf{w}(\phi \circ \alpha')$ for directed, edge-labelled multigraphs under the following assumptions:

(i) the interface A must be discrete (contains no edges),
(ii) the weighted type graph has only edge weights (no node weights).

As a consequence of the discrete interface A, all arrows involved in the pushout are edge-monic (Example 4.7).

Lemma 4.10 generalizes Lemma 2 of [8] in multiple directions:

(a) it works for arbitrary categories with some (strongly) traceable $dom(\mathbb{E})$,
(b) it does not require discreteness ($dom(\mathbb{E})$-freeness) of the interface, and
(c) it allows for α and α' that are not $dom(\mathbb{E})$-monic.

Both (a) and (b) are crucial to make the termination technique applicable in general categories, and (c) is important to allow for rewrite rules whose right morphism merges elements that we are weighing.

4.5   Weighing Pushout Objects

The following lemma (cf. [8, Lemma 1]) is a direct consequence of the universal property of pushouts. We will use this property for weighing pushout objects:

$$\begin{array}{ccc} A & \overset{\alpha}{\longrightarrow} & B \\ \beta \downarrow & \delta & \downarrow \beta' \\ C & \underset{\alpha'}{\longrightarrow} & D \end{array} \tag{2}$$

Lemma 4.12 (Pushout morphisms). *Let $T \in Ob(\mathbf{C})$ and consider a pushout square as in (2). For every morphism $t_A : A \to T$, there is a bijection θ from*

(a) the class U of morphisms $t_D : D \to T$ for which $t_A = t_D \circ \beta' \circ \alpha$, to
(b) the class P of pairs $(t_B : B \to T, t_C : C \to T)$ of morphisms for which $t_A = t_B \circ \alpha = t_C \circ \beta$,

such that for all $t_D \in U$ and pairs $(t_B, t_C) \in P$, $\theta(t_D) = (t_B, t_C) \iff t_B = t_D \circ \beta'$ and $t_C = t_D \circ \alpha'$.

Lemma 4.13 (Weighing pushout objects). *Let $\mathcal{T} = (T, \mathbb{E}, S, \mathbf{w})$ be a finitary weighted type graph. Consider an oriented pushout square δ as in (2). Define*

$$k = \bigoplus_{t_A : A \to T} \left(\bigoplus_{\substack{t_C : C \to T \\ t_A = t_C \circ \beta}} \mathbf{w}(t_C - \beta) \right) \odot \left(\underbrace{\bigoplus_{\substack{t_B : B \to T \\ t_A = t_B \circ \alpha}} \mathbf{w}(t_B)}_{= \mathbf{w}(\{ - \circ \alpha = t_A \})} \right)$$

Then the following holds:

(A) We have $\mathbf{w}_{\mathcal{T}}(D) = k$ if δ is weighable with $\mathcal{T}$.
(B) We have $\mathbf{w}_{\mathcal{T}}(D) \leq k$ if δ is bounded above by $\mathcal{T}$.

5   Termination via Weighted Type Graphs

In the previous section, we have developed techniques for weighing pushout objects. Our next goal is to apply these techniques to the pushout squares of double pushout diagrams in order to prove that rewrite steps reduce the weight.

As we have already noted in Remark 2.2, there exist many variations of the general concept of double pushout graph transformation. To keep our treatment as general as possible, we introduce in Sect. 5.1 an abstract concept of a double pushout framework $\mathfrak{F}$ which maps rules $\rho : L \leftarrow K \rightarrow R$ to classes $\mathfrak{F}(\rho)$ of double pushout diagrams with top span ρ.

In order to prove that rewrite steps decrease the weight, we need to ensure that every object that can be rewritten admits some morphism into the weighted type graph. To this end, we introduce the concept of context closures of rules in Sect. 5.2. In Sect. 5.3, we formulate criteria on rewrite rules (Definition 5.9) that ensure that the induced rewrite steps are decreasing (see [21, Theorem C.3]). Finally, we state our main theorem (Theorem 5.11) for proving (relative) termination.

5.1   Double Pushout Frameworks

Definition 5.1. *A* **double pushout diagram** δ *is a diagram of the form*

$$
\begin{array}{ccccc}
L & \leftarrow l- & K & -r\rightarrow & R \\
\downarrow m & \text{PO} & \downarrow u & \text{PO} & \downarrow w \\
G & \leftarrow l'- & C & -r'\rightarrow & H
\end{array}
$$

This diagram δ *is a* **witness** *for the* **rewrite step** $G \Rightarrow^{\delta} H$. *We use the following notation for the left and the right square of the diagram, respectively:*

$$
\mathit{left}(\delta) = \langle\, C \xleftarrow{u} K \xrightarrow{l} L; \; C \xrightarrow{l'} G \xleftarrow{m} L \,\rangle
$$

$$
\mathit{right}(\delta) = \langle\, C \xleftarrow{u} K \xrightarrow{r} R; \; C \xrightarrow{r'} H \xleftarrow{w} R \,\rangle
$$

Both $\mathit{left}(\delta)$ *and* $\mathit{right}(\delta)$ *are oriented pushout squares (where* u *is vertical). For a class* Δ *of double pushout diagrams, we define* $\mathit{left}(\Delta) = \{\, \mathit{left}(\delta) \mid \delta \in \Delta \,\}$ *and* $\mathit{right}(\Delta) = \{\, \mathit{right}(\delta) \mid \delta \in \Delta \,\}$.

Definition 5.2 (Double pushout framework). *A* **double pushout framework** $\mathfrak{F}$ *is a mapping of DPO rules to classes of DPO diagrams such that, for every DPO rule* ρ, $\mathfrak{F}(\rho)$ *is a class of DPO diagrams with top-span* ρ.

The rewrite relation $\Rightarrow_{\rho,\mathfrak{F}}$ *induced by a DPO rule* ρ *in* $\mathfrak{F}$ *is defined as follows:* $G \Rightarrow_{\rho,\mathfrak{F}} H$ *iff* $G \Rightarrow^{\delta} H$ *for some* $\delta \in \mathfrak{F}(\rho)$. *The rewrite relation* $\Rightarrow_{\mathfrak{S},\mathfrak{F}}$ *induced by a set* $\mathfrak{S}$ *of DPO rules is given by:* $G \Rightarrow_{\mathfrak{S},\mathfrak{F}} H$ *iff* $G \Rightarrow_{\rho,\mathfrak{F}} H$ *for some* $\rho \in \mathfrak{S}$. *When* $\mathfrak{F}$ *is clear from the context, we suppress* $\mathfrak{F}$ *and write* $\Rightarrow_{\rho}$ *and* $\Rightarrow_{\mathfrak{S}}$.

5.2   Context Closures

For proving termination, we need to establish that every rewrite step causes a decrease in the weight of the host graph. In particular, we need to ensure that every host graph admits some morphism into the weighted type graph. For this purpose, we introduce the concept of a context closure. Intuitively, a context closure for a rule $L \leftarrow K \rightarrow R$ is a morphism $c : L \rightarrow T$ into the type graph T in such a way that every match $L \rightarrow G$ can be mapped 'around c'.

Definition 5.3 (Context closure). *Let $\mathcal{A}$ be a class of morphisms. A **context closure** of an object $L \in Ob(\mathbf{C})$ for class $\mathcal{A}$ is an arrow $c : L \rightarrow T$ such that, for every arrow $\alpha : L \rightarrow G$ in $\mathcal{A}$, there exists a morphism $\beta : G \rightarrow T$ with $c = \beta \circ \alpha$. This can be depicted as*

$$
\begin{array}{ccc}
L & \xrightarrow{\ \alpha\ } & G \\
 & \searrow_{c} \ {=}\ \swarrow_{\beta} & \\
 & T &
\end{array}
$$

Remark 5.4 (Partial map classifiers as context closures). The notion of a context closure is strictly weaker than a partial map classifier. If category $\mathbf{C}$ has an $\mathcal{M}$-partial map classifier (T, η) (see [1, Definition 28.1] and [14, Definition 5]) for a stable class of monics $\mathcal{M}$, then the $\mathcal{M}$-partial map classifier arrow $\eta_X : X \hookrightarrow T(X)$ for an object X is a context closure of X for $\mathcal{M}$, because for any $\alpha : X \hookrightarrow Z$ in $\mathcal{M}$ there exists a unique $\beta : Z \rightarrow T(X)$ making the following square a pullback:

$$
\begin{array}{ccc}
X & = & X \\
\eta_X \downarrow & \text{PB} & \downarrow \alpha \\
T(X) & \xleftarrow{\ \beta\ } & Z
\end{array}
$$

Definition 5.5 (Context closure for a rule). *Let $\mathfrak{F}$ be a DPO framework and ρ a DPO rule $L \leftarrow K \rightarrow R$. A **context closure** for ρ and T (in $\mathfrak{F}$) is a context closure $c : L \rightarrow T$ for the class of match morphisms occurring in $\mathfrak{F}(\rho)$.*

Remark 5.6 (Context closures for unrestricted matching). Consider a DPO rule $\rho = L \leftarrow K \rightarrow R$ in an arbitrary DPO framework $\mathfrak{F}$ (in particular, matching can be unrestricted). Every morphism $c : L \rightarrow 1 \rightarrow T$, that factors through the terminal object 1, is a context closure for ρ and T in $\mathfrak{F}$.

This particular choice of a context closure generalizes the flower nodes from [8] which are defined specifically for the category of edge-labelled multi-graphs. A flower node [8] in a graph T is just a morphism $1 \rightarrow T$.

When employing such context closures, we establish termination with respect to unrestricted matching. This is bound to fail if termination depends on the matching restrictions.

Remark 5.7 (Increasing the power for monic matching). For rules $L \xleftarrow{l} K \xrightarrow{r} R$ with monic matching, it is typically fruitful to choose a type graph T and context closure $c : L \rightarrow T$ in such a way that c avoids collapsing $l(K)$ in L. In other words, the context closure c should be monic for $\{\, l \circ h \mid h \in \mathrm{Hom}(-, K)\,\}$. In the case that l is monic, this is equivalent to $c \circ l$ being monic.

For **Graph**, the intuition is as follows. When collapsing nodes of $l(K)$ along c, we actually prove a stronger termination property than intended. We then prove termination even if matches $m : L \to G$ are allowed to collapse those nodes.

Example 5.8 (Context closure for monic matching). Let ρ be the DPO rule

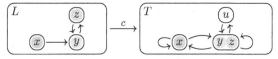

from [33, Example 55] in **Graph** with a framework $\mathfrak{F}$ with monic matching.

Following Remark 5.7, we construct a context closure $c : L \to T$ for ρ that avoids collapsing $l(K)$ of L. Since $l(K) \cong K$, this means that the complete graph around the nodes of K must be a subobject of T. We pick

Indeed, for every monic match $m : L \rightarrowtail G$ there exists a morphism $\alpha : G \to T$ such that $c = \alpha \circ m$. Thus c is a context closure for ρ and T in $\mathfrak{F}$.

5.3 Proving Termination

In this section we prove our main theorem (Theorem 5.11) for using weighted type graphs to prove termination of graph transformation systems.

First, we introduce different notions of decreasing rules (Definition 5.9).[1] Then, we state our technique for proving (relative) termination (Theorem 5.11).

The following definition introduces the notions of weakly, uniformly and closure decreasing rules. Weakly decreasing rules never increase the weight of the host graph. Uniformly decreasing rules reduce the weight of the host graph for every mapping of the host graph onto the type graph. If the underlying semiring is strictly monotonic, then closure decreasingness suffices: the weight of the host graph decreases for those mappings that correspond to the context closure of the rule (and is non-increasing for the other mappings).

Definition 5.9 (Decreasing rules). *Let $\mathfrak{F}$ be a DPO framework. Moreover, let $\mathcal{T} = (T, \mathbb{E}, S, \mathbf{w})$ be a finitary weighted type graph over a well-founded commutative semiring $\langle S, \oplus, \odot, \mathbf{0}, \mathbf{1}, \prec, \leq \rangle$. A DPO rule $\rho : L \xleftarrow{l} K \xrightarrow{r} R$ is called*

*(i) **weakly decreasing** (w.r.t. $\mathcal{T}$ in $\mathfrak{F}$) if*
$- \mathbf{w}(\{ - \circ l = t_K \}) \geq \mathbf{w}(\{ - \circ r = t_K \})$ *for every $t_K : K \to T$;*
*(ii) **uniformly decreasing** (w.r.t. $\mathcal{T}$ in $\mathfrak{F}$) if*
  - for every $t_K : K \to T$:
  $\mathbf{w}(\{ - \circ l = t_K \}) \succ \mathbf{w}(\{ - \circ r = t_K \})$ *or*
  $\{ - \circ l = t_K \} = \varnothing = \{ - \circ r = t_K \}$, *and*

---

[1] In the extended version [21], we prove that rewrite steps arising from these rules cause a decrease in the weight of the host graphs (see [21, Theorem C.3]).

  – *there exists a context closure $\mathfrak{c}_\rho$ for ρ and T in $\mathfrak{F}$;*

If semiring S is moreover strictly monotonic, we say that ρ is

(iii) **closure decreasing** *(w.r.t. T in $\mathfrak{F}$) if*
  – ρ *is weakly decreasing,*
  – *there exists a context closure $\mathfrak{c}_\rho$ for ρ and T in $\mathfrak{F}$, and*
  – $\mathbf{w}(\{-\circ l = t_K\}) \succ \mathbf{w}(\{-\circ r = t_K\})$ *for $t_K = \mathfrak{c}_\rho \circ l$.*

By saying that a rule is closure decreasing, we tacitly assume that the semiring is strictly monotonic (for otherwise the concept is not defined). Note that if S is strictly monotonic, uniformly decreasing implies closure decreasing.

Remark 5.10. Our notions of uniformly and closure decreasing rules generalize the corresponding concepts (strongly and strictly decreasing rules) in [8,10]. The techniques in [8] fail to generalize [10] as explained in [21, Remark A.2].

We now state our main theorem.

Theorem 5.11 (Proving termination). *Let $\mathfrak{F}$ be a DPO framework and let $\mathfrak{S}_1$ and $\mathfrak{S}_2$ be sets of double pushout rules. Let $\langle S, \oplus, \odot, \mathbf{0}, \mathbf{1}, \prec, \leq \rangle$ be a well-founded commutative semiring, and let $\mathcal{T} = (T, \mathbb{E}, S, \mathbf{w})$ be a finitary weighted type graph such that*

  – *left(δ) is weighable with $\mathcal{T}$, and*
  – *right(δ) is bounded above by $\mathcal{T}$*

for every rule $\rho \in (\mathfrak{S}_1 \cup \mathfrak{S}_2)$ and double pushout diagram $\delta \in \mathfrak{F}(\rho)$. If

1. ρ *is weakly decreasing for every $\rho \in \mathfrak{S}_2$, and*
2. ρ *is uniformly or closure decreasing for every $\rho \in \mathfrak{S}_1$,*

then $\mathfrak{S}_1$ is terminating relative to $\mathfrak{S}_2$.

The proof of this theorem may be found in [21, Appendix C].

6   Examples

Notation 6.1 (Visual Notation). *We use the visual notation as in [33]. The vertices of graphs are non-empty sets $\{x_1, \ldots, x_n\}$ depicted by boxes $\boxed{x_1 \cdots x_n}$. We will choose the vertices of the graphs in such a way that the homomorphisms $h : G \to H$ are fully determined by $S \subseteq h(S)$ for all $S \in V_G$. For instance, in Example 6.3 below, the morphism $\mathfrak{c}_\rho : L \to T$ maps nodes $\{x\}, \{y\}, \{z\} \in L$ as follows: $\mathfrak{c}_\rho(\{x\}) = \{x\}$ and $\mathfrak{c}_\rho(\{y\}) = \mathfrak{c}_\rho(\{z\}) = \{y, z\} \in T$.*

We start with two examples (Examples 6.2 and 6.3) for which the approaches from [8,10] fail, as the systems are not terminating with unrestricted matching.

Example 6.2 (Loop unfolding with monic matching). Let ρ be the DPO rule

$$L \; \boxed{x \; \circlearrowleft \; y} \; \leftarrow l - \boxed{K \; x \qquad y} \; - r \rightarrow \boxed{R \; x \longrightarrow y}$$

in **Graph**, and let matches m restrict to monic matches. We prove termination using the weighted type graph $\boxed{T \quad 2 \subset x \rightleftarrows y \circlearrowleft 2}$ over the (strictly monotonic) arithmetic semiring. The type graph T is unlabelled. The numbers along the edges indicate the weighted elements $\mathbb{E}$. Here $\mathbb{E}$ consists of 2 morphisms with domain $\bullet \longrightarrow \bullet$ (see further Remark 4.3) and their weight is the number along the edge they target. So $\mathbb{E} = \{\, e_x, e_y \,\}$ where e_x and e_y map $\bullet \longrightarrow \bullet$ onto the loop on $\{x\}$ and the loop on $\{y\}$ in T, respectively. The weights of e_x and e_y are $\mathbf{w}(e_x) = \mathbf{w}(e_y) = 2$.

As the context closure $\mathfrak{c}_\rho$ for ρ and T we use the inclusion $\mathfrak{c}_\rho : L \rightarrowtail T$. Clearly, for any monic match $L \rightarrowtail G$, there exists a map $\alpha : G \to T$ such that $\mathfrak{c}_\rho = \alpha \circ m$. For $t_K = \mathfrak{c}_\rho \circ l$ we have $\mathbf{w}(\{\, - \circ l = t_K \,\}) = 2 \succ 1 = \mathbf{w}(\{\, - \circ r = t_K \,\})$ since the empty product is 1. Moreover, $\mathbf{w}(\{\, - \circ l = t_K \,\}) \geq \mathbf{w}(\{\, - \circ r = t_K \,\})$ for all other $t_K : K \to T$. Thus ρ is closure decreasing, and consequently terminating by Theorem 5.11.

Example 6.3 (Reconfiguration). Continuing Example 5.8, we prove termination of ρ with matches m restrict to monic matches. We use the weighted type graph shown on the right, over the (strictly monotonic) arithmetic semiring. This is the type graph constructed in Example 5.8, enriched with weighted elements.

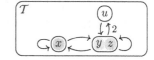

$\mathbb{E}$ consists of a single morphism with weight 2, mapping $\bullet \longrightarrow \bullet$ onto the edge from $\{y, z\}$ to $\{u\}$ in T.

As the context closure $\mathfrak{c}_\rho$ for ρ and T we use $\mathfrak{c}_\rho : L \to T$ as indicated by the node names ($\{y\}$ and $\{z\}$ are mapped onto $\{y, z\}$). For $t_K = \mathfrak{c}_\rho \circ l$ we then obtain $\mathbf{w}(\{\, - \circ l = t_K \,\}) = 1 + 1 + 2 \succ 1 + 1 = \mathbf{w}(\{\, - \circ r = t_K \,\})$ since the empty product is 1, and $\{z\}$ in L can be mapped onto $\{x\}$, $\{y, z\}$ and $\{u\}$, while $\{w\}$ in R can only be mapped onto $\{x\}$ and $\{y, z\}$ in T. Furthermore, $\mathbf{w}(\{\, - \circ l = t_K \,\}) \geq \mathbf{w}(\{\, - \circ r = t_K \,\})$ for all other $t_K : K \to T$. Thus ρ is closure decreasing, and hence terminating by Theorem 5.11.

We continue with an example (Example 6.4) of rewriting simple graphs. Here the techniques from [8, 10] are not applicable, because they are defined only for multigraphs. Even when considering the rule as a rewrite rule on multigraphs, the technique from [10] is not applicable due to non-monic r, and the technique from [8] fails because the rule is not terminating with respect to unrestricted matching.

Example 6.4 (Simple graphs and monic matching). Let ρ be the DPO rule

$$L \; \boxed{x \longrightarrow y} \; \leftarrow l - \boxed{K \; x \qquad y} \; - r \rightarrow \boxed{R \; x\,y \circlearrowleft \quad z \quad w}$$

in the category of simple graphs **SGraph**, and let matches m restrict to monic matches. This rule folds an edge into a loop, similar to [33, Example 49], but extended with the addition of two fresh nodes in the right-hand side. We prove termination using the weighted type graph $T = (T, \mathbb{E}, S, \mathbf{w})$ given by

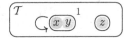

over the tropical semiring. So $\mathbb{E}$ consists of a single morphism e mapping $\bullet$ onto x in T, and $\mathbf{w}(e) = 1$. (See Remark 4.4 on strong traceability in **SGraph**.) As the context closure $\mathfrak{c}_\rho$ for ρ and T we use the morphism $\mathfrak{c}_\rho : L \to T$ as indicated by the node names ($\{x\}$ and $\{y\}$ are mapped onto $\{x, y\}$).

Then $\mathbf{w}(\{-\circ l = t_K\}) = 2 \succ 1 = 3 \oplus 2 \oplus 2 \oplus 1 = \mathbf{w}(\{-\circ r = t_K\})$ for $t_K = \mathfrak{c}_\rho \circ l$. For every other morphism $t_K : K \to T$ it holds that $\{-\circ l = t_K\} = \varnothing = \{-\circ r = t_K\}$. Thus ρ is uniformly decreasing, and Theorem 5.11 yields termination of ρ.

Example 6.5 (Reconfiguration on simple graphs). Example 6.3 can also be considered on **SGraph** with monic matching. Assume that we add in K an edge from $\{x\}$ to $\{y\}$ (this does not change the rewrite relation). Then l is regular monic, and we have strong traceability for the edges along the left pushout squares of rewrite steps. Hence all left squares are weighable, and all right squares are bounded above by the type graph in Example 6.3; so the same type graph proves also termination on **SGraph**.

Additional examples with unrestricted matching are in [21, Appendix D]:

- [21, Example D.1] treats [39, Example 3] and [36, Example 3.8] by Plump;
- [21, Example D.2] treats [8, Example 6] by Bruggink et al.;
- [21, Example D.3] shows that the technique also can be fruitfully applied for rules $\rho : L \leftarrow K \to R$ for which L is a subgraph of R, that is, $L \rightarrowtail R$.

Finally, [21, Example D.4] illustrates a limitation of our technique:

Open Question 6.6. *We conjecture that our technique cannot be applied for rules $\rho : L \xleftarrow{l} K \xrightarrow{r} R$ for which there exists an epimorphism $e : R \twoheadrightarrow L$ with $l = e \circ r$. Can the technique be enhanced to deal with such rules?*

In [21, Appendix E] we provide some benchmarks from a Scala implementation.

7   Related Work

The approaches [8,10] by Bruggink et al. are the most relevant to our paper. Both employ weighted type graphs for multigraphs to prove termination of graph transformation. We significantly strengthen the weighted type graphs technique for rewriting with monic matching and we generalize the techniques to arbitrary categories and different variants of DPO.

Plump has proposed two systematic termination criteria [36,39] for hypergraph rewriting with DPO. The approach in [36] is based on the concept of forward closures. The paper [39] introduces a technique for modular termination proofs based on sequential critical pairs. The modular approach can often simplify the termination arguments significantly. However, the smaller decomposed systems still require termination arguments. Therefore, our techniques and the approach from [39] complement each other perfectly. Our technique can handle all examples from the papers [36,39], except for one (see [21, Example D.4] for discussion of this system).

In [33], we have proposed a termination approach based on weighted element counting. Like the method presented in this paper, the method [33] is defined for general categories (more specifically, rm-adhesive quasitoposes), but for PBPO$^+$ [34,35] instead of DPO. PBPO$^+$ subsumes DPO in the quasitopos setting [34, Theorem 72] if both the left-hand morphism l of DPO rules and matches m are required to be regular monic. The weighted element counting approach defines the weight of an object G by means of counting weighted elements of the form $t : T \to G$, meaning it is roughly dual to the approach presented in this paper.

An example that the weighted type graph approach can prove, but the weighted element counting approach cannot prove, is [8, Example 6], which requires measuring a global property rather than a local one; see [21, Example D.2]. Conversely, [39, Example 6] is an example that seems unprovable using weighted type graphs, while the element counting approach can prove it rather straightforwardly [33, Example 39]; see further [21, Example D.4]. So it appears the two methods may be complementary even in settings where they are both defined.

The paper [30] proposes a criterion for termination of DPO rules with respect to injective matching. Roughly speaking, they show that a system is terminating if the size of repeated sequential compositions of rules tends to infinity. It remains to be seen if this criterion can be fruitfully automated.

The paper [6] introduces a framework for proving termination of high-level replacement units (these are systems with external control expressions). They show that their framework can be used for node and edge counting arguments which are subsumed by weighted type graphs (see further [8]).

There are also various techniques that generalize TRS methods to term graphs [32,37,38] and drags [17].

8   Future Work

There are several interesting directions for future research:

1. Can the weighted type graphs technique be extended to other algebraic graph transformation frameworks, such as SPO [31], SqPO [15], PBPO [13], AGREE [14] and PBPO$^+$ [34]?
2. Can the technique be enhanced to deal with examples like [21, Example D.4]? (See also Open Question 6.6.)
3. Can the technique be extended to deal with negative application conditions [25]?
4. Are there semirings other than arctic, tropical and arithmetic that can be fruitfully employed for termination proofs with the weighted type graph technique?

Concerning (1), an extension to PBPO$^+$ would be particularly interesting. In [34], it has been shown that, in the setting of quasitoposes, every rule of the other formalisms can be straightforwardly encoded as a PBPO$^+$ rule that generates the same rewrite relation[2]. However, our proofs (Lemma 4.13) crucially depend on a property specific to pushouts (Lemma 4.12). It will therefore be challenging to lift the reasoning to the pullbacks in PBPO$^+$ without (heavily) restricting the framework (to those pullbacks that are pushouts).

References

1. Adámek, J., Herrlich, H., Strecker, G.E.: Abstract and Concrete Categories: The Joy of Cats, vol. 17, pp. 1–507 (2006). http://www.tac.mta.ca/tac/reprints/articles/17/tr17.pdf
2. Apt, K.R., Pedreschi, D.: Reasoning about termination of pure prolog programs. Inf. Comput. **106**(1), 109–157 (1993). https://doi.org/10.1006/inco.1993.1051
3. Behr, N., Harmer, R., Krivine, J.: Concurrency theorems for non-linear rewriting theories. In: Gadducci, F., Kehrer, T. (eds.) ICGT 2021. LNCS, vol. 12741, pp. 3–21. Springer, Cham (2021). https://doi.org/10.1007/978-3-030-78946-6_1
4. Behr, N., Harmer, R., Krivine, J.: Fundamentals of compositional rewriting theory. CoRR abs/2204.07175 (2022). https://doi.org/10.48550/arXiv.2204.07175
5. Beyer, D.: Progress on software verification: SV-COMP 2022. In: Fisman, D., Rosu, G. (eds.) Tools and Algorithms for the Construction and Analysis of Systems, TACAS 2022. LNCS, vol. 13244, pp. 375–402. Springer, Cham (2022). https://doi.org/10.1007/978-3-030-99527-0_20
6. Bottoni, P., Hoffmann, K., Parisi-Presicce, F., Taentzer, G.: High-level replacement units and their termination properties. J. Vis. Lang. Comput. **16**(6), 485–507 (2005). https://doi.org/10.1016/j.jvlc.2005.07.001
7. Braatz, B., Golas, U., Soboll, T.: How to delete categorically - two pushout complement constructions. J. Symb. Comput. **46**(3), 246–271 (2011). https://doi.org/10.1016/j.jsc.2010.09.007
8. Bruggink, H.J.S., König, B., Nolte, D., Zantema, H.: Proving termination of graph transformation systems using weighted type graphs over semirings. In: Parisi-Presicce, F., Westfechtel, B. (eds.) ICGT 2015. LNCS, vol. 9151, pp. 52–68. Springer, Cham (2015). https://doi.org/10.1007/978-3-319-21145-9_4
9. Bruggink, H.J.S., König, B., Nolte, D., Zantema, H.: Proving termination of graph transformation systems using weighted type graphs over semirings (2023). arXiv:1505.01695v3
10. Bruggink, H.J.S., König, B., Zantema, H.: Termination analysis for graph transformation systems. In: Diaz, J., Lanese, I., Sangiorgi, D. (eds.) TCS 2014. LNCS, vol. 8705, pp. 179–194. Springer, Heidelberg (2014). https://doi.org/10.1007/978-3-662-44602-7_15
11. Cook, B., Podelski, A., Rybalchenko, A.: Termination proofs for systems code. In: Proceedings of Conference on Programming Language Design and Implementation, pp. 415–426. ACM (2006). https://doi.org/10.1145/1133981.1134029
12. Cook, B., Podelski, A., Rybalchenko, A.: Proving program termination. Commun. ACM **54**(5), 88–98 (2011). https://doi.org/10.1145/1941487.1941509

---

[2] The subsumption of SPO by PBPO$^+$ in quasitoposes remains a conjecture [34, Remark 23], but has been established for **Graph**.

13. Corradini, A., Duval, D., Echahed, R., Prost, F., Ribeiro, L.: The PBPO graph transformation approach. J. Log. Algebraic Methods Program. **103**, 213–231 (2019)
14. Corradini, A., Duval, D., Echahed, R., Prost, F., Ribeiro, L.: Algebraic graph rewriting with controlled embedding. Theor. Comput. Sci. **802**, 19–37 (2020). https://doi.org/10.1016/j.tcs.2019.06.004
15. Corradini, A., Heindel, T., Hermann, F., König, B.: Sesqui-pushout rewriting. In: Corradini, A., Ehrig, H., Montanari, U., Ribeiro, L., Rozenberg, G. (eds.) ICGT 2006. LNCS, vol. 4178, pp. 30–45. Springer, Heidelberg (2006). https://doi.org/10. 1007/11841883_4
16. Corradini, A., Montanari, U., Rossi, F., Ehrig, H., Heckel, R., Löwe, M.: Algebraic approaches to graph transformation - Part I: basic concepts and double pushout approach. In: Handbook of Graph Grammars and Computing by Graph Transformations, Volume 1: Foundations, pp. 163–246. World Scientific (1997)
17. Dershowitz, N., Jouannaud, J.: Graph path orderings. In: Proceedings of Conference on Logic for Programming, Artificial Intelligence and Reasoning (LPAR). EPiC Series in Computing, vol. 57, pp. 307–325. EasyChair (2018). https://doi. org/10.29007/6hkk
18. Ehrig, H., Ehrig, K., Prange, U., Taentzer, G.: Fundamentals of Algebraic Graph Transformation. MTCSAES, Springer, Heidelberg (2006). https://doi.org/ 10.1007/3-540-31188-2
19. Ehrig, H., Golas, U., Hermann, F.: Categorical frameworks for graph transformation and HLR systems based on the DPO approach. Bull. EATCS **102**, 111–121 (2010)
20. Ehrig, H., Pfender, M., Schneider, H.J.: Graph-grammars: an algebraic approach. In: Proceedings of Symposium on Switching and Automata Theory (SWAT), pp. 167–180. IEEE Computer Society (1973). https://doi.org/10.1109/SWAT.1973.11
21. Endrullis, J., Overbeek, R.: Generalized weighted type graphs for termination of graph transformation systems (2024). https://arxiv.org/abs/2307.07601v2
22. Giesl, J., et al.: Proving termination of programs automatically with AProVE. In: Demri, S., Kapur, D., Weidenbach, C. (eds.) IJCAR 2014. LNCS (LNAI), vol. 8562, pp. 184–191. Springer, Cham (2014). https://doi.org/10.1007/978-3-319-08587-6_13
23. Giesl, J., Raffelsieper, M., Schneider-Kamp, P., Swiderski, S., Thiemann, R.: Automated termination proofs for Haskell by term rewriting. ACM Trans. Program. Lang. Syst. **33**(2), 7:1–7:39 (2011). https://doi.org/10.1145/1890028.1890030
24. Giesl, J., Rubio, A., Sternagel, C., Waldmann, J., Yamada, A.: The termination and complexity competition. In: Beyer, D., Huisman, M., Kordon, F., Steffen, B. (eds.) TACAS 2019. LNCS, vol. 11429, pp. 156–166. Springer, Cham (2019). https://doi. org/10.1007/978-3-030-17502-3_10
25. Habel, A., Heckel, R., Taentzer, G.: Graph grammars with negative application conditions. Fundam. Informaticae **26**(3/4), 287–313 (1996). https://doi.org/10. 3233/FI-1996-263404
26. Habel, A., Müller, J., Plump, D.: Double-pushout graph transformation revisited. Math. Struct. Comput. Sci. **11**(5), 637–688 (2001). https://doi.org/10.1017/ S0960129501003425
27. König, B., Nolte, D., Padberg, J., Rensink, A.: A tutorial on graph transformation. In: Heckel, R., Taentzer, G. (eds.) Graph Transformation, Specifications, and Nets. LNCS, vol. 10800, pp. 83–104. Springer, Cham (2018). https://doi.org/10.1007/ 978-3-319-75396-6_5

28. Lack, S., Sobociński, P.: Adhesive categories. In: Walukiewicz, I. (ed.) FoSSaCS 2004. LNCS, vol. 2987, pp. 273–288. Springer, Heidelberg (2004). https://doi.org/10.1007/978-3-540-24727-2_20

29. Lee, C.S., Jones, N.D., Ben-Amram, A.M.: The size-change principle for program termination. In: Proceedings of Symposium on Principles of Programming Languages (POPL), pp. 81–92. ACM (2001). https://doi.org/10.1145/360204.360210

30. Levendovszky, T., Prange, U., Ehrig, H.: Termination criteria for DPO transformations with injective matches. In: Proceedings of Workshop on Graph Transformation for Concurrency and Verification. ENTCS, vol. 175, pp. 87–100. Elsevier (2006). https://doi.org/10.1016/J.ENTCS.2007.04.019

31. Löwe, M.: Algebraic approach to single-pushout graph transformation. Theor. Comput. Sci. **109**(1&2), 181–224 (1993). https://doi.org/10.1016/0304-3975(93)90068-5

32. Moser, G., Schett, M.A.: Kruskal's tree theorem for acyclic term graphs. In: Proceedings of Workshop on Computing with Terms and Graphs, TERMGRAPH. EPTCS, vol. 225, pp. 25–34 (2016). https://doi.org/10.4204/EPTCS.225.5

33. Overbeek, R., Endrullis, J.: Termination of graph transformation systems using weighted subgraph counting. CoRR abs/2303.07812 (2023). https://doi.org/10.48550/arXiv.2303.07812

34. Overbeek, R., Endrullis, J., Rosset, A.: Graph rewriting and relabeling with PBPO⁺: a unifying theory for quasitoposes. J. Logical Algebraic Methods Program. (2023). https://doi.org/10.1016/j.jlamp.2023.100873

35. Overbeek, R.: A unifying theory for graph transformation. Ph.D. thesis, Vrije Universiteit Amsterdam (2024). https://doi.org/10.5463/thesis.524

36. Plump, D.: On termination of graph rewriting. In: Nagl, M. (ed.) WG 1995. LNCS, vol. 1017, pp. 88–100. Springer, Heidelberg (1995). https://doi.org/10.1007/3-540-60618-1_68

37. Plump, D.: Simplification orders for term graph rewriting. In: Prívara, I., Ružička, P. (eds.) MFCS 1997. LNCS, vol. 1295, pp. 458–467. Springer, Heidelberg (1997). https://doi.org/10.1007/BFb0029989

38. Plump, D.: Term graph rewriting. In: Handbook of Graph Grammars and Computing by Graph Transformation: Volume 2: Applications, Languages and Tools, pp. 3–61 (1999). https://www-users.york.ac.uk/~djp10/Papers/tgr_survey.pdf

39. Plump, D.: Modular termination of graph transformation. In: Heckel, R., Taentzer, G. (eds.) Graph Transformation, Specifications, and Nets. LNCS, vol. 10800, pp. 231–244. Springer, Cham (2018). https://doi.org/10.1007/978-3-319-75396-6_13

40. Vigna, S.: A guided tour in the topos of graphs (2003). https://doi.org/10.48550/arxiv.math/0306394

41. Zantema, H.: Termination. In: Term Rewriting Systems, Cambridge Tracts in Theoretical Computer Science, vol. 55, pp. 181–259. Cambridge University Press (2003)

Extension and Restriction of Derivations in Adhesive Categories

Hans-Jörg Kreowski[1]([envelope]) [ID], Aaron Lye[2] [ID], and Aljoscha Windhorst[1,2] [ID]

[1] Department of Computer Science, University of Bremen, P.O.Box 33 04 40,
28334 Bremen, Germany
{kreo,windhorst}@uni-bremen.de
[2] Institute for the Protection of Maritime Infrastructures, German Aerospace Center
(DLR), Fischkai 1, 27572 Bremerhaven, Germany
{aaron.lye,aljoscha.windhorst}@dlr.de

Abstract. Extension and restriction of derivations are long-known operations in the framework of graph transformation. In this paper, we continue the study of extension and restriction on the higher level of the adhesive categories. A construction of extensions is provided by means of extension spans, and several properties of extensions are shown relating them, in particular, to restrictions.

1 Introduction

A good part of the research on graph transformation has been devoted to operations on derivations and their properties like, e.g., the local Church-Rosser theorems, parallelization, sequentialization, shift, canonical derivations, amalgamation and concurrent derivations. Extension and restriction are two further basic and long-known operations on derivations. In the special case of a derivation $w_0 \to w_1 \to \ldots \to w_n$ of strings in a Chomsky grammar, an extension adds a left and a right context x and y resp. to each string yielding $xw_0y \to xw_1y \to \ldots \to xw_ny$ while a restriction removes such common initial and final sections if they are not involved in any of the rule applications. In the special case of a rewrite sequence of terms or trees $\triangle_{t_0} \to \triangle_{t_1} \to \ldots \to \triangle_{t_n}$, extension adds an outer context yielding $\triangle_{t_0} \to \triangle_{t_1} \to \ldots \to \triangle_{t_n}$ while restriction removes such a common outer context. In the case of graphs, the situation is more complicated because there is no left and right, no outer and inner, and it is more difficult to avoid dangling edges. Given a graph-transformational derivation $G_0 \Longrightarrow G_1 \Longrightarrow \ldots \Longrightarrow G_n$ and a supergraph $\overline{G}_0$ of G_0, the derivation can be extended to a derivation $\overline{G}_0 \Longrightarrow \overline{G}_1 \Longrightarrow \ldots \Longrightarrow \overline{G}_n$ such that $G_i \subseteq \overline{G}_i$ and $\overline{G}_i \setminus G_i = \overline{G}_0 \setminus G_0$ for $i = 1, \ldots, n$ provided that the edges of the complement $\overline{G}_0 \setminus G_0$ are attached to vertices in $\overline{G}_0 \setminus G_0$ or to vertices of G_0 that are invariant in the derivation meaning that none of them is removed in any of the rule applications (see Ehrig, Pfender and Schneider [1], Ehrig and Kreowski [2,3] and Ehrig [4]). Conversely, the derivation can be restricted to a subgraph G_0' of G_0 if G_0' contains all vertices and edges of G_0 that are accessed by some matching

R. Harmer and J. Kosiol (Eds.): ICGT 2024, LNCS 14774, pp. 59–76, 2024.
https://doi.org/10.1007/978-3-031-64285-2_4

morphism of a given derivation (see Kreowski [5]). Beyond graph transformation in the narrow sense, Ehrig et al. [6,7] consider extension in the context of adhesive high-level replacement systems, restriction of a single rule application and restriction of an arbitrary derivation as inverse to extension. In [8], we introduce and study restrictions of derivations over adhesive categories including spines being minimal restrictions.

In this paper, we start a new attempt to investigate the concepts of extension and restriction of derivations in adhesive high-level replacement systems in a systematic and comprehensive way. After the preliminaries in Sect. 2, Sect. 3 is mainly devoted to a novel construction of extensions. It allows to choose an extension span that adds a larger context to the invariant part of a given derivation. The pushout of the extension span extends the invariant part, and a sequence of further pushouts extends the whole derivation. This is possible as the invariant part, which is defined as the intersection of all intermediate objects of the derivation, is embedded in all the derived and intermediate objects. At the end of the section, we show how an extension span induces the invariant part of the corresponding extension. In Sect. 4, we recall the constructions of the accessed part of a derivation, which covers the part of the start object that is needed for the matching of rules along the derivation, and of the restriction to a "subobject" (given by a monomorphism) of the start object provided that the accessed part factors through the "subobject". As a new result that is not yet known even for the category of graphs, we show that the invariant part and the accessed part of a derivation together cover the start object. Section 5 concludes the paper. Throughout the Sects. 2 to 4, a running example demonstrates the potentials of the interplay between extension and restriction in the redesign and remodeling of processes and computations from subprocesses and subcomputations of given processes and computations. The example is a colored variant of Sierpinski triangles taken from the area of pattern generation (see, e.g., [9,10]).

2   Preliminaries

In this section, the categorical prerequisites are provided. For the well-known categorical notions including initial objects, pullbacks, and pushouts and their basic properties confer, e.g., Adamek et al. [11] and Ehrig et al. [6,7]. The concepts and properties of adhesive categories as far as used in this paper are recalled explicitly. Many graph-like structures behave like the category of graphs. Capturing the required properties for rule-based transformation leads to the definition of adhesive categories.

A category $\mathbf{C}$ is adhesive if it has pullbacks and pushouts along monomorphisms (meaning that at least one of the two morphisms in the pushout span is a monomorphism) which are Van Kampen squares in addition. A Van Kampen square is a pushout of $A \leftarrow C \rightarrow B$ satisfying the following condition: Each commutative cube for which the pushout of $A \leftarrow C \rightarrow B$ forms the bottom face and the back faces are pullbacks, the front faces are pullbacks if and only if the top face is a pushout.

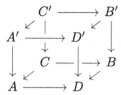

Besides the usual composition and decomposition properties of pullbacks and pushouts, some further nice properties hold in adhesive categories including the following facts (cf. [12]):

1. Monomorphisms are stable under pushouts.
2. Pushouts along monomorphisms are pullbacks.
3. The pushout complement of morphisms $m : K \to L$ and $f : L \to G$ is unique (up to isomorphism) whenever m is a monomorphism, i.e., there is an object Z and there are morphisms $z : K \to Z$ and $m' : Z \to G$ such that the square forms a pushout, and each other choice of Z, z, and m' differs only by an isomorphism.
4. The pushout-pullback decomposition property holds for a commuting diagram of the following shape:

$$
\begin{array}{ccc}
A & \longrightarrow B & \longrightarrow C \\
\downarrow & \downarrow & \downarrow \\
D & \longrightarrow E & \longrightarrow F
\end{array}
$$

   If the outer square is a pushout and the right square is a pullback, then the left square is a pushout.

   Pullbacks of monomorphisms and pushouts of their pullback morphisms have further very helpful properties. Consider the monomorphisms $a : A \to G$ and $b : B \to G$. The pullback $a' : A \cap B \to A$ and $b' : A \cap B \to B$ of a and b providing the morphism $a \cap b : A \cap B \to G$ with $a \cap b = a' \circ a = b' \circ b$ is called *intersection* of a and b. And the pushout $a'' : A \to G$ and $b'' : B \to G$ of a' and b' is also called *union* of a and b providing the morphism $a \cup b : A \cap B \to A \cup B$ with $a \cup b = a'' \circ a' = b'' \circ b'$. If there is a monomorphism $m : A \to B$ with $b \circ m = a$, then m is unique so that one may write $a \subseteq b$. The use of the intersection symbol $\cap$, the union symbol $\cup$ and the subobject symbol $\subseteq$ is justified as the pullbacks of monomorphisms and the pushouts of such pullbacks can be constructed by intersections and unions of subobjects respectively in many typical adhesive categories like **Sets, Graphs, Hypergraphs**, etc. Moreover, several properties hold that are typical for intersections, unions and subobjects using a further monomorphism $c : C \to G$:

(commutativity) $a \cap b = b \cap a$ and $a \cup b = b \cup a$,

(associativity) $a \cap (b \cap c) = (a \cap b) \cap c$ and $a \cup (b \cup c) = (a \cup b) \cup c$,

(distributivity) $a \cap (b \cup c) = (a \cap b) \cup (a \cap c)$ and $a \cup (b \cap c) = (a \cup b) \cap (a \cup c)$,

(subproperty) $a \subseteq b$ implies $a \cap b = a$ and $a \cup b = b$.

In all cases, the equality is up to isomorphism, i.e., $a = b$ if and only if there is an isomorphism i with $b \circ i = a$. As a denotational simplification, we may replace the names of the monomorphisms by the domains if the monomorphisms are clear from the context. This means for the properties above that a, b, c may be replaced by A, B, C.

The rewriting formalism for graphs and graph-like structures which we use throughout this paper is the double-pushout (DPO) approach as introduced by Ehrig, Pfender and Schneider in [1]. It was originally introduced for graphs. However, it is well-defined in many graph-like categories, more specifically, in adhesive categories.

A *rule* is a span of monomorphisms $p = (L \xleftarrow{l} K \xrightarrow{r} R)$. L and R are called *left-hand side* and *right-hand side*, K is called *gluing object*.

A rule application to some object G is defined wrt a morphism $g \colon L \to G$ which is called *(left) matching morphism*. G directly derives H if p and g extend to the diagram

$$
\begin{array}{ccccc}
L & \xleftarrow{\ l\ } & K & \xrightarrow{\ r\ } & R \\
{\scriptstyle g}\downarrow & (1) & {\scriptstyle z}\downarrow & (2) & {\scriptstyle h}\downarrow \\
G & \xleftarrow{m_{Z \to G}} & Z & \xrightarrow{m_{Z \to H}} & H
\end{array}
$$

such that both squares are pushouts. Z is called *intermediate object* and h is called *right matching morphism*.

The application of a rule p to G wrt g is called *direct derivation* and is denoted by $G \underset{p}{\Longrightarrow} X$ (where g is kept implicit). A *derivation* from G to X is a sequence of direct derivations $G_0 \underset{p_1}{\Longrightarrow} G_1 \underset{p_2}{\Longrightarrow} \cdots \underset{p_n}{\Longrightarrow} G_n$ with $G_0 = G$, $G_n = X$ and $n \geq 0$. If $p_1, \cdots, p_n \in P$, then the derivation is also denoted by $G \underset{P}{\overset{n}{\Longrightarrow}} X$. If the length of the derivation does not matter, we write $G \underset{P}{\overset{*}{\Longrightarrow}} X$. For a derivation d, the derived and intermediate objects together with the monic embeddings form the *derivation diagram diagr(d)*.

Example 1. Our running example demonstrates the concepts of extension and restriction as well as their interplay by means of derivations that allow to generate colored Sierpinski triangles. The underlying category, called **Trigraphs**, is the category of hypergraphs with hyperedges of type 3 that are called tripods. Trigraphs is defined quite similar to the category of graphs. A *trigraph* has a set of vertices and a set of tripods. Each tripod δ is attached to three vertices $one(\delta)$, $two(\delta)$ and $three(\delta)$ and has a label $l(\delta) \in \Sigma$. Morphisms are pairs of mappings between the sets of vertices and the sets of tripods resp. that are compatible with the attachments and labels. Trigraphs can be shown to be adhesive just as the category of graphs. In our examples, vertices are represented by bullets, and a tripod δ has the form $\triangle$ or $\triangle$ where Δ and $c \in \{red, yellow, blue\}$ are labels. The upper corner is $one(\delta)$, the left one $two(\delta)$ and the right one $three(\delta)$.

Consider the following rules.

| Tripod | Corners | 3 Tripods | Yellow | Red | Blue |

An application of the first rule is called *refinement* and an application of one of the other three rules a *coloring*. Two sample derivations are

$$nine = (\quad \Longrightarrow \quad \overset{3}{\Longrightarrow} \quad)$$

9 Tripods

$$red = (\quad \Longrightarrow \quad \Longrightarrow \quad \Longrightarrow \quad)$$

Analogously to *red*, two further derivations are

$$yellow = (\quad \overset{3}{\Longrightarrow} \quad) \quad blue = (\quad \overset{3}{\Longrightarrow} \quad)$$

3   Extension and Invariant Part

In this section, the extension and the invariant part of a derivation are considered. The idea of extending a derivation is to set it into a larger context. Formally, a derivation extends a given derivation if the double pushouts that establish their derivation steps differ by further pushouts in a certain way. An extension of a derivation can be constructed using the invariant part which is inductively defined as the intersection of all intermediate objects of the derivation. Intuitively, it represents the part of the start object that is not changed along the derivation. If one extends the invariant part by gluing it together with an extra context object, then this extension can be carried over to all derived and intermediate objects of the derivation yielding an extended derivation the invariant part of which is the extension of the invariant part of the given derivation. At the end of the section, the related work is discussed.

3.1   Extension

We start with the definition of extensions.

Definition 1. *1. A derivation* $\overline{d} = (\overline{G}_0 \underset{p_1}{\Longrightarrow} \overline{G}_1 \underset{p_2}{\Longrightarrow} \cdots \underset{p_n}{\Longrightarrow} \overline{G}_n)$ *is an* extension *of the derivation* $d = (G_0 \underset{p_1}{\Longrightarrow} G_1 \underset{p_2}{\Longrightarrow} \cdots \underset{p_n}{\Longrightarrow} G_n)$ *if, for* $i = 1, \ldots, n$, *the double pushouts defining the i-th derivation step are related as follows*

$$
\begin{array}{ccccc}
L_i & \xleftarrow{\quad l_i \quad} & K_i & \xrightarrow{\quad r_i \quad} & R_i \\
\downarrow{g_i} & (i_L) & \downarrow{z_i} & (i_R) & \downarrow{h_i} \\
\overline{g}_i \Big(G_{i-1} & \xleftarrow{m_{Z_i \to G_{i-1}}} & Z_i & \xrightarrow{m_{Z_i \to G_i}} & G_i \Big) \overline{h}_i \\
\downarrow{q_{G_{i-1} \to \overline{G}_{i-1}}} & (i'_L) & \downarrow{q_{Z_i \to \overline{Z}_i}} & (i'_R) & \downarrow{q_{G_i \to \overline{G}_i}} \\
\overline{G}_{i-1} & \xleftarrow{m_{\overline{Z}_i \to \overline{G}_{i-1}}} & \overline{Z}_i & \xrightarrow{m_{\overline{Z}_i \to \overline{G}_i}} & \overline{G}_i
\end{array}
$$

where (i'_L) *and* (i'_R) *are pushouts and* (i_L, i_R) *and* $((i_L) + (i'_L), (i_R) + (i'_R))$ *are the double pushouts of* d *and* $\overline{d}$, *respectively.*
2. *The morphism* $q_{G_0 \to \overline{G}_0}$ *is called* extension morphism *and the derivation* $\overline{d}$ *is denoted by* $d \uparrow q_{G_0 \to \overline{G}_0}$.

Remark 1. 1. It may be noted that the definition means in particular that, for $i = 1, \ldots, n - 1$, the i-th right matching morphism $\overline{h}_i$ and the left matching morphism $\overline{g}_{i+1}$ factor both through the same $q_{G_i \to \overline{G}_i}$.
2. As pushout complements of pushouts along monomorphisms are unique up to isomorphisms as pushouts anyway, an extension - provided it exists - is determined up to isomorphism by the extension morphism. This justifies the denotation $d \uparrow q_{G_0 \to \overline{G}_0}$.
3. Due to the composition property of pushouts, it is easy to see that extensions of extensions are extensions. More formally, let $d \uparrow q_{G_0 \to \overline{G}_0}$ and $(d \uparrow q_{G_0 \to \overline{G}_0}) \uparrow q_{\overline{G}_0 \to \overline{\overline{G}}_0}$ be extensions. Then the latter equals $d \uparrow (q_{\overline{G}_0 \to \overline{\overline{G}}_0} \circ q_{G_0 \to \overline{G}_0})$.

3.2 Invariant Part

The invariant part of a derivation is inductively defined as intersection of all intermediate objects.

Construction 1 (Invariant part).

1. The *invariant part* $inv(d) \colon INV(d) \to G$ of the derivation $d = (G \overset{*}{\underset{P}{\Longrightarrow}} X)$ is constructed by induction on the length n of d.
   Base $n = 0$: $inv(G \overset{0}{\underset{P}{\Longrightarrow}} X) = inv(G \overset{0}{\underset{P}{\Longrightarrow}} G) = (id_G \colon G \to G)$, and
   Step $n + 1$: $inv(G \underset{p}{\Longrightarrow} H \overset{n}{\underset{P}{\Longrightarrow}} X) \colon INV(d) \to G$ is given by the pullback (1) and the composition (2) in

$$
\begin{array}{ccccc}
 & & INV(d) = Z \cap INV(\hat{d}) & \xrightarrow{m_{INV(d)}} & INV(\hat{d}) \\
inv(d) \nearrow & (2) & inv(d)_Z \downarrow & (1) & \downarrow inv(\hat{d}) \\
G & \xleftarrow{\quad m_{Z \to G} \quad} & Z & \xrightarrow{\quad m_{Z \to H} \quad} & H
\end{array}
$$

where $inv(\hat{d}): INV(\hat{d}) \to H$ is given by induction hypothesis for the tail $\hat{d} = (H \overset{n}{\underset{P}{\Longrightarrow}} X)$ of d.

2. The invariant part is defined as a monomorphism into the start object of a derivation. But the construction gives rise to a cone of monomorphisms into all the derived and intermediate objects of the derivation.

Let $d = (G_0 \overset{n}{\underset{P}{\Longrightarrow}} G_n)$ be a derivation with the derivation diagram

$$G_0 \xleftarrow{\hspace{1cm}} Z_1 \xrightarrow{\hspace{1cm}} G_1 \xleftarrow{\hspace{1cm}} \cdots \xrightarrow{\hspace{1cm}} G_n.$$
$$m_{Z_1 \to G_0} \qquad m_{Z_1 \to G_1} \qquad m_{Z_2 \to G_1} \qquad m_{Z_n \to G_n}$$

Then two families of monomorphisms $\{inv(d)_{G_i}: INV(d) \to G_i\}_{i=0,\ldots,n}$ and $\{inv(d)_{Z_i}: INV(d) \to Z_i\}_{i=1,\ldots,n}$ such that, for $i = 1,\ldots,n$, $inv(d)_{G_{i-1}} = m_{Z_i \to G_{i-1}} \circ inv(d)_{Z_i}$ and $inv(d)_{G_i} = m_{Z_i \to G_i} \circ inv(d)_{Z_i}$ can be constructed by induction on n.

Base $n = 0$: The first family contains only id_{G_0} and the second one is empty. Step $n > 0$: Applying the induction hypothesis for $n - 1$ to the derivation $\hat{d} = (G_1 \overset{n-1}{\underset{P}{\Longrightarrow}} G_n)$, one gets two families of monomorphisms $\{inv(\hat{d})_{G_i}: INV(\hat{d}) \to G_i\}_{i=1,\ldots,n}$ and $\{inv(\hat{d})_{Z_i}: INV(\hat{d}) \to Z_i\}_{i=2,\ldots,n}$ such that $(*)$, for $i = 2,\ldots,n$, $inv(\hat{d})_{G_{i-1}} = m_{Z_i \to G_{i-1}} \circ inv(\hat{d})_{Z_i}$ and $inv(\hat{d})_{G_i} = m_{Z_i \to G_i} \circ inv(\hat{d})_{Z_i}$. Using these monomorphisms and the ones of the definition of the invariant part, the families for d can be defined by $inv(d)_{G_0} = inv(d), inv(d)_{Z_1}$ as above, and, for $i = 1,\ldots,n$, $inv(d)_{G_i} = m_{INV(\hat{d})} \circ inv(\hat{d})_{G_i}$ and, for $i = 2,\ldots,n$, $inv(d)_{Z_i} = m_{INV(\hat{d})} \circ inv(\hat{d})_{Z_i}$. Composing the morphisms in $(*)$ with $m_{INV(\hat{d})}$, one gets the required equations for the case of d for $i = 2,\ldots,n$. For $i = 1$, the definition of the invariant part yields

$$inv(d)_{G_0} = inv(d) = m_{Z_1 \to G_0} \circ inv(d)_{Z_1}$$

and

$$inv(d)_{G_1} = m_{INV(\hat{d})} \circ inv(\hat{d})_{G_1} = m_{INV(\hat{d})} \circ inv(\hat{d}) = m_{Z_1 \to G_1} \circ inv(d)_{Z_1}.$$

The whole situation looks in diagrammatic form as follows:

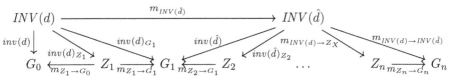

as an iteration of pullbacks of the embeddings of the intermediate objects in the respective derived objects is the limit of the derivation diagram.

3.3 Extension by Means of an Extension Span

By adding an enlarging context to the invariant part of a derivation, an extension of the derivation can be constructed. The enlargement is obtained by a pushout of a chosen extension span. If this construction is combined with the cone of monomorphisms that embed the invariant part into the derivation diagram by further pushouts, then one gets an extension.

Construction 2 (Extension). Let $d = (G_0 \underset{P}{\overset{n}{\Longrightarrow}} G_n)$ be a derivation with the derivation diagram

$$G_0 \xleftarrow{\ \ \ \ \ \ } Z_1 \xrightarrow{\ \ \ \ \ \ } G_1 \xleftarrow{\ \ \ \ \ } \cdots \xrightarrow{\ \ \ \ \ \ } G_n$$
$$\quad m_{Z_1 \to G_0} \quad m_{Z_1 \to G_1} \quad m_{Z_2 \to G_1} \quad m_{Z_n \to G_n}$$

and the two families of monomorphisms $\{inv(d)_{G_i} : INV(d) \to G_i\}_{i=0,\ldots,n}$ and $\{inv(d)_{Z_i} : INV(d) \to Z_i\}_{i=1,\ldots,n}$. Let $es = (C \xleftarrow{add} B \xrightarrow{glue} INV(d))$ be a span of morphisms called *extension span*.

The extension, denoted by $d \uparrow es$, is constructed by the following steps.

1. Construct the pushouts (a) as well as the pushouts (b) for $i = 1, \ldots, n$.

$$
\begin{array}{ccc}
B \xrightarrow{glue} INV(d) & INV(d) \xrightarrow{inv(d)_{G_0}} G_0 \\
\Big\downarrow{add} \ (E) \ \Big\downarrow{add'} & \Big\downarrow{add'} (E_0') \ \Big\downarrow{add_0'} \\
C \xrightarrow{glue'} A & A \xrightarrow{a_0'} \overline{G}_0 \\
(a) &
\end{array}
$$

$$
\begin{array}{cc}
INV(d) \xrightarrow{inv(d)_{Z_i}} Z_i & Z_i \xrightarrow{m_{Z_i \to G_i}} G_i \\
\Big\downarrow{add'} (E_i) \ \Big\downarrow{add_i} & \Big\downarrow{add_i}(E_i') \ \Big\downarrow{add_i'} \\
A \xrightarrow{a_i} \overline{Z}_i & \overline{Z}_i \xrightarrow{a_i'} \overline{G}_i \\
& (b)
\end{array}
$$

2. Consider the diagram

$$
\begin{array}{ccccc}
 & & inv(d) & & \\
 & & (1) & & \\
INV(d) & \xrightarrow{inv(d)_{Z_1}} & Z_1 & \xrightarrow{m_{Z_1 \to G_1}} & G_0 \\
\Big\downarrow{add'} & (E_1) & \Big\downarrow{add_1} & (3) & \Big\downarrow{add_0'} \\
A & \xrightarrow{a_1} & \overline{Z}_1 & \dashrightarrow{a_1''} & \overline{G}_0 \\
 & & (2) & & \\
 & & a_0' & &
\end{array}
$$

(1) is commutative due to the construction. The outer rectangle is commutative as the pushout (E_0'). Therefore, the universal pushout property of (E_1) induces the mediating morphism a_1'' such that (2) and (3) become commutative. As (E_1) and $(E_1) + (3) = (E_0')$ are pushouts, (3) is a pushout.

3. Consider, for $i = 2, \ldots, n$, the diagram

$$
\begin{array}{ccccccc}
 & & inv(d)_{Z_{i-1}} & & & & \\
 & & inv(d)_{G_{i-1}} & & & & \\
 & & (4_i) & & (5_i) & & \\
INV(d) & \xrightarrow{inv(d)_{Z_i}} & Z_i & \xrightarrow{m_{Z_i \to G_i}} & G_{i-1} \xleftarrow{m_{Z_i \to G_i}} & Z_{i-1} \\
\Big\downarrow{add'} & (E_i) & \Big\downarrow{add_i} & (7_i) & \Big\downarrow{add_{i-1}'} (E_{i-1}') & \Big\downarrow{add_{i-1}} \\
A & \xrightarrow{a_1} & \overline{Z}_i & \dashrightarrow{a_i''} & \overline{G}_{i-1} \xleftarrow{a_{i-1}'} & \overline{Z}_{i-1} \\
 & & (6_i) & & & & \\
 & & a_0' & & & &
\end{array}
$$

(4_i) and (5_i) are commutative due to the cone of monomorphisms associated to the invariant part. The outer rectangle (E_{i-1}) and (E'_{i-1}) are commutative as pushouts. Therefore, the universal property of (E_i) induces mediating morphisms a''_i such that (6_i) and (7_i) become commutative. As (E_i) and $(E_i) + (7_i) = (E_{i-1}) + (E'_{i-1})$ are pushouts, the diagrams (7_i) are pushouts, too.

4. Altogether, one gets, for $i = 1, \ldots, n$, the following double pushout

$$
\begin{array}{ccccc}
L_i & \xleftarrow{\ l_i\ } & K_i & \xrightarrow{\ r_i\ } & R_i \\
\downarrow{g_i} & (i_L) & \downarrow{z_i} & (i_R) & \downarrow{h_i} \\
G_{i-1} & \xleftarrow{\ m_{Z_i \to G_{i-1}}\ } & Z_i & \xrightarrow{\ m_{Z_i \to G_i}\ } & G_i \\
\downarrow{add'_{i-1}} & (7_i) & \downarrow{add_i} & (E'_i) & \downarrow{add'_i} \\
\overline{G}_{i-1} & \xleftarrow{\ m_{\overline{Z}_i \to \overline{G}_{i-1}}\ } & \overline{Z}_i & \xrightarrow{\ m_{\overline{Z}_i \to \overline{G}_i}\ } & \overline{G}_i
\end{array}
$$

where (i_L, i_R) is the double pushout of the i-th rule application of the derivation d. Obviously, the diagrams form an extension of d.

The result of this construction is summarized in the following theorem.

Theorem 1. *Let* $d = (G \overset{*}{\underset{P}{\Longrightarrow}} X)$ *be a derivation and* $C \xleftarrow{add} B \xrightarrow{glue} INV(d)$ *be an extension span. Let*

$$
\begin{array}{ccccc}
B & \xrightarrow{\ glue\ } & INV(d) & \xrightarrow{\ inv(d)\ } & G \\
\downarrow{add\ (E)} & & \downarrow{add'\ (E')} & & \downarrow{add''} \\
C & \xrightarrow{\ glue'\ } & A & \xrightarrow{\ a'\ } & \overline{G}
\end{array}
$$

be the pushouts of add and glue as well as of add' and inv(d). Then the extension of d with respect to the extension morphism add'' can be constructed.

Example 2. Consider the derivation *red* of Example 1 and the extension span

$$
es(red) = (\qquad \qquad \longleftarrow \qquad \longrightarrow \qquad)
$$

yielding the extension

$$
\overline{red} = red \uparrow es(red) = (\qquad \Longrightarrow \qquad \Longrightarrow \qquad \Longrightarrow \qquad)
$$

Analogously, the derivations *yellow* and *blue* can be extended by means of the extension spans

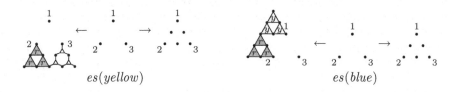

es(yellow)   es(blue)

yielding the extensions

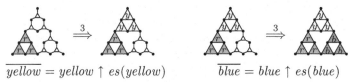

$$\overline{yellow} = yellow \uparrow es(yellow) \qquad \overline{blue} = blue \uparrow es(blue)$$

Starting with the derivation *nine*, the sequential composition

$$nine = \overline{red}\ \overline{yellow}\ \overline{blue} = (\ \overset{13}{\Longrightarrow}\ )$$

derives a new pattern from the single tripod.

3.4   Extension Span and the Invariant Part

The pushout of the extension span induces a mediating morphism from the pushout object into the start object of the corresponding extension. It can be shown that this morphism is the invariant part of the extension provided that the extension span is monic.

Theorem 2. *Let $d = (G \overset{*}{\underset{P}{\Rightarrow}} X)$ be a derivation with the invariant part $inv(d)$: $INV(d) \to G$ and $es = (C \leftarrow B \to INV(d))$ be a monic extension span. Let*

$$
\begin{array}{ccccc}
B & \overset{glue}{\longrightarrow} & INV(d) & \overset{inv(d)}{\longrightarrow} & G \\
\Big\downarrow {\scriptstyle add\ (E)} & & \Big\downarrow {\scriptstyle add'\ (E')} & & \Big\downarrow {\scriptstyle add''} \\
C & \underset{glue'}{\longrightarrow} & A & \underset{a'}{\longrightarrow} & \overline{G}
\end{array}
$$

be the corresponding pushouts of es and of $inv(d)$ and add' resp. Then $a' : A \to \overline{G}$ is the invariant part of $d \uparrow es$.

Sketch of Proof. The statement is proved by induction on the length n of d. The case $n = 0$ is obviously true. For $n+1$, some commutative diagrams provided by the constructions of the invariant part and the extension can be arranged into a Van Kampen cube in such a way that it implies that the diagram

$$
\begin{array}{ccc}
A & \longrightarrow & \widehat{A} \\
\downarrow & & \downarrow \\
\overline{Z} & \longrightarrow & \overline{G}
\end{array}
$$

is a pullback. Applying the induction hypothesis to the tail $\widehat{d}$ of d, one gets $\widehat{A} = INV(\widehat{d})$ such that $A = Z \cap INV(\widehat{d}) = INV(d)$.

Related Work. Extensions of derivations over adhesive categories are defined and discussed in Ehrig et al. [6,7]. Given a derivation $d = (G \underset{P}{\overset{*}{\Rightarrow}} X)$, the span $G \xleftarrow{\ inv(d)\ } INV(d) \xrightarrow{\ inv(d)_X\ } X$ is called derived span there, so that invariant parts and derived spans imply each other. In both books, one finds a construction of an extension with respect to a chosen extension morphism $q : G \to \overline{G}$ provided that q is consistent wrt d, i.e., the initial pushout

$$
\begin{array}{ccc}
B_0 & \xrightarrow{\ m\ } & G \\
{\scriptstyle add_0}\downarrow & (IPO) & \downarrow{\scriptstyle q} \\
C_0 & \xrightarrow{\ \overline{m}\ } & \overline{G}
\end{array}
$$

with a monomorphism m exists and there is a monomorphism $m' : B_0 \to INV(d)$ such that $m = inv(d) \circ m'$. In this case, $C_0 \xleftarrow{\ add_0\ } B_0 \xrightarrow{\ m'\ } INV(d)$ forms an extension span in such a way that the induced extension coincides with the extension constructed in [6,7]. Conversely, let $es = (C \xleftarrow{\ add\ } B \xrightarrow{\ glue\ } INV(d))$ be an extension span with a monic right-hand side and

$$
\begin{array}{ccccc}
B & \xrightarrow{\ glue\ } & INV(d) & \xrightarrow{\ inv(d)\ } & G \\
{\scriptstyle add}\downarrow & (E) & {\scriptstyle add'}\downarrow & (E') & \downarrow{\scriptstyle add''} \\
C & \xrightarrow{\ glue'\ } & A & \xrightarrow{\ a'\ } & \overline{G}
\end{array}
$$

be the pushouts of add and $glue$ as well as of add' and $inv(d)$. Let (IPO) be the initial pushout wrt $q = add''$. As the composition of (E) and (E') is a respective pushout, the initiality of (IPO) induces morphisms $b : B_0 \to B$ and $c : C_0 \to C$ such that $inv(d) \circ glue \circ b = m$ and $a' \circ glue' \circ c = \overline{m}$. The first equality means that $q = add''$ is consistent wrt d so that the extension by means of es can also be constructed according to [6,7]. We do not know whether this remains true if the right-hand side is not monomorphic. Moreover, our construction works independent of initial pushouts. We wonder whether there is a general construction or whether they exist in general at all.

The difference between the two constructions may be considered small. But using the extension span, there is always the simple choice of the boundary morphism $glue : B \to INV(d)$ as the identity of $INV(d)$. The early constructions of extensions of derivations in the context of graph transformation in Ehrig, Pfender and Schneider [1], Ehrig and Kreowski [2,3] and Ehrig [4] are covered by the more general approaches in adhesive categories.

4   Restrictions and Accessed Parts

While an extension embeds a derivation into a larger context, restriction is the inverse counterpart. It is known that a derivation can be restricted with respect to a monomorphism into the start object of the derivation if the monomorphism factors through the so-called accessed part. We recall the constructions as we relate them to extensions and invariant parts. In particular, we show that the start object of a derivation is the union of the invariant part and the accessed part of the given derivation. As a consequence, it turns out that the start object of a derivation is the union of the invariant part and the start object of a monomorphic restriction. As restriction and extension are inverse to each other, this means at the same time that the start object of a monomorphic extension is the union of the invariant part of the extension and the start object of the extended derivation.

Definition 2 (Restriction).

1. A derivation $d' = (G' \overset{*}{\underset{P}{\Rightarrow}} X')$ is a restriction of the derivation $d = (G \overset{*}{\underset{P}{\Rightarrow}} X)$ if d is an extension of d'.
2. The corresponding extension morphism $q_{G' \to G}$ is called restriction morphism, and the derivation d' is denoted by $d \downarrow q_{G' \to G}$.

Remark 2. 1. By definition, extension and restriction are inverse to each other, i.e., $d' = (d' \uparrow q_{G' \to G}) \downarrow q_{G' \to G}$ and $d = (d \downarrow q_{G' \to G}) \uparrow q_{G' \to G}$.
2. As an extension is uniquely determined by the extension morphism, a restriction is uniquely determined by the restriction morphism.
3. As extensions of extensions are extensions, restrictions of restrictions are restrictions, i.e., $((d \downarrow q_{G' \to G}) \downarrow q_{G'' \to G'}) = d \downarrow (q_{G' \to G} \circ q_{G'' \to G'})$.

4.1   Accessed Parts

In [8], it is shown that, under certain assumptions, a chosen monomorphism into the start object of a derivation induces a restriction if the accessed part of the derivation factors through the monomorphism. Intuitively, this means that the monomorphism covers the part of the start object that is accessed by some left matching morphism along the derivation.

Assumption and Notation. 1. We assume that the underlying adhesive category $\mathbf{C}$ has a strict initial object $\emptyset$, i.e., an initial object $\emptyset$ with the property that every morphism in $\mathbf{C}$ with codomain $\emptyset$ is an isomorphism. Moreover, we require that the initial morphism $\emptyset_G \colon \emptyset \to G$ for every object G is a monomorphism.
2. Every morphism in $\mathbf{C}$ has an epi-mono factorization, i.e., for every morphism f there is a factorization $f = m \circ e$ where e is an epimorphism and m is a monomorphism.
3. Let $d = (G \overset{*}{\underset{P}{\Rightarrow}} X)$ be a derivation with two cases $G \overset{0}{\underset{P}{\Rightarrow}} X$ if and only if $G = X$, and $(G \overset{n+1}{\underset{P}{\Rightarrow}} X) = (G \underset{p}{\Rightarrow} H \overset{n}{\underset{P}{\Rightarrow}} X)$ for $n \in \mathbb{N}$ where the first step is defined by the double pushout

$$L \xleftarrow{\quad l \quad} K \xrightarrow{\quad r \quad} R$$
$$g\downarrow \qquad (1) \qquad z\downarrow \qquad (2) \qquad h\downarrow$$
$$G \xleftarrow{\quad m_{Z\to G} \quad} Z \xrightarrow{\quad m_{Z\to H} \quad} H$$

with $p = (L \xleftarrow{l} K \xrightarrow{r} R) \in P$. The remaining derivation is denoted by $\hat{d} = tail(d) = (H \overset{n}{\underset{P}{\Rightarrow}} X)$.

4. Let $(L \xrightarrow{g} G) = (L \xrightarrow{\quad e_{L\to g(L)} \quad} g(L) \xrightarrow{\quad m_{g(L)\to G} \quad} G)$ be the epi-mono factorization of g.

5. Let $G' \xrightarrow{\quad m_{G'\to G} \quad} G$ be a monomorphism for some object G'.

Construction 3 (Accessed Part). The *accessed part* $acc(d) \colon ACC(d) \to G$ of d is constructed by induction on the length n of d.

Base $n = 0$: $acc(G \overset{0}{\underset{P}{\Rightarrow}} X) = (\emptyset_G \colon \emptyset \to G)$, and

Step $n + 1$: $acc(G \overset{n+1}{\underset{P}{\Rightarrow}} X) = acc(G \underset{p}{\Rightarrow} H \overset{n}{\underset{P}{\Rightarrow}} X) \colon ACC(d) \to G$ is constructed

in four steps where $acc(\hat{d}) = acc(tail(d)) \colon ACC(\hat{d}) \to H$ is the accessed part of $\hat{d}$ using the induction hypothesis for a derivation of length n. Moreover, the double pushout defining the first derivation step is used as given above.

1. Construct the pullbacks

$$Z \cap ACC(\hat{d}) \xrightarrow{\quad m_{ACC(\hat{d})} \quad} ACC(\hat{d})$$
$$m_Z\downarrow \qquad (A1) \quad acc(\hat{d})\downarrow$$
$$Z \xrightarrow{\quad m_{Z\to H} \quad} H$$

$$g(L) \cap (Z \cap ACC(\hat{d})) \xrightarrow{\quad m_{Z\cap ACC(\hat{d})} \quad} Z \cap ACC(\hat{d})$$
$$m_{g(L)}\downarrow \qquad (A2) \qquad m_Z\downarrow \quad Z$$
$$\qquad\qquad\qquad m_{Z\to G}\downarrow$$
$$g(L) \xrightarrow{\quad m_{g(L)\to G} \quad} G$$

2. Construct the pushout

$$g(L) \cap (Z \cap ACC(\hat{d})) \xrightarrow{\quad m_{Z\cap ACC(\hat{d})} \quad} Z \cap ACC(\hat{d})$$
$$m_{g(L)}\downarrow \qquad (A3) \qquad \overline{m}_{Z\cap ACC(\hat{d})}\downarrow$$
$$g(L) \xrightarrow{\quad \overline{m}_{g(L)} \quad} g(L) \cup (Z \cap ACC(\hat{d}))$$

3. Using the commutativity of the right pullback in Step 1, the pushout in Step 2 induces a mediating morphism $acc(d) \colon ACC(d) \to G$ with $ACC(d) = g(L) \cup (Z \cap ACC(\hat{d}))$ and the commutative diagrams

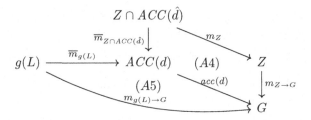

4.2   Restrictions

In the category of graphs, it is known that the derivation d can be restricted to every subgraph G' of G with $ACC(d) \subseteq G'$ by clipping off vertices and edges outside of G' (or their isomorphic counterparts) from all graphs of d keeping all matches invariant in this way. The construction can be generalized to a monomorphism from G' to G if $acc(d)\colon ACC(d) \to G$ factors through it on the level of adhesive categories. The restriction is constructed by induction on the length of the derivation d. The construction is divided into a restriction of a direct derivation and a continuation of the restriction after the first derivation step.

Construction 4 (Restriction)

1. Let $g'\colon L \to G'$ be a morphism and $m_{G'\to G}\colon G' \to G$ be a monomorphism for some object G' such that $g = m_{G'\to G} \circ g'$. Then the double pushout defining $G \underset{p}{\Longrightarrow} H$ can be decomposed into

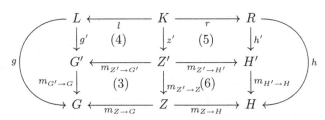

such that $z = m_{Z'\to Z} \circ z'$, $h = m_{H'\to H} \circ h'$, (4) and (5) are pushouts and (3) and (6) are pullbacks. In particular, (4) and (5) define a direct derivation $G' \underset{p}{\Longrightarrow} H'$. The derivation $G' \underset{p}{\Longrightarrow} H'$ is called *single-step restriction* of $G \underset{p}{\Longrightarrow} H$ wrt $m_{G'\to G}\colon G' \to G$, denoted by $(G \underset{p}{\Longrightarrow} H) \downarrow m_{G'\to G}$.

2. Let $acc(d)$ factor through $m_{G'\to G}$. Then the *restriction* of d wrt $m_{G'\to G}$, denoted by $d \downarrow m_{G'\to G}$, can be constructed by induction on the length n of d:

   Base $n = 0$: $((G \overset{0}{\underset{P}{\Longrightarrow}} X) \downarrow m_{G'\to G}) = (G' \overset{0}{\underset{P}{\Longrightarrow}} G')$, and

   Step $n + 1$: $((G \overset{n+1}{\underset{P}{\Longrightarrow}} X) \downarrow m_{G'\to G}) = ((G \underset{p}{\Longrightarrow} H \overset{n}{\underset{P}{\Longrightarrow}} X) \downarrow m_{G'\to G}) = (G' \underset{p}{\Longrightarrow} H' \overset{n}{\underset{P}{\Longrightarrow}} X')$ where $G' \underset{p}{\Longrightarrow} H'$ is the single-step restriction of $G \underset{p}{\Longrightarrow} H$ wrt

$m_{G'\to G}$ and $(H' \underset{P}{\overset{n}{\Longrightarrow}} X') = (H \underset{P}{\overset{n}{\Longrightarrow}} X) \downarrow m_{H'\to H}$ is given by the induction hypothesis for derivations of length n.

Example 3. Applying the rules of Example 1, one obtains a derivation of the form

Obviously, a tripod must be created before it can be colored, and the colorings of two different tripods are independent of each other. Therefore, one can assume without loss of generality that the derivation above starts with all refinements followed by all red colorings followed by all yellow colorings ending with all blue colorings such that the derivation can be decomposed into the subderivation *nine* from Example 1 and the subderivations $\overline{red}$, $\overline{yellow}$, and $\overline{blue}$ from Example 2.

The accessed part of $\overline{red}$ is the substructure *3Tripods* at the left corner, the accessed part of $\overline{yellow}$ the *3Tripods* at the upper corner, and the accessed part of $\overline{blue}$ the *3Tripods* at the right corner. Therefore, the three derivations can be restricted to their accessed parts yielding the derivations *red*, *yellow*, and *blue* from Example 1 respectively.

As demonstrated for the sample pattern *9Tripods* in Example 2, the restrictions *red*, *yellow* and *blue* can be extended to every pattern that contains *3Tripods* as a subpattern. If one removes the subpattern up to the corners, then one gets a context that provides an extension span using the corners as boundary. If one proceeds with the derived pattern of this extension in the same way, then one can color a sequence of *3Tripods* yielding arbitrary complex patterns. An example of this kind is the following derivation

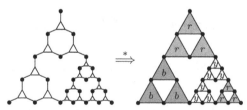

4.3 Monomorphic Restriction Morphisms and Accessed Parts

As shown above, a derivation can be restricted with respect to a monomorphism if the accessed part factors through it. Conversely, it can be shown that the accessed part of a derivation factors through the restriction morphism of a restriction if the restriction morphism is a monomorphism.

Theorem 3. *Let* $d = (G \underset{P}{\overset{*}{\Longrightarrow}} X)$ *be a derivation and* $d' = d \downarrow q_{G'\to G}$ *be a restriction of* d *with a monomorphic restriction morphism* $q_{G'\to G}$*. Then there is a morphism* $q \colon ACC(d) \to G'$ *such that* $q_{G'\to G} \circ q = acc(d)$*.*

Sketch of Proof. The statement is proved by induction on the length n of d. The case $n = 0$ is obviously true. For $n+1$, $acc(d) = m_{g(L) \to G} \cup (m_{Z \to H} \cap acc(\widehat{d}))$ where $g : L \to G$ is the first matching morphism, Z the first intermediate object, and $\widehat{d}$ the tail of d. With $g, m_{g(L) \to G}$ factors through $q_{G' \to G}$. That $m_{Z \to H} \cap acc(\widehat{d})$ factors through $q_{G' \to G}$ follows from the induction hypothesis for $\widehat{d}$ and the properties of intersections. With both components of the union, the union itself factors through $q_{G' \to G}$.

4.4 Accessed Part and Invariant Part are Complementary

Intuitively, it is clear that the start object of a derivation consists of the invariant part kept unchanged throughout the derivation and the part that is removed by some of the rule applications. As the accessed part covers what is removed, one should expect that the union of the invariant part and the accessed part covers the start object. This is stated and proved in the following theorem.

Theorem 4. *Let* $d = (G \overset{*}{\underset{P}{\Longrightarrow}} X)$ *be a derivation. Then* $acc(d) \cup inv(d) = id_G$.

Proof.

$$
\begin{aligned}
ACC(d) \cup INV(d) &= (g(L) \cup (Z \cap ACC(\widehat{d})) \cup (Z \cap INV(\widehat{d}))) && \text{(definitions)} \\
&= g(L) \cup ((Z \cap ACC(\widehat{d})) \cup (Z \cap INV(\widehat{d}))) && \text{(associativity } \cup) \\
&= g(L) \cup ((Z \cap (ACC(\widehat{d})) \cup INV(\widehat{d}))) && \text{(distributivity)} \\
&= g(L) \cup (Z \cap H) && \text{(ind. hypothesis)} \\
&= g(L) \cup Z && \text{(subproperty)} \\
&= G && \text{(known fact)}
\end{aligned}
$$

Corollary 1. *Let* $d = (G \overset{*}{\underset{P}{\Longrightarrow}} X)$ *be a derivation and* $d \downarrow q_{G' \to G}$ *be a restriction of* d. *Then* $acc(d) \cup q_{G' \to G} = id_G$.

Proof. As the union preserves monomorphisms, $q : ACC(d) \to G'$ with $g' \circ q = g$ yields a monomorphism $q \cup INV(d) : G = ACC(d) \cup INV(d) \to G' \cup INV(d)$. As G' and $INV(d)$ are monomorphically embedded into G, there is a monomorphism $q' : G' \cup INV(d) \to G$. Both are inverse to each other so that $G = G' \cup INV(d)$.

Related Work. Restrictions of derivations over adhesive categories are considered in some detail in [8] while they are only mentioned as inverse to extensions in Ehrig et al. [6,7]. In Ehrig et al. [6] the single-step restriction is introduced in addition. The early investigations by Kreowski [5] and Plump [13] provide constructions of restrictions for graphs and hypergraphs respectively and prove some properties concerning the iteration of restrictions and their relation to extensions. For the category of graphs, a more detailed investigation of restrictions can be found [14]. In the latter four references, the constructions are not fully categorical as specific properties of graphs and hypergraphs are used, too.

5   Conclusion

In this paper, we have done some steps of an attempt to investigate extensions and restrictions of derivations in an adhesive category systematically and comprehensively. This includes a new construction of extensions based on a choice of an extension span and several properties of extensions, restrictions and their relations in connection with the accessed parts and the invariant parts of derivations. For example, we have shown that the union of the accessed part and the invariant part of a derivation covers its start object.

But various questions are still not answered like the following ones.

1. It is easy to see that extensions of extensions are extensions. But if the two given extensions are constructed by means of extension spans, then it would be interesting to know whether the composed extension can be constructed by an induced extension span.
2. If one cuts an extension of a derivation into sequential pieces, then the pieces are obviously extensions of the corresponding pieces of the underlying derivation. But what about the other way around? Under which conditions do extensions of sequential pieces of a derivation compose properly?
3. How do extension and restriction behave with respect to other operations on derivations like sequentialization, parallelization, shift, amalgamation, etc.?
4. In particular, the Concurrency Theorem in [6] states that a derivation of two steps induces a concurrent derivation with the same start and result objects resp. applying a concurrent production that can be constructed from the two given rules and vice versa. The concurrent production and its matches combine the two rules and their matches in such a way that this plays a similar role as the accessed part of a derivation. Therefore, it may be of interest to investigate the relation between concurrent productions and accessed parts more closely.
5. The interplay of extension and restriction as demonstrated in our running example of pattern generation follows a general principle. Given a derivation that represents a complex process of some kind, it can be cut into sequential pieces (maybe after some rearrangement by interchanging sequentially independent derivation steps). Restrictions of the pieces isolate subprocesses, and their proper extensions allow to compose new processes. It may be interesting to conduct case studies to demonstrate the usefulness of this procedure in the context of business processes, chemical processes, construction plans, schedules, and such.

Acknowledgement. We are grateful to the anonymous reviewers for their valuable comments that led to various improvements.

References

1. Ehrig, H., Pfender, M., Schneider, H.-J.: Graph grammars: an algebraic approach. In: IEEE Conference on Automata and Switching Theory, Iowa City, pp. 167–180 (1973)
2. Ehrig, H., Kreowski, H.-J.: Categorical approach to graphic systems and graph grammars. In: Marchesini, G., Mitter, S.K. (eds.) Mathematical Systems Theory, pp. 323–351. Springer, Heidelberg (1976). https://doi.org/10.1007/978-3-642-48895-5_23
3. Ehrig, H., Kreowski, H.-J.: Pushout-properties: an analysis of gluing constructions for graphs. Math. Nachr. **91**, 135–149 (1979)
4. Ehrig, H.: Embedding theorems in the algebraic theory of graph grammars. In: Karpiński, M. (ed.) FCT 1977. LNCS, vol. 56, pp. 245–255. Springer, Heidelberg (1977). https://doi.org/10.1007/3-540-08442-8_91
5. Kreowski, H.-J.: Manipulationen von Graphmanipulationen. Ph.D. thesis, Technische Universität Berlin (1978)
6. Ehrig, H., Ehrig, K., Prange, U., Taentzer, G.: Fundamentals of Algebraic Graph Transformation. Monographs in Theoretical Computer Science. An EATCS Series, Springer, Heidelberg (2006). https://doi.org/10.1007/3-540-31188-2
7. Ehrig, H., Ermel, C., Golas, U., Hermann, F.: Graph and Model Transformation - General Framework and Applications. Monographs in Theoretical Computer Science. An EATCS Series, Springer, Heidelberg (2015). https://doi.org/10.1007/978-3-662-47980-3
8. Kreowski, H.-J., Lye, A., Windhorst, A.: Moving a derivation along a derivation preserves the spine in adhesive high-level replacement systems (2024, submitted)
9. Hutchinson, J.E.: Fractals and self similarity. Indiana Univ. Math. J. **30**(5), 713–747 (1981)
10. Drewes, F., Kreowski, H.-J.: Picture generation by collage grammars. In: Ehrig, H., Engels, G., Kreowski, H.-J., Rozenberg, G. (eds.) Handbook Of Graph Grammars And Computing By Graph Transformation, vol. 2, chap. 11, pp. 397–457. World Scientific (1999)
11. Adámek, J., Herrlich, H., Strecker, G.E.: Abstract and Concrete Categories - The Joy of Cats. Dover Publications, Mineola (2009)
12. Lack, S., Sobociński, P.: Adhesive categories. In: Walukiewicz, I. (ed.) FoSSaCS 2004. LNCS, vol. 2987, pp. 273–288. Springer, Heidelberg (2004). https://doi.org/10.1007/978-3-540-24727-2_20
13. Plump, D.: Computing by graph rewriting. Habilitation thesis, University of Bremen (1999)
14. Kreowski, H.-J., Kuske, S., Lye, A., Windhorst, A.: Moving a derivation along a derivation preserves the spine. In: Fernández, M., Poskitt, C.M. (eds.) ICGT 2023. LNCS, vol. 13961, pp. 64–80. Springer, Cham (2023). https://doi.org/10.1007/978-3-031-36709-0_4

Causal Graph Dynamics and Kan Extensions

Luidnel Maignan$^{(\boxtimes)}$ and Antoine Spicher$^{(\boxtimes)}$

Univ Paris Est Creteil, LACL, 94000 Creteil, France
{luidnel.maignan,antoine.spicher}@u-pec.fr

Abstract. On the one side, the formalism of Global Transformations comes with the claim of capturing any transformation of space that is local, synchronous and deterministic. The claim has been proven for different classes of models such as mesh refinements from computer graphics, Lindenmayer systems from morphogenesis modeling, and cellular automata from biological, physical and parallel computation modeling. The Global Transformation formalism achieves this by using category theory for its genericity, and more precisely the notion of Kan extension to determine the global behaviors based on the local ones. On the other side, Causal Graph Dynamics describe the transformation of port graphs in a synchronous and deterministic way. In this paper, we show the precise sense in which the claim of Global Transformations holds for them as well. This is done by showing different ways in which they can be expressed as Kan extensions, each of them highlighting different features of Causal Graph Dynamics. Along the way, this work uncovers the interesting class of Monotonic Causal Graph Dynamics and their universality among General Causal Graph Dynamics.

1 Introduction

Initial Motivation. This work started as an effort to understand the framework of Causal Graph Dynamics (CGDs) from the point of view of Global Transformations (GTs), both frameworks expanding on Cellular Automata (CAs) with the similar goal of handling dynamical spaces, but with two different answers. Indeed, we have on the one hand CGDs that have been introduced in 2012 in [1] as a way to describe synchronous and local evolutions of labeled port graphs whose structures also evolve. Since then, the framework has evolved to incorporate many considerations such as stochasticity, reversibility [3] and quantumness [2] On the other hand, we have GTs that have been proposed in 2015 in [6] as a way to describe synchronous local evolution of any spatial structure whose structure also evolves. This genericity over arbitrary kind of space is obtained using the language of category theory. It should therefore be the case that CGDs are special cases of GTs in which the spatial structure happens to be labeled port graphs. So the initial motivation is to make this relationship precise.

Initial Plan. Initially, we expected this to be a very straightforward work. First, because of the technical features of CGDs (recalled in the next section), it is

© The Author(s), under exclusive license to Springer Nature Switzerland AG 2024
R. Harmer and J. Kosiol (Eds.): ICGT 2024, LNCS 14774, pp. 77–96, 2024.
https://doi.org/10.1007/978-3-031-64285-2_5

appropriate to study them in the same way that CAs had been studied in [4]. By this, we mean that although GTs use the language of category theory, the categorical considerations all simplify into considerations from order theory when an absolute position system is used, as is the case in CAs and in CGDs. Secondly, CGDs come directly with an order from the notion of sub-graph used implicitly throughout. The initial plan was to unfold the formalism on this basis to make sure that everything is as straightforward as expected, and then proceed to the next step: quotienting absolute positions out to only keep relative ones, thus actually using categorical features and not only order-theoretic ones, as done in [4] for CAs.

Section 2.2 recalls how CGDs are defined and work. Section 2.3 recalls how Kan extensions and GTs are defined and simplifies in the particular case of order theory. Section 3.1 makes clear the strong relation between the two concepts and how this relation leads to the initial ambition.

Actual Plan. It turns out that the initial ambition falls short, but the precise way in which it does reveals something interesting about CGDs. The expected relationship is in fact partial and only allows to accommodate Kan extensions with CGDs which happen to be monotonic. This led us to change our initial plan to investigate the role played by monotonic CGDs within the framework. Doing so, we uncover the universality of monotonic CGDs among general CGDs, thus implying the initially wanted result: all CGDs are GTs.

Sections 3.2 and 3.3 make clear the relationship between GTs and monotonic CGDs, while Sect. 4 shows the universality of monotonic CGDs by proposing an encoding transforming any CGD into a monotonic one. Finally, Sect. 5 discusses this interesting property of CGDs. All proofs are included, auxiliary ones being deferred to the appendix to ease the reading.

2   Preliminaries

2.1   Notations

The definitions in this article mostly use common notations from set theory. The set operations symbols, especially set inclusion $\subseteq$ and union $\cup$, are heavily overloaded, but the context always allows to recover the right semantics. Two first overloads concern the inclusion and the union of partial functions, which are to be understood as the inclusion and union of the graphs of the functions respectively, *i.e.*, their sets of input-output pairs. A partial function f from a set A to a set B is indicated as $f : A \rightharpoonup B$, meaning that f is defined for some elements of A. The restriction of a function $f : A \rightharpoonup B$ to a subset $A' \subseteq A$ is denoted $f \restriction A' : A' \rightharpoonup B$.

2.2   Causal Graph Dynamics

Our work strongly relies on the objects defined in [1]. We recall here the definitions required to understand the rest of the paper. Particularly, [1] establishes the equivalence of the so-called *causal dynamics* and *localisable functions*. For the present work, we focus on the latter. Our method is to stick to the original notations in order to make clear our relationship with the previous work. However, some slight differences exist but are locally justified.

We consider an uncountable infinite set $\mathcal{V}$ of symbols for naming vertices.

Definition 1 (Labeled Graphs with Ports). *Let Σ and Δ be two sets, and π a finite set. A graph G with states in Σ and Δ, and ports in π is the data of:*

- *a countable subset $V(G) \subset \mathcal{V}$ whose elements are the* vertices *of G,*
- *a set $E(G)$ of non-intersecting two-element subsets of $V(G) \times \pi$, whose elements are the* edges *of G, and are denoted $\{u : i, v : j\}$,*
- *a partial function $\sigma(G) : V(G) \rightharpoonup \Sigma$ labeling vertices of G with states,*
- *a partial function $\delta(G) : E(G) \rightharpoonup \Delta$ labeling edges of G with states.*

The set of graphs with states in Σ and Δ, and ports π is written $\mathcal{G}_{\Sigma,\Delta,\pi}$.

A pointed graph (G, v) is a graph $G \in \mathcal{G}_{\Sigma,\Delta,\pi}$ with a selected vertex $v \in V(G)$ called pointer.

In this definition, the fact that edges are non-intersecting means that each $u : i \in V(G) \times \pi$ appears in at most one element of $E(G)$. The vertices of the graph are therefore of degree at most $|\pi|$, the size of the finite set π. Let us note right now that the particular elements of $\mathcal{V}$ used to build $V(G)$ in a graph G should ultimately be irrelevant, and only the structure and the labels should matter, as made precisely in Definition 6. This is the reason of the definition of the following object.

Definition 2 (Isomorphism). *An isomorphism is the data of a bijection $R : \mathcal{V} \to \mathcal{V}$. Its action on vertices is straightforwardly extended to any edge by $R(\{u : i, v : j\}) := \{R(u) : i, R(v) : j\}$, to any graph $G \in \mathcal{G}$ by*

$$V(R(G)) := \{R(v) \mid v \in V(G)\}, \qquad E(R(G)) := \{R(e) \mid e \in E(G)\},$$
$$\sigma(R(G)) := \sigma(G) \circ R^{-1}, \qquad\qquad \delta(R(G)) := \delta(G) \circ R^{-1},$$

and to any pointed graph (G, v) by $R(G, v) := (R(G), R(v))$.

Operations are defined to manipulate graphs in a set-like fashion.

Definition 3 (Consistency, Union, Intersection). *Two graphs G and H are* consistent *when*

- $E(G) \cup E(H)$ *is a set of non-intersecting two-element sets;*
- $\sigma(G)$ *and $\sigma(H)$ agree where they are both defined;*
- $\delta(G)$ *and $\delta(H)$ agree where they are both defined.*

In this case, the union *$G \cup H$ of G and H is defined by*

$$V(G \cup H) := V(G) \cup V(H), \qquad E(G \cup H) := E(G) \cup E(H),$$
$$\sigma(G \cup H) := \sigma(G) \cup \sigma(H), \qquad \delta(G \cup H) := \delta(G) \cup \delta(H).$$

The intersection $G \cap H$ *of G and H is always defined and given by*

$$V(G \cap H) := V(G) \cap V(H), \qquad E(G \cap H) := E(G) \cap E(H),$$
$$\sigma(G \cap H) := \sigma(G) \cap \sigma(H), \qquad \delta(G \cap H) := \delta(G) \cap \delta(H).$$

The empty graph *with no vertex is designed $\emptyset$.*

In this definition, the unions of functions seen as relations (that result in a non-functional relation in general) are here guaranteed to be functional by the consistency condition. The intersection of (partial) functions simply gives a (partial) function which is defined only on inputs on which both functions agree.

   To describe the evolution of such graphs in the CGD framework, we first need to make precise the notion of locality, which is captured by how a graph is pruned to a local view: a disk. We consider the usual distance between two vertices in graphs, *i.e.*, the minimal number of edges of the paths between them, and denote by $B_G(c, d)$ the set of vertices at distance at most d from c in the graph G.

Definition 4 (Disk). *Let G be a graph, $c \in V(G)$ a vertex, and r a non-negative integer. The* disk of radius r and center c *is the pointed graph $G_c^r = (H, c)$ with H given by*

$$V(H) := B_G(c, r + 1), \qquad\qquad\qquad \sigma(H) := \sigma(G) \restriction B_G(c, r),$$
$$E(H) := \{\{u{:}i, v{:}j\} \in E(G) \mid \{u, v\} \cap B_G(c, r) \neq \emptyset\}, \quad \delta(H) := \delta(G) \restriction E(H).$$

We denote by $\mathcal{D}_{\Sigma,\Delta,\pi}^r$ the set $\{G_c^r \mid G \in \mathcal{G}_{\Sigma,\Delta,\pi}, c \in V(G)\}$ of all disks of radius r, and by $\mathcal{D}_{\Sigma,\Delta,\pi}$ the set of all disks of any radius. When the disk notation is used with a set C of vertices as subscript, we mean

$$G_C^r := \bigcup_{c \in C} G_c^r. \tag{1}$$

   The CGD dynamics relies on a local evolution describing how local views generates local outputs consistently.

Definition 5 (Local Rule). *A* local rule of radius r *is a function $f : \mathcal{D}_{\Sigma,\Delta,\pi}^r \to \mathcal{G}_{\Sigma,\Delta,\pi}$ such that*

1. *for any isomorphism R, there is another isomorphism R', called the* conjugate *of R, with $f \circ R = R' \circ f$,*
2. *for any family $\{(H_i, v_i)\} \subset \mathcal{D}_{\Sigma,\Delta,\pi}^r$, $\bigcap_i H_i = \emptyset \implies \bigcap_i f((H_i, v_i)) = \emptyset$,*
3. *there exists b such that for all $D \in \mathcal{D}_{\Sigma,\Delta,\pi}^r$, $|V(f(D))| \leq b$,*
4. *for any $G \in \mathcal{G}_{\Sigma,\Delta,\pi}$, $u, v \in V(G)$, $f(G_v^r)$ and $f(G_u^r)$ are consistent.*

   In the second condition, note that a set of graphs have the empty graph as intersection iff their sets of vertices are disjoint. In the original work, functions respecting the two first conditions, the third condition, and the fourth condition are called respectively dynamics, bounded functions, and consistent functions. A local rule is therefore a bounded consistent dynamics.

The main result of [1] is the proof that *causal graph dynamics* are *localisable functions*, the concepts coming from the paper. We rely on this result in the following definition since we use the formal definition of the latter with the name of the former.

Definition 6 (Causal Graph Dynamics (CGD)). *A function* $F : \mathcal{G}_{\Sigma,\Delta,\pi} \to \mathcal{G}_{\Sigma,\Delta,\pi}$ *is a* causal graph dynamics *(CGD) if there exists a radius r and a local rule f of radius r such that*

$$F(G) = \bigcup_{v \in V(G)} f(G_v^r). \tag{2}$$

2.3   Global Transformations and Kan Extensions

In category theory [5], Kan extensions are a construction allowing to extend a functor along another one in a universal way. In the context of this article, we restrict ourselves to the case of pointwise left Kan extensions involving only categories which are posets. In this case, their definition simplifies as follows.

Definition 7 (Pointwise Left Kan Extension for Posets). *Given three posets A, B and C, and two monotonic functions $i : A \to B$, $f : A \to C$, the function $\Phi : B \to C$ given by*

$$\Phi(b) = \sup \{ f(a) \in C \mid a \in A \ s.t. \ i(a) \preceq b \} \tag{3}$$

is called the pointwise left Kan extension of f along i *when it is well-defined (some suprema might not exist), and in which case it is necessarily monotonic.*

Global Transformations (GTs) make use of left Kan extensions to tackle the question of the synchronous deterministic local transformation of *arbitrary* kind of spaces. It is a categorical framework, but in the restricted case of posets, it works as follows. While B and C capture as posets the local-to-global relationship between the spatial elements to be handled (inputs and outputs respectively), A specifies a poset of local transformation rules. The monotonic functions i and f give respectively the left-hand-side and right-hand-side of the rules in A. Glancing at Eq. (3), the transformation mechanism works as follows. Consider an input spatial object $b \in B$ to be transformed. The associated output $\Phi(b)$ is obtained by gathering (thanks to the supremum in C) all the right-hand-side $f(a)$ of the rules $a \in A$ with a left-hand-side $i(a)$ occurring in b. The occurrence relationship is captured by the respective partial orders: $i(a) \preceq b$ in B for the left-hand-side, and $f(a) \preceq \Phi(b)$ in C for the right-hand-side.

The monotonicity of Φ is the formal expression of a major property of a GT: if an input b is a subpart of an input b' (*i.e.*, $b \preceq b'$ in B), the output $\Phi(b)$ has to occur as a subpart of the output $\Phi(b')$ (*i.e.*, $\Phi(b) \preceq \Phi(b')$ in C). This property gives to the orders of B and C a particular semantics for GTs which will play an important role in the present work.

Remark 1. Elements of B are understood as information about the input. So, when $b \preceq b'$, b' provides a richer information than b about the input that Φ uses to produce output $\Phi(b')$, itself richer than output $\Phi(b)$. However, Φ cannot deduce the falsety of a property about the input from the fact this property is not included in b; otherwise the output $\Phi(b)$ might be incompatible with $\Phi(b')$.

At the categorical level, the whole GT formalism relies on the key ingredient that the collection A of rules is also a category. Arrows in A are called *rule inclusions*. They guide the construction of the output and allow overlapping rules to be applied all together avoiding the well-known issue of concurrent rules application [6].

In cases where Φ captures the evolution function of a (discrete time) dynamical system (so particularly for the present work where we want to compare Φ to a CGD), we consider the input and output categories/posets B and C to be the same category/poset, making Φ an endo-functor/function.

3 Unifying Causal Graph Dynamics and Kan Extensions

The starting point of our study is that Eq. (2) in the definition of CGDs has almost the same form as Eq. (3). Indeed, if we take Eq. (3) and set $A = \mathcal{D}^r_{\Sigma, \Delta, \pi}$, $B = C = \mathcal{G}_{\Sigma, \Delta, \pi}$, the function i to be the projection function from discs to graphs that drops their centers (*i.e.*, $i((H, c)) = H$) and the function f to be the local function from discs to graphs, we obtain an equation for Φ of the form

$$\Phi(G) = \sup \{ f(D) \in \mathcal{G}_{\Sigma, \Delta, \pi} \mid D \in \mathcal{D}^r_{\Sigma, \Delta, \pi} \text{ s.t. } i(D) \preceq G \} \tag{4}$$

which is close to Eq. (2) rewritten

$$F(G) = \bigcup \{ f(D) \in \mathcal{G}_{\Sigma, \Delta, \pi} \mid D = G^r_v, \, v \in V(G) \}.$$

This brings many questions. What is the partial order involved in Eq. (4)? Is the union of Eq. (2) given by the suprema of this partial order? Is it the case that $i(D) \preceq G$ implies $D = G^r_v$ for some v in this order? Are f and F of Eq. (2) monotonic functions for this order? We tackle these questions in the following sections.

3.1 The Underlying Partial Order

Considering the two first questions, there is a partial order which is forced on us. Indeed, we need this partial order to imply that suprema are unions of graphs. But partial orders can be defined from their binary suprema since $A \preceq B \iff \sup \{A, B\} = B$. Let us give explicitly the partial order, since it is very natural, and prove afterward that it is the one given by the previous procedure.

Definition 8 (Subgraph). *Given two graphs G and H, G is a subgraph of H, denoted $G \subseteq H$, when*

$$V(G) \subseteq V(H) \; \wedge \; E(G) \subseteq E(H) \; \wedge \; \sigma(G) \subseteq \sigma(H) \; \wedge \; \delta(G) \subseteq \delta(H).$$

This defines a partial order $- \subseteq -$ on $\mathcal{G}_{\Sigma,\Delta,\pi}$ called the subgraph order. *The subgraph order is extended to pointed graphs by*

$$(G, v) \subseteq (H, u) \quad :\Longleftrightarrow \quad G \subseteq H \wedge v = u.$$

In this definition, the relational condition $\sigma(G) \subseteq \sigma(H)$, where these two functions are taken as sets of input-output pairs, means that $\sigma(G)(v)$ is either undefined or equal to $\sigma(H)(v)$, for any vertex $v \in V(G)$. The same holds for the condition $\delta(G) \subseteq \delta(H)$.

Let us now state that the subgraph order has the correct relation with unions of graphs. It similarly encodes consistency and intersections of pairs of graphs.

Proposition 1. *Two graphs G and H are consistent precisely when they admit an upper bound in $(\mathcal{G}_{\Sigma,\Delta,\pi}, \subseteq)$. The union of G and H is exactly their supremum (least upper bound) in $(\mathcal{G}_{\Sigma,\Delta,\pi}, \subseteq)$. The intersection of G and H is exactly their infimum (greatest lower bound) in $(\mathcal{G}_{\Sigma,\Delta,\pi}, \subseteq)$.*

The two first questions being answered positively, let us rewrite Eq. (4) as

$$\Phi(G) = \bigcup \{ f(D) \mid D \in \mathcal{D}^r_{\Sigma,\Delta,\pi} \text{ s.t. } i(D) \subseteq G \}. \tag{5}$$

3.2   Comparing Disks and Subgraphs

Let us embark on the third question: is it the case that $i(D) \subseteq G$ implies $D = G^r_v$ for some v? Making the long story short, the answer is no. But it is crucial to understand precisely why. Fix some vertex $v \in V(G)$. Clearly, in Eq. (2), the only considered disk centered on v is G^r_v. Let us determine now what are exactly the disks centered on v involved in Eq. (5), that is, the set $I_v := \{(H, v) \in \mathcal{D}^r_{\Sigma,\Delta,\pi} \mid H \subseteq G \}$. Firstly, G^r_v is one of them of course.

Lemma 1. *For any vertex $v \in V(G)$, $G^r_v \in I_v$.*

The concern is that G^r_v is generally not the only one disk in I_v as expected by Eq. (2). However, G^r_v is the maximal one in the following sense.

Lemma 2. *For any vertex $v \in V(G)$, consider the disk $G^r_v = (H, v)$. Then for any disk $(H', v) \in I_v$, we have $H' \subseteq H$.*

In some sense, the converse of the previous proposition holds: it is roughly enough to be smaller than G^r_v to be a disk of I_v, as characterized by the following two propositions.

Lemma 3. *The set of disks is the set of pointed connected finite graphs.*

Proposition 2. *For any vertex $v \in V(G)$, I_v is a principal downward closed set in the poset of graphs restricted to connected finite graphs containing v.*

We now know that the union of Eq. (5) receives a bigger set of local outputs to merge than the union of Eq. (2). But we cannot conclude anything yet. Indeed, it might be the case that all additional local outputs do not contribute anything more. This is in particular the case if disks $D \in I_v$ are such that $f(D) \subseteq f(G^r_v)$. This is related to the fourth and last question.

3.3   Monotonic and General Causal Graph Dynamics

The last remark invites us to consider the case where the CGD is monotonic. We deal with the general case afterward.

Monotonic CGDs as Kan Extensions. As just evoked, things seem to go well if the local rule f happens to be monotonic. All the ingredients have been already given and the proposition can be made formal straightforwardly.

Proposition 3. *Let* $F : \mathcal{G}_{\Sigma,\Delta,\pi} \rightarrow \mathcal{G}_{\Sigma,\Delta,\pi}$ *be a CGD with local rule* $f :$ $\mathcal{D}^r_{\Sigma,\Delta,\pi} \rightarrow \mathcal{G}_{\Sigma,\Delta,\pi}$ *of radius* r. *If* f *is monotonic, then* F *is the pointwise left Kan extension of* f *along* $i : \mathcal{D}^r_{\Sigma,\Delta,\pi} \rightarrow \mathcal{G}_{\Sigma,\Delta,\pi}$, *the projection of discs to graphs.*

Proof. The proposition is equivalent to show that $F(G) = \Phi(G)$ for all G. Summarizing our journey up to here, we now know that Eq. (5) and Eq. (2) are similar except that the former iterates over the set of disks $I = \bigcup_{v \in V(G)} I_v$ while the latter iterates over $J = \{G^r_v \mid v \in V(G)\}$, with $J \subseteq I$ by Prop. 1. We get $F(G) \subseteq \Phi(G)$. Moreover, by Prop. 2, for any $D \in I_v$ for any v, $D \subseteq G^r_v$, and by monotonicity of f, $f(D) \subseteq f(G^r_v)$. So $f(D) \subseteq F(G)$ for all D and $\Phi(G) \subseteq F(G)$, leading to the expected equality. □

The class of CGDs having such a monotonic local rule is easily characterizable: they correspond to CGDs that are monotonic themselves as stated by the following proposition.

Proposition 4. *A CGD* F *is monotonic iff* F *admits a monotonic local rule.*

Corollary 1. *A CGD is monotonic iff it is a left Kan extension.*

The Non-monotonic Case. The previous result brings us close to our initial goal: encoding any CGD as a GT. The job would be considered done only if there is no non-monotonic CGDs, or if those CGDs are degenerate cases. However, it is clearly not the case and most of the examples in the literature are of this kind, as we can see with the following example, inspired from [3].

Consider for example the modeling of a particle going left and right on a linear graph by bouncing at the extremities. The linear structure is coded using two ports on each vertex, say l and r for left and right respectively, while the particle is represented with the presence of a label on some vertex, with two possible values indicating its direction. See Fig. 1 for illustrations of such graphs. The dynamics of the particle is captured by F as follows. Suppose that the particle is located at some vertex v (in green in Fig. 1), and wants to go to the right. If there is an outgoing edge to the right to an unlabeled vertex u (in red in Fig. 1), the label representing the particle is moved from v to u (second row of Fig. 1). If there is no outgoing edge to the right (the port r of v is free), it bounces by becoming a left-going particle (third row of Fig. 1). F works symmetrically for a left-going particle.

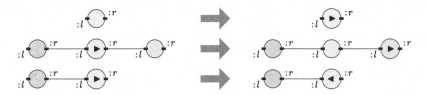

Fig. 1. Moving particle CGD - non-monotonic behavior. Each row represents an example of evolution with a graph G on the left and $F(G)$ on the right. Colors correspond to vertex names.

The behavior of F is non-monotonic since the latter situation is a sub-graph of the former, while the particle behaviors in the two cases are clearly incompatible. On Fig. 1, the right hand sides of the second and third rows are comparable by inclusion, but the left hand sides are not. This non-monotonicity involves a missing edge but missing labels may induce non-monotonicity as well. Suppose for the sake of the argument that F generates a right-going particle at any unlabeled isolated vertex (first row of Fig. 1). The unlabeled one-vertex graph is clearly a subgraph of any other one where the same vertex is labeled by a right-going particle and has some unlabeled right neighbor. In the former case, the dynamics puts a label on the vertex, while it removes it on the latter case. The new configurations are no longer comparable. See the first and second rows of Fig. 1 for an illustration.

Proposition 5. *CGDs are not necessarily monotonic.*

From Corollary 1, we conclude that all non-monotonic CGDs are not left Kan extensions as we have developed so far, *i.e.*, based on the subgraph relationship of Def 8. Analyzing the particle CGD in the light of Remark 1 tells us why. Indeed, the subgraph ordering is able to compare a place without any right neighbor with a place with some (left-hand-side of rows 3 and 2 in Fig. 1 for instance). Following Remark 1, in the GT setting, the former situation has less information than the latter: in the former, there is no clue whether the place has a neighbor or not; the dynamics should not be able to specify any behavior for a particle at that place. But clearly, for the corresponding CGD, both situations are totally different: the former is an extremity while the latter is not, and the dynamics specifies two different behaviors accordingly for a particle at that place.

4 Universality of Monotonic Causal Graph Dynamics

We have proven that the set of all CGDs is strictly bigger than the set of monotonic ones. However, we prove now that it is not more expressive. By this, we mean that we can simulate any CGD by a monotonic one, *i.e.*, monotonic CGDs are universal among general CGDs.

More precisely, given a *general* CGD F, a monotonic simulation of F consists of an encoding, call it $\omega(G)$, of each graph G, and a *monotonic* CGD F' such that

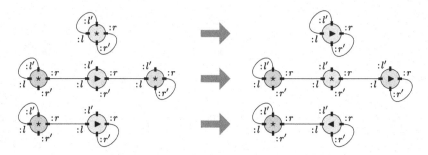

Fig. 2. Moving particle CGD - monotonic behavior. Compared to Fig. 1, the vertices have two additional ports (l' and r') and unlabeled vertices are now labeled by $\star$.

whenever $F(G) = H$ on the *general* side, $F'(\omega(G)) = \omega(H)$ on the *monotonic* side. Substituting H in the latter equation using the former equation, we get $F'(\omega(G)) = \omega(F(G))$, the exact property of the expected simulation: for any F, we want some ω and F' such that $F' \circ \omega = \omega \circ F$.

4.1 Key Ideas of the Simulation

In this section, we aim at introducing the key elements of the simulation informally and by the mean of the moving particle example.

Encoding the Original Graphs: The Moving Particle Case. Let us design a monotonic simulation of the particle dynamics. The original dynamics can be made monotonic by replacing the two missing edges at the extremities by easily identifiable loopback edges, making incomparable the two originally comparable situations. For such loopback edges to exist, we need an additional port for each original port, in our case say l' and r' for instance. For the case of non-monotonicity with labels, vertices where there is no particle (so originally unlabeled) are marked with a special label, say $\star$. Figure 2 depicts the same evolutions as Fig. 1 after those transformations.

This will be the exact role of the encoding function ω: the key idea to design a monotonic simulation is to make uncomparable the initially comparable situations. Any missing entity (edge or label) composing an original graph G needs to be replaced by a special entity (loopback edge or $\star$ respectively) in $\omega(G)$ indicating it was originally missing. At the end of the day, for any $G \subsetneq H$, we get $\omega(G)$ and $\omega(H)$ no longer comparable.

An Extended Universe of Graphs. All that remains is to design F' such that $F' \circ \omega = \omega \circ F$. It is important to note however that the universe of graphs

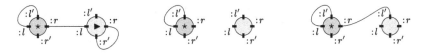

Fig. 3. Different classes of graphs: from left to right, a total graph, a coherent partial graph, an incoherent graph.

targeted by ω, which is the domain of F', is by design much larger than the original universe. Indeed, each vertex has now a doubled number of ports and the additional label $\star$ is available. So to be completely done with the task, we need F' to be able to work not only with graphs generated by the encoding, but also with all the other graphs.

Let us classify the various cases. Firstly, notice that the monotonic counterpart $\omega(G)$ of any graph G is "total" in the following sense: all vertices have labels, all original ports have edges, and all edges have labels. However, there also exists partial graphs with free ports and unlabeled vertices in this monotonic universe of graphs. Secondly, $\omega(G)$ uses the additional ports strictly for the encoding of missing edges with loopback edges. But, there are also graphs making arbitrary use of those ports and which are not "coherent" with respect to the encoding. Figure 3 illustrates the three identified classes: the middle graph is a coherent subgraph of the total graph on the right, and of an incoherent graph on the left. Notice that incoherent edges can always be dropped away to get the largest coherent subgraph of any graph. This is the case on the figure.

Disks in the Extended Universe. In order to design a local rule f' of the monotonic CGD F', we need to handle disks after encoding. The universe of graphs being bigger than initially, this also holds for disks, and the behavior of f' will depend on the class of the disk.

Let us first identify the monotonic counterpart of the original disks. They are called "total" disks and correspond to disks of some $\omega(G)$. For those disks, the injectivity of ω allows f' to simply retrieve the corresponding disk of the original graph G and invoke the original local rule f on it.

An arbitrary disk D may not be "total" and exhibit some partiality (free ports, unlabeled vertices or edges). Since f' is required to be monotonic, it needs to output a subgraph of the original local rule output. More precisely, we need to have $f'(D) \subseteq f'(D')$ for any total disk D' with $D \subseteq D'$. The easiest way to do that is by outputting the empty graph: $f'(D) = \emptyset$. (This solution corresponds to the so-called *coarse extension* proposed for CAs in [4]; *finer extensions* are also considered there).

Last, but not least, an arbitrary disk D may make an incoherent use of the additional ports with respect to ω. In that case, all incoherent information may be ignored by considering the largest coherent subdisk $D' \subseteq D$. The behavior of f' on D is then aligned with its behavior on D': $f'(D) = f'(D')$.

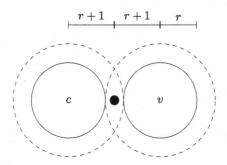

Fig. 4. We need to determine for each missing entity attached to a vertex or edge of $f(G_c^r)$, whether all other $f(G_v^r)$ agree to consider this entity as missing. The property 2 of local rules (Definition 5) tells us that only disks G_v^r that intersect G_c^r need to be checked. The furthest such v are at distance $2r + 2$. From there, we need to ask for radius r (so radius $r' = 3r + 2$ from c) to include the entirety of G_v^r (including its border at $r + 1$ from v, and therefore $3r + 3$ at most from c).

A Larger Radius. The last parameter of f' to tune is its radius. It will of course depend on the radius of the local rule of F. But it is worth noting that we are actually trying to build more information locally than the original local rule was trying to. Indeed, given a vertex of the original global output, many local rule applications may contribute concurrently to its definition. All of these contributions are consistent with each other of course, but it is possible for some local outputs to indicate some features of that vertex (label, edges) while others do not. If even one of them puts such a label for instance, then there is a label in the global output. It is only if none of them put a label that the global output will not have any label on it. The same holds similarly for ports: a port is free in the global output if it is so on all local outputs. Because the monotonic counterpart needs to specify locally if none of the local rule applications puts such a feature to a vertex, the radius of the monotonic local rule needs to be big enough to include all those local rule input disks. It turns out that the required radius for f' is $r' = 3r + 2$, if r is the radius of the original local rule f, as explained in Fig. 4.

4.2   Formal Definition of the Simulation

Now that we have all the components of the solution, let us make them precise.

The Encoding Function ω. The encoding function aims at embedding the graphs of $\mathcal{G}_{\Sigma,\Delta,\pi}$ in a universe where ports are doubled and labels are extended with an extra symbol. Set $\Sigma' := \Sigma \uplus \{\star\}$, $\Delta' := \Delta \uplus \{\star\}$, and $\pi' = \pi \times \{0, 1\}$. Ports $(a, 0) \in \pi'$ are considered to be semantically the same as their counterpart $a \in \pi$ while ports $(a, 1)$ are there for loopback edges. For simplicity, let us define the short-hands $\mathcal{G}$ and $\mathcal{G}'$ for $\mathcal{G}_{\Sigma,\Delta,\pi}$ and $\mathcal{G}_{\Sigma',\Delta',\pi'}$ respectively for the remainder of this section. Since we need to complete partial functions, we introduce the following notation: for any partial function $f : X \rightharpoonup Y$ and total function $z : X \to Z$, we define the total function $f!z : X \to Y \cup Z$ by

$$(f!z)(x) = \begin{cases} f(x) & \text{if } f(x) \text{ is defined,} \\ z(x) & \text{otherwise.} \end{cases}$$

For simplicity, we write $f!\star$ for $f!(x \mapsto \star)$ where x runs over X. This allows us to define the encoding $\omega(G) \in \mathcal{G}'$ of any $G \in \mathcal{G}$ as follows.

$$V(\omega(G)) := V(G)$$
$$\sigma(\omega(G)) := \sigma(G)!\star$$
$$E(\omega(G)) := \{\, \{u:(i,0), \mathrm{trg}_G^\star(u:i)\} \mid u:i \in V(G):\pi \,\}$$
$$\delta(\omega(G))(\{u:(i,a), v:(j,b)\}) := \begin{cases} (\delta(G)!\star)(\{u:0, v:j\}) & \text{if } b = 0 \\ \star & \text{if } u:i = v:j \end{cases}$$

The definition of $E(\omega(G))$ is written to make clear that all original ports are indeed occupied by an edge. The total function $\mathrm{trg}_G^\star : V(G) \times \pi \to V(G) \times \pi'$ is defined based on the partial function $\mathrm{trg}_G : V(G) \times \pi \rightharpoonup V(G) \times \pi$ as follows.

$$\mathrm{trg}_G(u:i) := v:j \text{ iff } \{u:i, v:j\} \in E(G)$$
$$\mathrm{trg}_G^\star := (t_0 \circ \mathrm{trg}_G)!t_1 \text{ where } t_m(u:i) := u:(i,m)$$

The definition of $\delta(\omega(G))$ deals with original edges for the first case, and with loopback edges for the second, which are the only two possibilities with respect to $\mathrm{trg}_G^\star$.

The Disk Encoding Function ω_r. Let us now define the shorthands $\mathcal{D}_r$ and $\mathcal{D}'_r$ for $\mathcal{D}^r_{\Sigma,\Delta,\pi}$ and $\mathcal{D}^r_{\Sigma',\Delta',\pi'}$ for any radius r. We put the radius as a subscript to make readable the four combinations $\mathcal{D}_r$, $\mathcal{D}'_r$, $\mathcal{D}_{r'}$, and $\mathcal{D}'_{r'}$. The function $\omega_r : \mathcal{D}_r \to \mathcal{D}'_r$ aims at encoding original disks as total disks. It is defined as

$$\omega_r((D, v)) = \omega(D)^r_v. \tag{6}$$

The reason we apply $(-)^r_v : \mathcal{G}' \to \mathcal{D}'_r$ on the raw result of ω is that the behavior of ω (adding loopback edges on each unused port and labels on unlabeled vertices and edges) is not desired at distance $r + 1$ for the result to be a disk of radius r. More conceptually, ω_r is the unique function having the following commutation property, which expresses that ω_r is the disk counterpart of ω.

Proposition 6. *For any graph $G \in \mathcal{G}$, $\omega_r(G^r_v) = \omega(G)^r_v$.*

Coherent Subgraph Function coh. As discussed informally, some graphs in $\mathcal{G}'$ may be ill-formed from the perspective of the encoding by using doubled ports arbitrarily. We define the function coh : $\mathcal{G}' \to \mathcal{G}'$ that removes the incoherencies.

$$V(\text{coh}(G')) := V(G')$$
$$\sigma(\text{coh}(G')) := \sigma(G')$$
$$E(\text{coh}(G')) := \{\{u\!:\!(i,0), v\!:\!(j,b)\} \in E(G') \mid$$
$$b = 1 \Rightarrow (v\!:\!j = u\!:\!i \ \wedge \ \delta(G')(\{u\!:\!(i,0), u\!:\!(i,1)\}) = \star)\}$$
$$\delta(\text{coh}(G')) := \delta(G') \restriction E(\text{coh}(G'))$$

Since coh only removes incoherent edges, and since they are stable by inclusion, we trivially have that

Lemma 4. coh *is monotonic.*

The Monotonic Local Rule. The monotonic local rule f' is defined in two stages. The first stage is to take a disk in $\mathcal{D}'_{r'}$ and transform it into something that the original local rule can work with. To do this, we remove any incoherencies with the function coh, and if the result is total, we restrict it to the correct radius and retrieve its counterpart in the original universe (which is possible because ω_r is injective). The original local function can be called on this counterpart. As discussed before, if the coherent sub-disk is not total, we simply output an empty graph. The signature of this function is $\phi : \mathcal{D}'_{r'} \to \mathcal{G}$.

$$\phi(D', c) := \begin{cases} \emptyset & \text{if } \text{coh}(D') \notin \text{Im}(\omega_{r'}) \\ f(\omega_r^{-1}(\text{coh}(D')^r_c)) & \text{otherwise} \end{cases} \tag{7}$$

Turning this result in $\mathcal{G}$ into its counterpart in $\mathcal{G}'$ is more complicated than simply invoking ω since one have to check that missing entities are really missing, as discussed in Fig. 4. This is the purpose of the second stage leading to the definition of the monotonous local rule $f' : \mathcal{D}'_{r'} \to \mathcal{G}'$ which, thanks to its radius $r' = 3r + 2$, is able to inspect all r-disks at distance $2r + 2$ from c.

$$f'((D', c)) = \omega \left(\bigcup_{v \in D'^{2r+1}_c} \phi(D', v) \right)^0_{\phi(D', c)} \tag{8}$$

In this equation, the union of all local results that can contribute to entities attached to $\phi(D', c)$ is built. In this way, they are all given a chance to say if what was missing in $\phi(D', c)$ is actually missing. Finally ω is applied and its result restricted to the sole vertices of $\phi(D', c)$ and their edges, using the notation $(-)^0_X$ of Eq. (1). Those vertices and edges are the only one for which all the possibly contributing disks have been inspected.

Proposition 7. f' *is a monotonic local rule.*

The Monotonic Simulation. We finally have the wanted CGD F' of local rule f':

$$F'(G') := \bigcup_{v \in V(G')} f'(G'^{r'}_v).$$

It remains then to show that F' is indeed a monotonic simulation of F, which is achieved with the two next propositions.

Proposition 8. *F' is monotonic.*

Proof. Since f' is monotonic by Proposition 7, F' is monotonic as well by Proposition 4. □

Proposition 9. *F' simulates F via the ω encoding, i.e., $F' \circ \omega = \omega \circ F$.*

Proof. Suppose $G' = \omega(G)$ for some $G \in \mathcal{G}$. Since all involved disks in $F'(\omega(G))$ are total and thanks to Proposition 6, the expression of $F'(\omega(G))$ simplifies drastically.

$$F'(\omega(G)) = \bigcup_{c \in V(G)} \omega \left(\bigcup_{v \in G_c^{2r+1}} f(G_v^r) \right)^0_{f(G_c^r)} . \tag{9}$$

Clearly, $F'(\omega(G))$ does not exhibit any incoherencies and $\omega(F(G))$ is total, so it is enough to show that $\omega(F(G)) \subseteq F'(\omega(G))$ to have the equality. This can be done entity by entity. Take a vertex $u \in \omega(F(G))$. It comes from some $f(G_c^r)$. So it belongs to the inner union of Eq. (9). It is preserved by ω then by $(-)^0_{f(G_c^r)}$, so it belongs to $F'(\omega(G))$. Consider now its label $\ell := \delta(\omega(F(G)))(u)$. If $\ell \neq \star$, there is some $v \in G_c^{2r+1}$ with $\ell = \delta(f(G_v^r))(u)$. So the inner union of Eq. (9) labels u by ℓ. The label is preserved by ω then by $(-)^0_{f(G_c^r)}$ since $u \in f(G_c^r)$. So $\delta(F'(\omega(G)))(u) = \ell$ as well. If $\ell = \star$, this means that none of the $f(G_v^r)$ put a label on u, so u is unlabeled in the inner union of Eq. (9). By definition, ω completes the labeling by $\star$ at u, and $(-)^0_{f(G_c^r)}$ preserves this label. So $\delta(F'(\omega(G)))(u) = \star$ as well. The proof continues similarly for edges and their labels, with an additional care for dealing with loopback edges. □

5   Conclusion

In this article, we planned to compare CGD and GT frameworks. The very particular route we have chosen for this task led us to identify the class of Monotonic CGDs which are both CGDs and GTs, and happen to be universal among all CGDs.

This work was guided by the formal similarities between the two frameworks leading to a list of four questions. The chosen strategy to cope with these questions was to tackle the first one "what is the order?" with the goal of answering positively to the second question "is the union of CGDs the supremum of this order?". This journey led us to identify the subgraph order to structure the set of port graphs instead of considering them independent (more precisely related by the disks only) as the original framework does. This pushes forward the idea of using graph inclusion to express gain of information as proposed by the GT framework, opening a new direction when designing CGDs by respecting the order with monotonicity.

One interesting thing to note is that a slight adaptation of the original definition of port graphs allows to represent general CGDs as Kan extensions without any encoding. Indeed, the encoding considered in this article leads to a universe of graphs where most of them are ill-formed. By adding explicitly to graphs additional features to represent the positive information that some port is not occupied or some label is missing, it is possible to get a universe where all objects make sense.

An alternative route to answer the four-question list is possible by tackling the first question with the goal of answering the third one (almost) positively, thus falsifying the second one. An order closely related to this route is the "induced subgraph" order stating that G is lower than H if $G = H^0_{V(G)}$. This order is stronger than the usual subgraph order but is also very interesting. Fewer graphs are "inducedly consistent" and one may ask if the subclass of CGDs definable with this stronger notion of consistency is universal. This might give an idea on whether the result of this article is isolated or, on the contrary, if it follows a common pattern shared with many instances.

Finally, let us not forget the next steps sketched in the introduction, in particular the quotienting of vertex names. In this paper, the notions of isomorphism and renaming-invariance of CGDs play no real role. Because GTs use a categorical language, this renaming-invariance can be handled by quotienting the objects of the poset, but not the arrows between the objects. This means that there would be now typically many arrows between two objects that the GT framework is able to cope with. Indeed, the formalism of GT was designed to work with objects like abstract graphs where no notion of names or positioning exist, but only their (possible multiple) occurrences in each other. A similar line of thought was already explored for cellular automata in [4]. Indeed, cellular automata are typically defined with a global positioning system, but many studies actually only care about the relative positioning of the information. A similar treatment for CGDs is left for future work.

Acknowledgments. This publication was made possible through the support of the ID #62312 grant from the John Templeton Foundation, as part of the 'The Quantum Information Structure of Spacetime' Project (QISS). The opinions expressed in this project/publication are those of the author(s) and do not necessarily reflect the views of the John Templeton Foundation. We also acknowledge the support of INRIA Saclay through the funding of two years of "delegation" of the first author to work on these topics.

Appendix

Proposition 1. *Two graphs G and H are consistent precisely when they admit an upper bound in $(\mathcal{G}_{\Sigma,\Delta,\pi}, \subseteq)$. The union of G and H is exactly their supremum (least upper bound) in $(\mathcal{G}_{\Sigma,\Delta,\pi}, \subseteq)$. The intersection of G and H is exactly their infimum (greatest lower bound) in $(\mathcal{G}_{\Sigma,\Delta,\pi}, \subseteq)$.*

Proof. Admitting an upper bound in $\mathcal{G}_{\Sigma,\Delta,\pi}$ means there is $K \in \mathcal{G}_{\Sigma,\Delta,\pi}$ such that $G \subseteq K$ and $H \subseteq K$. Since the subgraph order is defined componentwise, we consider the union of G and H as quadruplets of sets:

$$(V(G) \cup V(H), E(G) \cup E(H), \sigma(G) \cup \sigma(H), \delta(G) \cup \delta(H)),$$

which always exists and is the least upper bound in the poset of quadruplets of sets with the natural order. For this object to be a graph, it is enough to check $E(G) \cup E(H)$ is a non-intersecting two-element set, and that $\sigma(G)$ and $\sigma(H)$ (resp. $\delta(G)$ and $\delta(H)$) coincide on their common domain. This is the case when G and H admit an upper bound as required in the definition. Conversely, when the condition holds, the union is itself an upper bound.

For intersection, consider

$$(V(G) \cap V(H), E(G) \cap E(H), \sigma(G) \cap \sigma(H), \delta(G) \cap \delta(H))$$

which is clearly the greatest lower bound and is always a graph.   □

Lemma 1. *For any vertex $v \in V(G)$, $G_v^r \in I_v$.*

Proof. In the Definition 4, the graph component H of the pointed graph G_v^r is explicitly defined by taking a subset for each of the four components of G, as required in Definition 8 of subgraph.   □

Lemma 2. *For any vertex $v \in V(G)$, consider the disk $G_v^r = (H, v)$. Then for any disk $(H', v) \in I_v$, we have $H' \subseteq H$.*

Proof. By Definition 8 of the subgraph order, we need to prove four inclusions. For the inclusion of vertices, consider an arbitrary vertex $w \in V(H')$. By definition of disks of radius r, there is a path in H' from v to w of length at most $r + 1$. But since $H' \subseteq G$, we have $w \in V(G)$ and this path itself is also in G. So $w \in V(G)$ respects the defining property of the set $V(G_v^r)$ and therefore belongs to it. The three other inclusions ($E(H') \subseteq E(H)$, $\sigma(H') \subseteq \sigma(H)$ and $\delta(H') \subseteq \delta(H)$) are proved similarly, by using the definition of disks, then the fact $H' \subseteq G$, and finally the definition of G_v^r.   □

Lemma 3. *The set of disks is the set of pointed connected finite graphs.*

Proof. Indeed, for any disk $(H, v) \in \mathcal{D}_{\Sigma,\Delta,\pi}$, H is connected since all vertices are connected to v, and H is finite since all vertices have at most $|\pi|$ neighbors, so a rough bound is $|\pi|^r$ with r the radius of the disk. Conversely, for any pointed connected finite graphs (G, v), we have $(G, v) = G_v^r$ for any $r \geq |V(G)| - 1$, the length of the longest possible path.   □

Proposition 2. *For any vertex $v \in V(G)$, I_v is a principal downward closed set in the poset of graphs restricted to connected finite graphs containing v.*

Proof. Indeed, consider any disk (H, v) such that $H \subseteq G$. Now, take H' a connected graph containing v and such that $H' \subseteq H$. Since H is finite, so is H'. By Proposition 3, (H', v) is also a disk. By transitivity, $H' \subseteq H \subseteq G$. This proves that we have a downward closed set. This is moreover a principal one because of Propositions 1 and 2.   □

Proposition 4. *A CGD F is monotonic iff F admits a monotonic local rule.*

Proof. If F has a monotonic local rule f, F is a left Kan extension by Proposition 3 and is monotonic as recalled in Sect. 2.3.

Conversely, suppose F monotonic. By Definition 6 of CGDs, there is a local rule f', not necessarily monotonic, of radius r generating F. Consider f defined by $f((H,c)) = F(H)$ for any disk (H,c). f is monotonic by monotonicity of F. f is a local rule, which is checked easily using that f' is itself a local rule and that $f((H,c)) = \bigcup_{v \in H} f'(H_v^r)$. f generates F. For any graph G:

$$\bigcup_{c \in V(G)} f(G_c^r) = \bigcup_{c \in V(G)} \bigcup_{v \in G_c^r} f'(i(G_c^r)_v^r).$$

But $f'(i(G_c^r)_v^r) \subseteq F(i(G_c^r)) \subseteq F(G)$, the last inclusion coming from the monotonicity of F. So $\bigcup_{c \in V(G)} f(G_c^r) \subseteq F(G)$. Moreover, for any c, $f'(G_c^r) = f'(i(G_c^r)_c^r)$, so $f'(G_c^r) \subseteq \bigcup_{c \in V(G)} f(G_c^r)$ and $F(G) \subseteq \bigcup_{c \in V(G)} f(G_c^r)$. □

Proposition 5. *CGDs are not necessarily monotonic.*

Proof. Take $|\pi| = 1$, $\Sigma = \Delta = \emptyset$. In a graph G, we have isolated vertices and pairwise connected vertices. For the local rule, consider $r = 1$, so the two possible disks (modulo renaming) are the isolated vertex and a pair of connected vertices. There is no graph G such that the two disks appear together (otherwise G would ask the vertex to be connected and disconnected at the same time). So we can define f such that it acts inconsistently on the two disks, since property 4 of local rules does not apply here. But the isolated vertex is included in the connected vertices in the sense of $\subseteq$, but the image by f, then by F are not. □

Proposition 6. *For any graph $G \in \mathcal{G}$, $\omega_r(G_v^r) = \omega(G)_v^r$.*

Proof. Take $G_v^r = (D, v)$. After inlining the definition of ω_r, we are left to show $\omega(D)_v^r = \omega(G)_v^r$. For any entity (vertices, labels, edges) at distance at most r of v, $\omega(D)$ and $\omega(G)$ coincide. At distance $r + 1$, things may differ for the special label $\star$ and loopback edges, and beyond distance $r + 1$, $\omega(D)$ has no entities at all. However, all these differences are exactly what is removed by $(-)_v^r$. So, we get the expected equality. □

Proposition 7. *f' is a monotonic local rule.*

Proof. We first show that f' is a local rule, that is, it checks the four properties of Definition 5.

We need to show that for any isomorphism R, there is a conjugate R' such that $f' \circ R = R' \circ f'$. We show that the conjugate R' of R for f works. The result is obvious for $\mathrm{coh}(D') \in \mathrm{Im}(\omega_{r'})$. In the other case, take $E' := \mathrm{coh}(D')$:

$$f'(R(D',c)) = f'((R(D'),R(c)))$$

$$= \omega\left(\bigcup_{v \in R(D')_{R(c)}^{2r+1}} f(\omega_r^{-1}(R(E')_v^r)))\right)_{f(\omega_r^{-1}(R(E')_{R(c)}^r))}$$

$$= \omega\left(\bigcup_{v \in R(D')_{R(c)}^{2r+1}} R'(f(\omega_r^{-1}(E_{R'(v)}'^r))))\right)_{R'(f(\omega_r^{-1}(E_c'^r)))}$$

$$= R'\left(\omega\left(\bigcup_{v \in D_c'^{2r+1}} f(\omega_r^{-1}(E_v'^r))))\right)_{R'(f(\omega_r^{-1}(E_c'^r)))}\right)$$

$$= R'[\omega\left(\bigcup_{v \in D_c'^{2r+1}} f(\omega_r^{-1}(E_v'^r)))\right)_{f(\omega_r^{-1}(E_c'^r))}]$$

$$= R'(f(D',c)).$$

Consider a family of disks $\{(D_i',c_i)\}_{i \in I} \subset \mathcal{D}_{r'}'$ such that $\bigcap_i D_i' = \emptyset$. We need to show that $\bigcap_i f'(D_i',c_i) = \emptyset$. We suppose that for all i, $E_i' := \mathrm{coh}(D_i') \in \mathrm{Im}(\omega_{r'})$ (otherwise the result is trivial). First notice that for any family $\{v_i \in (D_i')_{c_i}^{2r+1}\}$, we have $\bigcap_i \omega_r^{-1}((E_i')_{v_i}^r) = \emptyset$, so $\bigcap_i f(\omega_r^{-1}((E_i')_{v_i}^r)) = \emptyset$. Suppose now some vertex u in $\bigcap_i f'(D_i',c_i)$. So, for each i there is some v_i such that $u \in f(\omega_r^{-1}((E_i')_{v_i}^r))$ which is impossible.

We need a bound b such that for all $(D',c) \in \mathcal{D}_{r'}'$, $|V(f'(D',c))| \leq b$. We consider the bound b given by f and show that it works. Indeed the last step of the computation of f' is precisely a restriction to the vertices of $f(\omega_r^{-1}(\mathrm{coh}(D')_c^r))$. So $|V(f'(D',c))| \leq |V(f(\omega_r^{-1}(\mathrm{coh}(D')_c^r)))| \leq b$.

Consider $G' \in \mathcal{G}'$ and $c,d \in V(G')$. We need $f'(G_c'^{r'})$ and $f'(G_d'^{r'})$ to be consistent. Once again, we suppose that $\mathrm{coh}(G_c'^{r'}) \in \mathrm{Im}(\omega_{r'})$ and $\mathrm{coh}(G_d'^{r'}) \in \mathrm{Im}(\omega_{r'})$ (otherwise the result is trivial). In order to use the consistency property of f, we build $G = \omega_r^{-1}(\mathrm{coh}(G_c'^{r'})) \cup \omega_r^{-1}(\mathrm{coh}(G_d'^{r'}))$. So for any pair of r-disks (D_1,D_2) of G, $f(D_1)$ and $f(D_2)$ are consistent. In other words, all the original r-disks involved in $f'(G_c'^{r'})$ and $f'(G_d'^{r'})$ are consistent with each other. It is particularly true for $\phi(G_c'^{r'})$ and $\phi(G_d'^{r'})$. Edges of $f'(G_c'^{r'})$ and $f'(G_d'^{r'})$ are consistent because either they come from $f(G_c^r)$ and $f(G_d^r)$ which are consistent, or they are loopback edges added by ω. Labels are also consistent for the same kind of reason, making $f'(G_c'^{r'})$ and $f'(G_d'^{r'})$ consistent.

We finally show that f' is monotonic. Take two disks D_1' and D_2' such that $D_1' \subseteq D_2'$. As f' is ultimately defined by cases, let us consider first the case where $\mathrm{coh}(D_1') \notin \mathrm{Im}(\omega_{r'})$. In this case $f'(D_1') = \emptyset$ so we necessarily have $f'(D_1') \subseteq f'(D_2')$. We are left with the case $\mathrm{coh}(D_1') \in \mathrm{Im}(\omega_{r'})$, meaning that $\mathrm{coh}(D_1')$ is a total disk. Since coh is monotonic by Lemma 4, $\mathrm{coh}(D_1') \subseteq \mathrm{coh}(D_2')$. And as a total disk, nothing but incoherences can be added to $\mathrm{coh}(D_1')$. So $\mathrm{coh}(D_1') = \mathrm{coh}(D_2')$ and therefore $f'(D_1') = f'(D_2')$. So the order is preserved in all cases, and f' is monotonic. □

References

1. Arrighi, P., Dowek, G.: Causal graph dynamics. In: Czumaj, A., Mehlhorn, K., Pitts, A.M., Wattenhofer, R. (eds.) Automata, Languages, and Programming - 39th International Colloquium, ICALP 2012, Warwick, UK, 9–13 July 2012, Proceedings, Part II. LNCS, vol. 7392, pp. 54–66. Springer, Cham (2012). https://doi.org/10.1007/978-3-642-31585-5_9

2. Arrighi, P., Martiel, S.: Quantum causal graph dynamics. Phys. Rev. D **96**(2), 024026 (2017)

3. Arrighi, P., Martiel, S., Perdrix, S.: Reversible causal graph dynamics: invertibility, block representation, vertex-preservation. Nat. Comput. **19**(1), 157–178 (2020). https://doi.org/10.1007/S11047-019-09768-0

4. Fernandez, A., Maignan, L., Spicher, A.: Cellular automata and Kan extensions. Nat. Comput. **22**(3), 493–507 (2023). https://doi.org/10.1007/S11047-022-09931-0

5. MacLane, S.: Categories for the Working Mathematician. Graduate Texts in Mathematics, Springer, New York (2013). https://doi.org/10.1007/978-1-4757-4721-8

6. Maignan, L., Spicher, A.: Global graph transformations. In: Plump, D. (ed.) Proceedings of the 6th International Workshop on Graph Computation Models Co-located with the 8th International Conference on Graph Transformation (ICGT 2015) Part of the Software Technologies: Applications and Foundations (STAF 2015) federation of conferences, L'Aquila, Italy, 20 July 2015. CEUR Workshop Proceedings, vol. 1403, pp. 34–49. CEUR-WS.org (2015). http://ceur-ws.org/Vol-1403/paper4.pdf

Application Domains

The 'Causality' Quagmire for Formalised Bond Graphs

Richard Banach[1]([⊠])[iD] and John Baugh[2][iD]

[1] Department of Computer Science, University of Manchester, Oxford Road, Manchester M13 9PL, UK
richard.banach@manchester.ac.uk
[2] Department of Civil, Construction, and Environmental Engineering, North Carolina State University, Raleigh North Carolina, 27695-7908, USA
jwb@ncsu.edu

Abstract. Bond graphs represent the structure and functionality of mechatronic systems from a power flow perspective. Unfortunately, presentations of bond graphs are replete with ambiguity, significantly impeding understanding. We extend the formalisation in preceding work to address the phenomenon of 'causality', intended to help formulate solution strategies for bond graphs, but usually presented in such vague terms that the claims made are easily shown to be false. We show that 'causality' only works as advertised in the simplest cases, where it mimics the mathematical definition of bond graph semantics. Counterexamples severely limit the applicability of the notion.

Keywords: Bond Graph · Formalisation · Bond Graph 'Causality' · Bond Graph Solution Strategy

1 Introduction

Bond graphs were introduced in the work of Paynter in 1959 [16, 17]. These days the most authoritative presentation is [12]. From the large related literature we cite [4, 7, 12, 14, 18].

Even the best presentations, though, are replete with ambiguity, often arising from a non-standard use of language that leaves the reader who is more used to conventional parlance in physical and engineering terminology feeling insecure and confused. In [3] we introduced a formalisation of bond graphs, allowing results to be presented with precision, and we discussed bond graph transformation, abstraction and refinement in that framework. In the present paper, we use the same framework to tackle so-called bond graph 'causality'. This has nothing to do with the normal physical notion of causality. Instead it is a game (in the formal mathematical sense of the word), for decorating a bond graph (usually in a locally progressive manner). The aim of this game is that its outcomes provide insight into how the equations of bond graph semantics (which are indeed equations, and thus undirected) may be transformed into a more algorithmically directed system, which can be used to actually solve for the behaviour of the system that the bond graph represents.

© The Author(s), under exclusive license to Springer Nature Switzerland AG 2024
R. Harmer and J. Kosiol (Eds.): ICGT 2024, LNCS 14774, pp. 99–117, 2024.
https://doi.org/10.1007/978-3-031-64285-2_6

The bottom line is that the game only works in the most simple of cases, and then it just follows the semantics, interpreted in the simplest possible way. When the game doesn't work out that way, it indicates that something more sophisticated is needed, though it gives little clue about *what* might be needed, without deeper (and often *ad hoc*) analysis. For such cases, more powerful generic techniques have been available for quite some time, and so, the case for the continued adherence to the 'causality' notion, with the enthusiasm seen in the textbook literature, is severely weakened.

The rest of this paper is as follows. Sections 2 and 3 recapitulate, tersely, the essentials of bond graphs as presented in [3]. Section 4 constructs bond graph 'causality', which we prefer to call dependency. Section 5 attempts to use the dependency notion to construct solutions to bond graph systems, but instead comes up with many counterexamples that show that the original idea doesn't work out. Lack of space forces many details to be omitted here. Section 6 presents two key theorems that explain why the 'nice' examples succeed, but the others don't. Section 7 indicates much more mathematically robust ways of approaching the solving of bond graphs, and Sect. 8 concludes.

2   Classical Physical Theories for Classical Engineering

We tersely summarise the physical theories of [3].

[**PT.1**] A system consists of interconnected devices, and operates within an environment from which it is separated by a notional boundary. A system can input or output energy from the environment through specific devices. Aside from this, the system is assumed to be isolated from the environment.

[**PT.2**] The classical physics relevant to bond graphs is captured, in general, by a system of second order ordinary differential equations (ODE) of the form:

$$\Phi(q'', q', q) = e \tag{1}$$

Of most interest is the case where Φ is a linear constant coefficients (LCC) ODE:

$$L \frac{d^2 q}{dt^2} + R \frac{dq}{dt} + K q = e \tag{2}$$

The system (1) or (2) concerns the behaviour of one (or more) generalised displacement(s), referred to as **gendis** with typical symbol q (*mech*: displacement; *elec*: charge). The gendis time derivative q' is called the **flow**, with typical symbol f (*mech*: velocity; *elec*: current). The gendis second time derivative q'' is called the generalised acceleration **genacc**, with typical symbol a (*mech*: acceleration; *elec*: induction). These all occur in the LHS of (1)-(2). On the RHS of (1)-(2) is the **effort**, typical symbol e (*mech*: force; *elec*: voltage).

[**PT.3**] Of particular importance among the variables mentioned is the product of effort and flow, because $e \times f$ is **power**, i.e., the rate at which energy is processed. The transfer and processing of power is crucial for the majority of engineered systems. According to [**PT.1**], energy can only enter or exit a system through specific kinds of device. Therefore, all other devices conserve energy within the system.

[PT.4] Engineered systems are made by connecting relatively simple devices.

Dissipator: R-device (*mech*: dashpot; *elec*: resistor)	$R f = e$	(3)
Compliant: C-device (*mech*: spring; *elec*: capacitor)	$K q = e$	(4)
Inertor: L-device (*mech*: mass; *elec*: inductor)	$L a = e$	(5)

A dissipator is a device that can output energy into the environment in the form of heat. Compliants and inertors are devices that store energy. Specifically, the power they receive is accumulated within the device as stored energy, to be released back into the rest of the system later.

Sources input power to/from the system of interest in predefined ways.

Effort source: SE-device (*mech*: force; *elec*: voltage)	$e = \Phi_E(t)$	(6)
Flow source: SF-device (*mech*: velocity; *elec*: current)	$f = \Phi_F(t)$	(7)

Note that the power input and output to/from each of these cases is not determined by Eqs. (6)–(7) alone (since the other variable is not specified), but by the behaviour of the rest of the system that they are connected to.

Transformers and gyrators are devices that are connected to two power connections (two efforts and two flows), and allow non-trivial tradeoffs between the effort and the flow in the two connections.

Transformer: TR-device (*mech*: lever; *elec*: transformer)
$$e_1 = h e_2 \quad \text{and} \quad h f_1 = f_2 \tag{8}$$
Gyrator: GY-device (*mech*: gyroscope; *elec*: transducer)
$$e_1 = g f_2 \quad \text{and} \quad g f_1 = e_2 \tag{9}$$

Junctions are devices that distribute power among several power connections $1 \ldots n$ (each with its own effort and flow), while neither storing nor dissipating energy. Aside from transformers and gyrators just discussed, the only remaining cases that arise are the common effort and common flow cases.

Common effort: E-device

(*mech*: common force; *elec*: common voltage, Kirchoff's Current Law)
$$e_1 = e_2 = \ldots = e_n \quad \text{and} \quad f_1 + f_2 + f_3 + \ldots + f_n = 0 \tag{10}$$
Common flow: F-device

(*mech*: common velocity; *elec*: common current, Kirchoff's Voltage Law)
$$e_1 + e_2 + e_3 + \ldots + e_n = 0 \quad \text{and} \quad f_1 = f_2 = \ldots = f_n \tag{11}$$

Noting that n is not fixed, **E** and **F** devices for different n are different devices.

[PT.5] From the bond graph perspective, the individual power connections to a device are conceptualised as power **port**s, through which power flows into or out of the device. Dissipators, compliants and inertors are therefore **one port** devices. Power sources are also one port devices. Transformers and gyrators are **two port** devices, while junctions are **three (or more) port** devices. For each category of device, all of its ports are individually labelled.

[**PT.6**] Since power is the product of an effort variable and a flow variable, each port is associated with an (effort, flow) variable pair whose values at any point in time define the power flowing through it.

[**PT.7**] All the variables involved in the description of a system are typed using a consistent system of dimensions and units. It is assumed that this typing is sufficiently finegrained that variables from different physical domains cannot have the same type. We do not have space to elaborate details, but since the only property of dimensions and units that we use is whether two instances are the same or not, it is sufficient to assume a set $\mathcal{DT} \times \mathcal{UT}$ of (dimension, unit) terms, that type the variables we need.

[**PT.8**] We refer to the elements of a system using a hierarchical naming convention. Thus, if Z-devices have ports p, then Z.p names the p ports of Z-devices. And if the effort variables of those ports are called e, then Z.p.e names those effort variables. Analogously, Z.p.f would name the flow variables corresponding to Z.p.e. Z.p.e.DU names the dimensions and units of Z.p.e, while Z.p.f.DU names the dimensions and units of Z.p.f.

[**PT.9**] For every (effort, flow) variable pair in a system (belonging to a port p of device Z say), for example (Z.p.e, Z.p.f), there is a **directional indication** (determined by the physics of the device in question and the equations used to quantify its behaviour). This indicates whether the power given by the product Z.p.e×Z.p.f is flowing **in**to or **out**of the port when the value of the product is ***positive***.

For the devices spoken of in [**PT.4**], there is a standard assignment of **in/out** indicators to its ports. Thus, for **R, C, L** devices, the standard assignment to their single port is **in**. For **SE, SF** devices, the standard assignment to their single port is **out**. For **TR, GY** devices, the standard assignment is **in** for one port and **out** for the other, depicting positive power flowing through the device. For the **E** and **F** devices, we standardise on a symmetric **in** assignment to all the ports.

3   Bond Graph Basics

Bond graphs are graphs which codify the physical considerations listed above.

[**UNDGR**] An **undirected graph** is a pair (V, E) where V is a set of vertices, and E is a set of edges. There is a map ends : $E \rightarrow \mathbb{P}(V)$, where ($\forall edg \in E \bullet$ card(ends(edg)) = 2) holds, identifying the pair of *distinct* elements of V that any edge edg connects. When necessary, we identify the individual ends of an edge edg, where ends(edg) = $\{a, b\}$ using (a, edg) and (b, edg). If ends(edg) = $\{a, b\}$, then we say that edg is incident on a and b.

Our formulation of conventional power level bond graphs (DPLBGs, directed power level bond graphs) starts with PLBGs, which are undirected labelled graphs. PLBGs are assembled out of the following ingredients. Figure 1 illustrates the process.

[**PLBG.1**] There is an alphabet $\mathcal{VL} = \mathcal{BVL} \cup \mathcal{CVL}$ of vertex labels, with basic vertex labels $\mathcal{BVL} = \{\mathbf{R, C, L, SE, SF, TR, GY, E, F}\}$, and user defined labels $\mathcal{CVL}$.

[**PLBG.2**] There is an alphabet $\mathcal{PL}$ of port labels and a map lab2pts : $\mathcal{VL} \rightarrow \mathbb{P}(\mathcal{PL})$, which maps each vertex element label to a set of port labels. (Below, we just say port, instead of port label, for brevity).

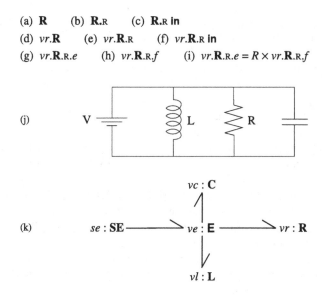

(a) **R**   (b) **R**.R   (c) **R**.R **in**

(d) *vr*.**R**   (e) *vr*.**R**.R   (f) *vr*.**R**.R **in**

(g) *vr*.**R**.R.*e*   (h) *vr*.**R**.R.*f*   (i) *vr*.**R**.R.*e* = $R \times vr$.**R**.R.*f*

(j)

(k)

Fig. 1. Stages in bond graph construction: (a) a vertex label (for a dissipator); (b) adding a port; (c) adding a directional indicator; (d)-(f) assigning attributes (a)-(c) to a vertex *vr*; (g) *vr*'s effort variable; (h) *vr*'s flow variable; (i) *vr*'s constitutive equation; (j) a simple electrical circuit embodying a dissipator (among other components); (k) a bond graph of the circuit in (j). Dimensions and units are not shown.

[PLBG.3] There are partial maps labpt2effDU , labpt2floDU : $\mathcal{VL} \times \mathcal{PL} \rightharpoonup$ $\mathcal{DT} \times \mathcal{UT}$ mapping each (vertex label, port) pair to the dimensions and units (not elaborated here) of the (forthcoming) effort and flow variables. Details are formalised in [3].

[PLBG.4] There is an alphabet $\mathcal{IO} = \{\textbf{in}, \textbf{out}\}$ of standard directional indicators, and a partial map labpt2stdio : $\mathcal{VL} \times \mathcal{PL} \rightharpoonup \mathcal{IO}$. Details are formalised in [3].

The above clauses capture properties of PLBGs that are common to all vertices sharing the same label. Other properties are defined per vertex. PLBGs can now be constructed.

[PLBG.5] A power level bond graph PLBG is based on an undirected graph $BG = (V, E)$ as in **[UNDGR]**, together with additional machinery as follows.

[PLBG.6] There is a map ver2lab : $V \rightarrow \mathcal{VL}$, assigning each vertex a label.

When map ver2lab is composed with lab2pts, yielding map ver2pts = ver2lab ⨾ lab2pts : $V \rightarrow \mathbb{P}(\mathcal{PL})$, each vertex acquires a set of port labels.

When map ver2lab, with a choice of port, is composed with maps labpt2effDU and labpt2floDU, yielding maps verpt2effDU = ver2lab × **Id** ⨾ labpt2effDU : $V \times \mathcal{PL} \rightharpoonup$ $\mathcal{DT} \times \mathcal{UT}$ and verpt2floDU = ver2lab × **Id** ⨾ labpt2floDU : $V \times \mathcal{PL} \rightharpoonup \mathcal{DT} \times \mathcal{UT}$, each (vertex, port) pair acquires dimensions and units for its effort and flow variables.

When map ver2lab, with a choice of port, is composed with map labpt2stdio, yielding partial map verpt2stdio = ver2lab × **Id** ⨾ labpt2io : $V \times \mathcal{PL} \rightharpoonup \mathcal{IO}$, each (vertex, port) pair acquires its *standard* directional indicator.

[PLBG.7] In practice, and especially for **E**, **F** devices, directional indicators are often assigned per (vertex, port) pair rather than generically per (vertex label, port). Thus there is a partial map verpt2io : $V \times \mathcal{PL} \nrightarrow \mathcal{IO}$, and verpt2io$(ver, pt)$ may, or may not, be the same as verpt2stdio(ver, pt). Details are formalised in [3].

There is a partial injective map verpt2eff : $V \times \mathcal{PL} \nrightarrow \mathcal{PV}$ giving each (vertex, port) pair (ver, pt) where $pt \in$ ver2pts(ver), an effort variable with dimensions and units verpt2effDU(ver, pt). Similarly, verpt2flo : $V \times \mathcal{PL} \nrightarrow \mathcal{PV}$ gives each (ver, pt) a flow variable with dimensions and units verpt2floDU(ver, pt). Also, we must have ran(verpt2eff) $\cap$ ran(verpt2flo) $= \varnothing$. These variables are referred to by extending the hierarchical convention of **[PT.8]**. Thus $v.Z.pt.e$ refers to vertex v, with label Z, having port pt, and so $v.Z.pt.e$ is the relevant effort variable, etc.

There is a map ver2defs : $V \rightarrow$ physdefs, which yields, for each vertex ver, a set of constitutive equations and/or other properties, that define the physical behaviour of the device corresponding to the vertex ver. Additionally, the properties in ver2defs(ver) can depend on generic parameters (from a set $\mathcal{PP}$ say), so there is a map ver2pars : $V \rightarrow \mathcal{PP}$. Additionally, the properties in ver2defs(ver) can depend on some bound variables. When necessary, we refer to such variables using $\mathbf{BV}$(ver2defs(ver)). Details are formalised in [3].

[PLBG.8] There is a bijection Eend2verpt : $V \times E \nrightarrow V \times \mathcal{PL}$, from edge ends in BG, to port occurrences, and for each edge $edg \in E$, where ends$(edg) = \{a, b\}$, the effort and flow variables at the ends of edg, have the same dimensions and units. Details are formalised in [3].

[PLBG.9] There is a map edge2dir : $E \rightarrow$ physdir, where physdir is a set of equalities and antiequalities between effort and flow variables, recording whether the power flow conventions of the two devices at the ends of each edge are compatible or not, and if not, adjusting suitably. Details are formalised in [3].

[SEMANTICS] The dynamics specified by a PLBG is the family of solutions to the collection of constraints specified by ver2defs (and ver2pars, edge2dir).

[PLBG.10] A PLBG is a DPLBG (directed PLBG, as in the literature) iff for each $edg \in E$, the power flow conventions are compatible. In such cases, edges become harpoons (half-arrows), showing the direction of positive power flow. In any case, the edges are called **bonds**.

A consequence of a unidirectional convention for variables along edges is that it permits the use of directed (rather than undirected) graphs as the underlying formalism. Although this makes the handling of edge ends a little easier, the impediments to bottom up bond graph construction that it imposes dissuaded us from following this approach.

4   Bond Graph 'Causality' / Dependency, and Its Assignment

In the bond graph literature the word/s 'causality'/'causal' appear a lot. It is a most unfortunate choice of terminology, and the quotes we use indicate our dissatisfaction.

In the context of physical systems, the concept of causality has a well understood meaning. It is connected with the idea that certain physical phenomena or situations force others to be the case. We do not have the space to discuss how this works properly, but briefly, the 'motive force' that causes a physical state of affairs to persist, or that

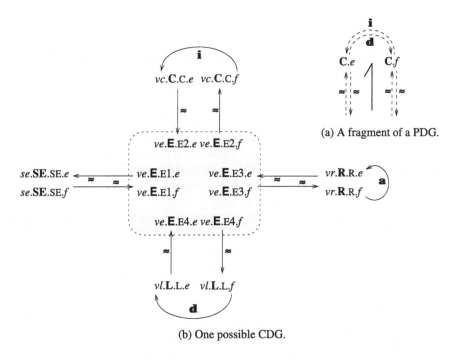

(a) A fragment of a PDG.

(b) One possible CDG.

Fig. 2. A PDG fragment and CDG option for Fig. 1.

causes one physical state of affairs to change to another, is primarily thermodynamic. In the physical context (and physical systems of a particular kind, are after all, what bond graphs are for), the firmly based concept of causality is a topic for discussion in thermodynamics and statistical physics.

By contrast, bond graph 'causality' is much more concerned with strategies for solving the system of equations that a bond graph defines. Worse, the connection between it and the necessities of actually solving these equation systems is tenuous, often to the point of invisibility. In this paper we prefer the word 'dependency', and in this section, we construct dependency, postponing thought about how it might be used till later.

Figure 2a shows a fragment of a **pre-dependency graph (PDG)**, for the bond graph of Fig. 1. Dependency concerns the variables of the bond graph, thus is easiest to depict using different, *directed* graphs, that follow the structure of the underlying bond graph. The idea is that the variable at the head of the arc *depends* (in some way not yet specified) on the variable at the tail of the arc. Since arcs are included in both directions for each variable, no commitment has yet been made as to what depends on what; thus it is a 'pre-' graph, indicated by making the arcs dashed. The $\approx$ label refers to the (anti-) equalities between device variables at the two ends of a bond. The **i** indicates that if the effort variable $\mathbf{C}.e$ depends on the flow variable $\mathbf{C}.f$, then **i**ntegration of $\mathbf{C}.f$ is needed (according to the constitutive equation (4)). If dependency is the other way round, then **d**ifferentiation of $\mathbf{C}.e$ is needed (again by (4)). Algebraic dependency (for dissipators) is indicated by **a**.

In a PDG, any two variables that are connected by an arc, are connected by two arcs, one in each direction, representing potential for, but indifference to, dependency between them. The indifference is removed in a **candidate dependency graph** (**CDG**), in which, for each such pair, one arc is deleted provided: (1) for each bond, the surviving arcs for effort and flow variables are oppositely directed; (2) the surviving arc for any physical device is co-aligned with the effort and flow arcs on the bond that joins it to the rest of the bond graph. Figure 2b shows a (somewhat perversely chosen) CDG for the Fig. 1 example. Solid arcs indicate commitment to a choice. All variables are shown in full detail. The only things missing are the device arcs for the **E** junction (all efforts strongly connected, similarly for flows) which would massively clutter the figure.

Note that we have not said what dependency *means*. The rest of this paper shows that this is not an easy question to answer.

Since all bond arcs come in oppositely directed pairs, any CDG can be represented more efficiently, by decorating the bonds. We use decorations ▶, ▷, >, which *always* point in the direction of the *effort arcs* of the underlying CDG, and work as follows.

Firstly, the decorations are determined by the devices at the ends of a bond, so there can be clashes. These are resolved by ▶ overriding ▷ overriding >, as needed. Then: **SE** has ▶ on its bond, pointing away from it; **SF** has ▶ on its bond, pointing towards it; **E** has one bond with ▷ pointing towards it (the **dominant bond**) and all other bonds (the **dependent bonds**) have > pointing away from it; **F** has one bond with ▷ pointing away from it (the **dominant bond**) and all other bonds (the **dependent bonds**) have > pointing towards it; **TR** has > on its bonds, one towards and one away from it; **GY** has > on its bonds, either both towards or both away from it; **L**, **C**, **R**, can have any decoration on their bond, but **L** *prefers*

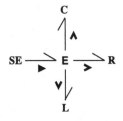

Fig. 3. A DBG for the example in Fig. 1.

towards and **C** *prefers* away from (and **R** is indifferent). A CDG whose dependencies conform to the restrictions on direction implicit in the preceding is called a **normal dependency graph** (**NDG**), and a bond graph that encodes these dependency decorations on its bonds in the manner just described is called a **dependency bond graph** (**DBG**). It is clear that the CDG in Fig. 2b breaks all the rules that we have been imposing for an NDG. Figure 3 shows the (only possible) DBG corresponding to Fig. 1. To make Fig. 2b into an NDG, we would have to reverse all the dependency arcs shown in the figure.

The DBG definition just defined specifies how the DBG decorations should look. The traditional bond graph literature (e.g. [4,7,12,14,18]) presents what is termed a sequential 'causality' assignment procedure (SCAP) to achieve an acceptable decoration. In reality, this is a greedy algorithm executing a rule based system involving priorities for the assignment, that features, potentially, a lot of concurrency that is implicit in the original formulation. Figure 4 summarises the rule based form.

It is clear that Fig. 4 is a greedy algorithm that depends on making choices at various points, and thus, for bond graphs with highly tangled junction structures, can involve a superpolynomial number of choice configurations to try, in order to determine whether a DBG set of decorations for a bond graph exists. In [15] the authors undertake a more incisive analysis, which shows that the job can be done in polynomial time.

Input: An undecorated bond graph
Output: A fully decorated bond graph, or **abort**

$Pri := 1$

Assign DBF decorations in order of priority Pri ;
   If inconsistency found at any point **Then abort Fi**
1. Assign to *all* **SE** and **SF** bonds ; $Pri := 2$
2. Assign to *all* **E** and **F** *not fully assigned but with already assigned dominant bond* ; $Pri := 3$
3. Assign to *some* **L** or **C** *able to take its preferred direction* ;
   If successful **Then** $Pri := 2$ **Else** $Pri := 4$ **Fi**
4. Assign to *some* **R**, **E**, **F**, **TR**, **GY** arbitrarily (but consistent with its definition) ; $Pri := 2$

Fig. 4. Traditional Dependency Assignment Procedure

5   Dependency Bond Graph Solution Strategies and Problems

The basic hypothesis of the traditional 'causality' approach of the bond graph literature, is that by decorating the bond graph with suitable markings, a strategy can be developed that reduces the solution of system behaviour to a set of oriented explicit ODEs and/or assignments, involving a minimal set of variables.

> *It has to be said that the hypothesis fails, and the stated objective is unachievable.*

The rule based process outlined in Fig. 4 was the main step of the first attempt to achieve the stated goal, and on benign examples it succeeds. But it was soon noticed that it does not succeed on all bond graphs. The rather extensive literature on the potential foundations of bond graphs that was produced in the '70 s, '80 s and '90 s (surveyed comprehensively in [4]) does not result in establishing a procedure that always produces the desired outcome. Despite this, all the standard texts unfailingly discuss 'causality' in terms that raise no qualms about the relevance and validity of the notion, or about what caveats might need to hold for its validity.

The first thing to do is to show how the ideal outcome fails, which happens in many ways. Let us assume, for now, that dependency means that once the thing at the tail of a dependency arc is 'known', the thing at the head becomes 'knowable' too. To start with, we consider Fig. 2b, which is out of scope really, since it breaks all the rules. The most 'known' thing there is **SE**.f, since it depends on nothing else. But it is *not* known, since only **SE**.e is defined for an effort source, so we have to guess the **SE**.f value. Once guessed, we would have to guess how that flow is partitioned between **L**, **C**, **R**. Once guessed, we would solve algebraically for **R**, differentiate for **L**, integrate for **C**. Miraculously, we would come up with the same effort in each case at **E**, which agreed with the predefined value of **SE**.e. Of course it is crazy to try to solve the system this way, but it is not logically inconsistent, being simply an angelically derived solution.

Within the rules is Fig. 3, which reverses the dependencies. Now we start with the truly known **SE**.e, shared via **E** to each of **L**, **C**, **R**. This is integrated to solve for **L**, differentiated for **C**, scaled for **R**. Thus, sticking to the rules is some help here as the solution process is deterministic. There is one niggle though. While integration is used for **L**, in line with **L**'s preferred dependency, differentiation is used for **C**, in line with **C**'s *unpreferred* dependency. While this is no problem mathematically, differentiation is highly deprecated in engineering terms as it generates jolts and spikes in variables' behaviours. The 'wrong' dependency for **C** heralds this kind of issue, but we see that merely having a DBG for Fig. 1 does not guarantee the desired explicit ODE solution.

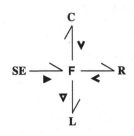

Fig. 5. A DBG for Fig. 1 with **F** instead of **E**.

Figure 5 replaces **E** with **F** in Fig. 1. This makes things more complicated. There are now three DBGs for Fig. 5 depending on which bond of **F** is chosen dominant. What is shown is the one that gives both **L** and **C** their preferred dependency. The traditional solution approach to more complex examples is **tracking back**, to derive the ODEs that solve the system. We start with the variable of an **L** or **C** that would appear in the LHS of an explicit ODE, equate it to the other side of the constitutive equation, and successively substitute for the variable there using the equations of the junction structure until source or physical variables are encountered. In **optimised tracking back** we do not stop at dissipators, but, recognising that their constitutive equations are linear algebraic, continue through those as if they were junctions. This causes **layer switching**, changing focus between efforts and flows, recognising that junction structures only couple efforts to efforts and flows to flows (**GY**s also layer switch).

Tracking back from **C**.e to **C**.f and then through the **equality layer** of **F** leads to **L**.f and to **C**.$e' = \frac{1}{C}$ **L**.f. Tracking back from **L**.f to **L**.e and then through the **sum layer** of **F** is more complicated as it branches out to **R**, **C** and **SE**. Eventually we derive **L**.$f' = \frac{1}{L}$ (**SE**.e − **C**.e − R**L**.f). We have a coupled pair of first order ODEs, just as promised by traditional 'causality'. Solving them does however require the derivative of **SE**.e — **C** is 'too close' to an effort source for any jolt (e.g. switching on a constant non-zero effort) to be smoothed out by the system.

A surprising number of examples of awkward behaviour in bond graphs do not require the involvement of inertors or compliants—dissipators are quite capable of generating them without help. Figure 6 shows a simple example of several dissipators connected to an effort source via a common effort junction. Using electrical language, we have resistors, composed in parallel. If we track back from the flow of the i'th dissipator, we get: $\mathbf{R}_i.f \times R_i = I_i R_i = V = \mathbf{R}_i.e = \mathbf{E}.e_i = \mathbf{E}.e_0 = \mathbf{SE}.e$, where $\mathbf{E}.e_0$ is the effort variable of **E** on the bond to **SE**. The choice of flow variable to track back from

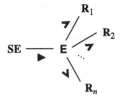

Fig. 6. Dissipators, composed in parallel.

was crucial to arriving at the *known* variable of **SE** rather than the unknown one. Assuming the natural current orientations for the dependent variable **SE**.f, given by: $\mathbf{SE}.f = \mathbf{E}.f_0 = \sum_{i=1}^{n} \mathbf{R}_i.f$ quickly leads to the familiar law $R^{-1} = \sum_{i=1}^{n} R_i^{-1}$, where R is the effective resistance of the parallel composition. Note that paying attention to the source unknown variable is indispensable for this derivation. Surprisingly, it is often stated in the bond graph literature that the unknown variables of sources are of no interest at all.

Figure 7 is the series counterpart of Fig. 6, i.e. several dissipators connected to an effort source via a common flow junction. In electrical language, we have resistors in series. The most immediate thing we notice is that the obvious symmetry of the physical system has been destroyed, in the bond graph representation, by the DBG dependency decorations. The most sensible arrangement would have been to have the dominant bond of **F** pointing towards **SE**. But since the decoration on the **SE** bond *must* be away from **SE**, and the **F** junction *must* have a dominant bond pointing outwards, one

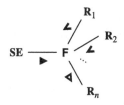

Fig. 7. Dissipators, composed in series.

dissipator must acquire the dominant decoration of the **F** junction, and R_n has been chosen randomly. From a computational perspective, although the effort is assumed known from **SE**, the unique value of the flow shared by the whole system has to be derived by aggregating the relevant constitutive equations for the dissipators. The source of the difficulty is that, in the sum layer of a junction, even if the value on the dominant bond is known precisely, this only supplies partial information to the dependent bonds.

In more detail, tracking the dependencies back to the source from the flow on the n'th dissipator via its effort, we find: $R_n.e = SE.e - \sum_{i=1}^{n-1} R_i.e$. Rearranging, we get: $SE.e = \sum_{i=1}^{n} R_i.e = \sum_{i=1}^{n} R_i.f \times R_i = SE.f \times \sum_{i=1}^{n} R_i$. After this, the value of the **SE** flow, and the familiar law for series composition $R = \sum_{i=1}^{n} R_i$, can be derived. For $i \neq n$, the DBG dependencies point in the opposite direction, but we have little option but to pursue the same calculational route. Similar issues arise if we replace parallel or series compositions of dissipators with compositions of inertors or of compliants, except that we have some integration or differentiation to do.

Figure 8 shows an example in which the tracking back ansatz leads to both known and unknown variables of sources appearing as inhomogeneous terms in the differential equations that are generated. It is a given, though invariably unstated in the bond graph literature, that when we derive the requirement that a source variable is needed for the value of some physical variable, i.e. it appears as an inhomogeneous term in an explicit ODE for the variable, it is the *known* variable that appears and not the unknown one. But bond graphs do not guarantee that that is necessarily the case.

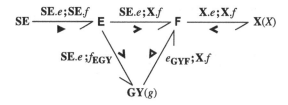

Fig. 8. A DBG showing tracking back to both known and unknown source variables.

In Fig. 8, the first gyrator equation is $\mathbf{SE}.e = g\,\mathbf{X}.f$, which coincides with tracking back from variable $\mathbf{X}.f$ to known source $\mathbf{SE}.e$. Tracking back from $\mathbf{X}.e$ however, requires both gyrator equations and leads to $\mathbf{X}.e = \mathbf{SE}.e(1 - \frac{1}{g}) + g\,\mathbf{SE}.f$, in which both known $\mathbf{SE}.e$ and unknown $\mathbf{SE}.f$ appear.

What happens next depends on what $\mathbf{X}$ is. If $\mathbf{X}$ is a dissipator, then $\mathbf{X}.e = X\mathbf{X}.f = \frac{X}{g}\mathbf{SE}.e$, solving for the physical device by simple algebra. If $\mathbf{X}$ is a compliant, we have $\mathbf{X}.e' = \frac{1}{C}\mathbf{X}.f = \frac{X}{g}\mathbf{SE}.e$ so we just have to integrate $\mathbf{SE}.e$ to solve for the physical device. But if $\mathbf{X}$ is an inertor, we get:

$$\mathbf{X}.f' = \tfrac{1}{L}\mathbf{X}.e = \tfrac{1}{L}\left(\mathbf{SE}.e(1 - \tfrac{1}{g}) + g\,\mathbf{SE}.f\right) \tag{12}$$

which cannot be solved immediately, given the unknown $\mathbf{SE}.f$. This has to be reconciled with the first gyrator equation $\mathbf{SE}.e = g\,\mathbf{X}.f$, which yields $\mathbf{SE}.e' = g\,\mathbf{X}.f'$. $\mathbf{X}.f'$ can now be eliminated to solve for $\mathbf{SE}.f$ algebraically. What we have is a simple example of a non-trivial differential algebraic system.

In some of the bond graph literature, tracking back can be interpreted as uncritically following the dependency decorations when there is a choice (in the equality layer of a junction). Figure 9 shows the inherent risks. If we track the effort variable back from dissipator $\mathbf{R}_x$, we encounter $\mathbf{E}_2$, and following the dependency decorations, $\mathbf{E}_1$, $\mathbf{E}_3$, $\mathbf{E}_2$, and so on indefinitely—a case of dependency capture. This is particularly bad considering that the $\mathbf{E}_1$-$\mathbf{E}_2$-$\mathbf{E}_3$ triangle can be reduced to a single $\mathbf{E}$ junction, using the bond graph transformation theory of [3], which eliminates the endless tracking back loop, and does not change the physical behaviour of the system. Moreover, if we consider the $\mathbf{E}_1$-$\mathbf{E}_2$-$\mathbf{E}_3$ triangle as the base of a tetrahedron, adding another $\mathbf{E}$ junction as apex, connected to the base in the obvious way, we cannot even assign consistent NDG/DBG decorations to the resulting bond graph. This shows that existence and properties of dependency assignments are not invariant under bond graph transformations that are correct according to the standard correctness notion.

Figure 10 shows a problem inherent in naively tracking back from physical variables on an individual basis. There are two dissipators, $\mathbf{R}_1$ and $\mathbf{R}_2$, both joined to an $\mathbf{E}$ junction, which is also joined to two further portions of the bond graph. The $\mathbf{E}$ junction is assumed to be a cut vertex of the graph. Tracking back from $\mathbf{R}_1$ may choose to go into 'the rest of the bond graph A'. This forces the equality of the two occurrences of e_1 shown. Tracking back from $\mathbf{R}_2$ may choose to go into 'the rest of the bond graph B'. This forces the equality of the two occurrences of e_2 shown. But the tracking back process itself does not force $e_1 = e_2$, which is required by the semantics, and will not be detected if the two tracking back instances terminate in A and B respectively. So naive tracking back can easily be unsound.

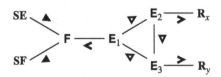

Fig. 9. A simple DBG showing dependency capture.

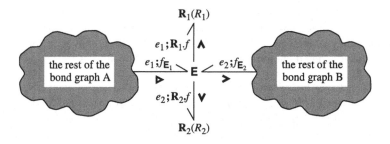

Fig. 10. A bond graph for which naive tracking back yields an incorrect solution.

It is therefore important to supplement naive tracking back with a policy that ensures that *all* equalities implied by junction equality layer semantics are taken care of. The easiest way of ensuring this is to nominate a distinguished bond of each junction that is *always* tracked back through at each tracking back instance involving that junction's equality layer, *and also* insisting that *all* bonds of each junction get involved in equality layer tracking back. The obvious candidate for the distinguished bond role is the junction's dominant bond.

Figure 11 shows an example, involving the earlier E_1-E_2-E_3 triangle, in which the preceding considerations have some impact. Naive tracking back of efforts from R may explore 'the rest of the bond graph' and terminate there, and thereby miss catching some of the effort equalities of E_1, and thereby those of E_2 and E_3. Exploring flows reaches further into the triangle and reveals that an unconstrained flow of arbitrary value may circulate round the triangle, but the details of the solution can differ from the standard semantics of the system.

On the other hand, if we follow the dominant bond tracking back policy, we get stuck in the triangle. Altogether, we have a choice between either bad semantics or nontermination.

6   Two Key Theorems

The preceding examples indicate that traditional bond graph 'causality' and the tracking back processes it is associated with are highly questionable notions, and placing naive faith in them is misplaced as far as the derivation of solutions to system behaviour is

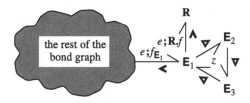

Fig. 11. An example for which tracking back yields equations different from those of the original bond graph, and following dominant bonds leads to nontermination.

concerned. Nevertheless, the examples treated in the standard texts referred to in the Introduction seem to work well enough under this approach. This is a phenomenon that begs an explanation. In this section we present two theorems that go a long way to explaining the apparent dichotomy. Their proofs are in Appendix A.

Theorem 1. *Let BG be a bond graph endowed with DBG dependency decorations. Suppose given a tracking back process for BG that always uses the dominant bond at junction equality layers. Then, for every physical device* **X** *for which tracking back encounters no gyrators:*

1. *If e is the effort variable of* **X** *and e is connected* either *to an* **E** *junction via a dependent bond of* **E***, or to an* **F** *junction via the dominant bond of* **F***, then the tracking back procedure tracks back only through bonds whose DBG dependency decorations point towards e.*
2. *If f is the flow variable of* **X** *and f is connected* either *to an* **F** *junction via a dependent bond of* **F***, or to an* **E** *junction via the dominant bond of* **E***, then the tracking back procedure tracks back only through bonds whose DBG dependency decorations point away from f.*

In both of these cases, the bond mentioned may have one or more transformers interposed without affecting the result.

Theorem 1 highlights the essential reason why, under the right conditions, knowing something about a variable of device **X** permits knowledge about distant parts of the bond graph that might be of interest. This can be the only justification for entertaining notions like dependency decorations and the rules for their allocation — otherwise they serve no purpose. Theorem 2 embellishes Theorem 1 with additional contextual conditions that enable its use in deriving solution strategies for actual bond graphs.

Theorem 2. *Let BG be a bond graph endowed with DBG dependency decorations. Assume given a choice of core variables, consisting of one variable for every inertor or compliant, each with integral dependency. Assume each is tracked back through the device's constitutive equation and then through the equations of the bond graph, in an optimised and terminating process, until inertor or compliant or source variables are reached. Then, the process yields a set of explicit ODEs for the system, each of the form* $v' = \Theta_i(vs)$. *The solution to these is semantically complete at peripheral variables, but need not be semantically sound at peripheral variables without the addition of algebraic equations.*

The structure of the proof of Theorem 1 reveals that the case analysis depends on: (1) whether an effort or a flow variable is being tracked; (2) whether the bond in question is dominant or not; (3) whether the bond is connected to an **E** junction or an **F** junction. So there are eight cases. Unfortunately, as we have seen, the theorem handles only four of them, leaving another four unconsidered. So the theorem far from exhausts the situations that are expressible using bond graphs, even when attention is restricted to bond graphs describing systems that are 'natural' from an engineering point of view. For bond graphs featuring these alternative cases, any connection with DBG dependency markings may appear tenuous indeed, and to a large extent, the plethora of awkward examples seen earlier can be seen to be rooted in these four cases.

In effect, Theorem 1 says that the nomenclature of 'following the causality', seen in the literature, is justified in the cases that the theorem handles, as it shows that the DBG decoration tree that is rooted at a physical device's variable, coincides with the parse tree that the bond graph's equations demand when discovering what expression the value of the relevant variable depends on, mathematically.

When tracking back enters the equality layer of a junction via a dependent bond b_{dep} and exits via the dominant bond b_{dom}, the traditional literature sometimes attempts to explain this by saying that the exit is via b_{dom} because 'b_{dep} is *caused by* b_{dom}'. Similar remarks can be seen when the sum layer is entered via the dominant bond and exited via all the dependent bonds. The limited remit of Theorem 1, and the impact of the preceding theory and counterexamples, show just how potentially unhelpful such phraseology can be.

We observe that the vast majority of textbook examples have tree shaped bond graphs, with a unique source device, and a tracking back process that overwhelmingly uses the dominant bond wherever possible at equality layer variables. Moreover, gyrators are typically cut points of the bond graph, even in cases that are not tree shaped as a whole. So the conditions of Theorems 1 and 2 are often naturally fulfilled, thus leading to the good behaviour seen for 'natural', integral dependency examples. This all additionally aligns with the fact that some of the counterexamples explored earlier contained loops, which, in bond graphs, are always an indication that caution is required.

7   Robust Solution Strategies for Bond Graphs

The most straightforward (and negative) interpretation of the previous section is that dependency bond graphs (so called 'causal bond graphs') are a deeply questionable notion intrinsically,[1] and are of little use in determining system solution strategies in the general case. We observed that the traditional derivation of systems of equations of the typical form desired for solving bond graph systems was a surprisingly deterministic process, with any possible role for dependency decorations being confined to the choice element of tracking back at equality layers. In this section we briefly overview a system solution strategy approach that does not depend on dependency decorations at all, but rather treats the constituent equations of bond graphs that the standard semantics in **[SEMANTICS]** is defined with respect to, at face value, resulting in a system of differential algebraic equations (DAEs). It is worth observing that in general, no completely satisfactory strategy for arbitrary DAE systems is known. The following are standard references: [5, 8–11, 13], of which [13] is the most pertinent for our purposes.

The approaches we outline can be applied to the raw bond graph equations given by the device definitions. This generates a large number of trivial linear equations. We can optimise to an extent, by substituting physical device variables into the junction structure as far as possible, taking care to preserve correctness by using **unification** rather than tracking back from individual variables. If the bond graph is tree structured, this will eliminate all junction variables. If not, residual junction variables will remain.

---

[1] We are by no means the first to say it; see e.g. [6], cited in [4].

Like ODEs, DAEs can be classified into general systems, linear systems, and LCC systems. Less like ODEs, non-CC linear systems are much harder than LCC ones. The best treatment known to us of all the DAE cases is [13].

LCC DAE Systems. The general structure studied in the LCC case is of the form $Ex' = Ax + b$, where x is the vector of variables of interest, x' is the vector of its derivatives, E and A are matrices, and b is an inhomogeneous vector term. Evidently this covers all the LCC bond graph cases we have covered. The focus is then on the *matrix pencil* $\mathbf{M} \equiv (\lambda E - A)$. The general approach reduces the pencil to Kronecker canonical form via matrix equivalence arguments, which covers all the singular cases. Of more interest practically is when E and A are square, whereupon, if $\det(\mathbf{M})(\lambda) \not\equiv 0$ the pencil is *regular*, and the Jordan normal form of $(A - \lambda_0 E)^{-1}E$, where λ_0 is a value that makes the determinant nonzero, enables the pencil to be brought into Weierstrass canonical form. This form neatly decomposes the original system into a conventional inhomogeneous ODE problem, and the application of successive powers of a nilpotent operator to correspondingly higher derivatives of an inhomogeneous term, with the inhomogeneities derived from the arbitrary (but sufficiently smooth) b. The nilpotency of the operator guarantees that the series has only a finite number of terms. It is the nilpotent part of the solution that brings in, for example, the derivative required of $\mathbf{SE}.e$ in the discussion of Fig. 5 above. It is clear that even in the regular LCC case, somewhat nontrivial mathematics is needed to handle all the exceptional circumstances that can be thrown up, and that our discourse on bond graphs had no difficulty in displaying.

In the case of Fig. 5 just mentioned, if $\mathbf{SE}.e$ is a step function that switches on a non-zero effort at $t = 0$, then b above is non-smooth, and, strictly speaking, we are outside the theory just described. Fortunately, there is a fairly straightforward extension of the smooth theory to generalised functions of a restricted form (piecewise smooth enough functions embellished with finite order derivatives of impulses at the join points (which themselves have no accumulations)). This enables such cases to be handled properly.

Linear Non-CC (LNCC) DAE Systems. The equivalence classes that yield the classification used in the LCC case are based on conventional matrix similarity notions. Unfortunately, if coefficients are not constant, linear combination does not commute with differentiation. So the classification for LNCC cases is much more complicated than for LCC, as [13] shows (in stark contrast to pure ODE systems).

Nonlinear (NL) DAE Systems. Aside from special cases that can be solved analytically, and that can always be created by reverse engineering the prospective solution, the only approach that is generally applicable for NL DAEs is numerical. The best known incarnation of this is the backward differentiation formula approach. In this, the derivative of the solution vector x' at time $n + 1$ is approximated by a linear expression in $x(n + 1), x(n) \ldots x(n - k)$. The DAE system then yields a system of non-linear algebraic equations in $x(n + 1), x(n) \ldots x(n - k)$, which is solved using a non-linear approach such as the Newton-Raphson method, and the time index is incremented. For much more on approaches to the NL case see [1,8], as well as [13] and elsewhere.

8   Conclusions

In the preceding sections we recalled, rather briefly, the essentials of formalised bond graphs, as per [3] (which made precise the more straightforward elements of bond graphs, as well as their rule based transformation). We then dived into the much murkier world of bond graph dependency (eschewing the highly misleading 'causality' word). Counterexamples proved ludicrously easy to find — once one overcame the inclination to accept the vague assertions regarding dependency's utility for solution derivation found in the bond graph textbook literature. This occupied a couple of sections above.

Nevertheless, the examples of bond graph 'causality' found in the bond graph textbook literature do seem to work. Why? In Sect. 6 we provided the answer. There is a case analysis for which half the cases assure 'nice behaviour' and the other half don't. By and large, the nice cases explain the nice textbook examples, and the other cases account for the plethora of trivially simple counterexamples we presented. To the best of our knowledge, no comparable result about 'causality' exists in the literature. We followed this by briefly covering mathematical approaches that deal robustly and reliably with the kind of equation systems that bond graphs generate. For lack of space, the whole of this paper is rather terse. A better illustrated and more detailed account of everything we have discussed is anticipated in [2].

A   Two Proofs

Proof of Theorem 1: To start with, we assume no transformer is encountered during tracking back (as well as no gyrator). For case 1 of the theorem, if e is connected to an **E** junction via a dependent bond $b_\mathbf{X}$ of **E**, then the DBG dependency decoration on $b_\mathbf{X}$ points towards **X**, thus towards e. Because BG has DBG decorations, the dominant bond, $b_\mathbf{E}$ of **E** points the same way, and the tracking back strategy selects it as the bond to track back along. The device at the other end of $b_\mathbf{E}$ is either a peripheral device, in which case we are done, or another **E** junction for which $b_\mathbf{E}$ is a dependent bond, or an **F** junction, for which $b_\mathbf{E}$ is the dominant bond, pointing away from **F**.

If it is an **E** junction with $b_\mathbf{E}$ a dependent bond, the argument just given repeats. If it is an **F** junction with $b_\mathbf{E}$ the dominant bond, then all other bonds of **F** are dependent bonds, and have DBG dependency decorations pointing towards **F**, and thus towards **X** again. The branching in the tracking back process that takes place at **F** is therefore all along dependent bonds. Let $b_\mathbf{F}$ be such a dependent bond of **F**. The device at the other end of $b_\mathbf{F}$ is either a peripheral device, in which case we are done, or another **F** junction for which $b_\mathbf{F}$ is the dominant bond, or another **E** junction, for which $b_\mathbf{F}$ is a dependent bond, pointing away from the **E** junction. The latter two cases have been dealt with already, so the arguments now repeat as often as needed until the entire parse tree that *SeekDependencyExp*(e) visits has been treated.

If, instead, e is connected to an **F** junction via a dominant bond of **F**, then the argument is as in the preceding paragraph, so need not be repeated. The argument for case 2 of the theorem is dual.

The general case chooses to interpose one or more transformers into one or more of the bonds processed during tracking back. For a given bond b, say there are n transformers. The process replaces b and its DBG dependency decoration, by $n + 1$ bonds, $b_0, b_1, \ldots b_n$ interleaving the n transformers, each having a DBG dependency decoration in the same direction as the one on b. Each one at the extreme ends of the sequence, b_0 and b_n, may become a dominant bond if the junction it attaches to demands it according to its direction. We are done. $\square$

Proof of Theorem 2: It is assumed that each inertor or compliant has integral dependency. In that case, the derivative of its core variable v is proportional to its complementary variable, and the tracking back from the complementary variable is covered by one of the four cases of Theorem 1.

So tracking back from an inertor or compliant variable v will align with the dependency markings until one of four situations is encountered. (1) A source or drain is encountered. Then, because the tracking back aligns with the dependency markings, and the DBG dependency markings point away from an **SE** or **DE** device and towards an **SF** or **DF** device, the variable encountered will be the effort variable for an **SE** or **DE** and will be the flow variable for an **SF** or **DF**, i.e. it is the known variable that is encountered, which will then appear as a conventional inhomogeneous term in the RHS(s) of relevant explicit ODE(s). (2) A core variable w of an inertor or compliant device is encountered. This terminates that branch of the tracking back process. Variable w is the LHS variable of the ODE for that device, and so w will occur in the RHS of the explicit ODE equation for v. (3) A dissipator is encountered. This would terminate that branch of an unoptimised process, but an optimised process continues via the complementary layer. This reverses the direction of tracking of the DBG markings, but since the continuation happens in the complementary layer, it simply interchanges the clause 1 and clause 2 cases of Theorem 1, and tracking back again becomes aligned with the DBG markings. (4) A gyrator is encountered. This is similar to the dissipator case except that continuation is along the other bond of the gyrator, rather than reversing back along the same bond after the layer switch. Since each instance of tracking back from a core variable terminates by assumption, one of the four cases discussed applies for each tracking back path in the parse tree of v. Since the derived ODEs are clearly satisfied by any behaviour of the system described by BG, we have semantic completeness. But, and not least because the statement of the theorem is applicable to pure dissipator bond graphs, we do not necessarily have semantic soundness without considering further algebraic equations (even if we disregard unknown variables of sources). We are done.   ☐

References

1. Ascher, U., Petzold, L.: Computer Methods for Ordinary Differential Equations and Differential-Algebraic Equations. SIAM (1998)
2. Banach, R.: Bond graphs: an abstract formulation (2024). (in preparation)
3. Banach, R., Baugh, J.: Formalisation, abstraction and refinement of bond graphs. In: Fernandez, M., Poskitt, C.M. (eds.) Proceedings ICGT-23. LNCS, vol. 13961, pp. 145–162. Springer, Heidelberg (2023). https://doi.org/10.1007/978-3-031-36709-0_8
4. Borutzky, W.: Bond Graph Methodology. Springer, Heidelberg (2010). https://doi.org/10.1007/978-1-84882-882-7
5. Brenan, K., Campbell, S., Petzold, L.: Numerical Solution of Initial-Value Problems in Differential-Algebraic Equations. SIAM (1996)
6. Cellier, F., Otter, M., Elmqvist, H.: Bond graph modeling of variable structure systems. In: Proceedings of ICBGM-95. Simulation Series, vol. 27, pp. 49–55. SCS Publishing (1995)
7. Damic, V., Montgomery, J.: Mechatronics by Bond Graphs. Springer, Heidelberg (2015). https://doi.org/10.1007/978-3-662-49004-4

8. Gerdts, M.: Optimal Control of ODEs and DAEs. De Gruyter (2012)
9. Hairer, E., Norsett, S., Wanner, G.: Solving Ordinary Differential Equations I Nonstiff Problems. Springer, Heidelberg (1993). https://doi.org/10.1007/978-3-540-78862-1
10. Hairer, E., Wanner, G.: Solving Ordinary Differential Equations II Stiff and Differential-Algebraic Problems. Springer, Heidelberg (1996). https://doi.org/10.1007/978-3-642-05221-7
11. Ilchmann, A., Reis, T. (eds.): Surveys in Differential-Algebraic Equations (3 vols.). Springer, Heidelberg (2013, 2015)
12. Karnopp, D., Margolis, D., Rosenberg, R.: System Dynamics: Modeling, Simulation, and Control of Mechatronic Systems, 5th edn. Wiley, Hoboken (2012)
13. Kunkel, P., Mehrmann, V.: Differential-Algebraic Equations: Analysis and Numerical Solution. European Mathematical Society (2006)
14. Kypuros, J.: Dynamics and Control with Bond Graph Modeling. CRC, Boca Raton (2013)
15. Lamb, J., Woodall, D., Asher, G.: Bond graphs II: causality and singularity. Disc. Appl. Math. **73**, 143–173 (1997). following on from Bond Graphs I, in vol. **72**, 261-293 (1997)
16. Paynter, H.: Analysis and Design of Engineering Systems. MIT Press, Cambridge (1961)
17. Paynter, H.: An Epistemic Prehistory of Bond Graphs. Elsevier, Amsterdam (1992)
18. Tenreiro Machado, J., Cunha, V.: An Introduction to Bond Graph Modeling with Applications. CRC, Boca Raton (2021)

Localized RETE for Incremental Graph Queries

Matthias Barkowsky$^{(\boxtimes)}$ and Holger Giese

Hasso Plattner Institute at the University of Potsdam, Prof.-Dr.-Helmert-Str. 2-3,
14482 Potsdam, Germany
{matthias.barkowsky,holger.giese}@hpi.de

Abstract. The growing size of graph-based modeling artifacts in model-driven engineering calls for techniques that enable efficient execution of graph queries. Incremental approaches based on the RETE algorithm provide an adequate solution in many scenarios, but are generally designed to search for query results over the entire graph. However, in certain situations, a user may only be interested in query results for a subgraph, for instance when a developer is working on a large model of which only a part is loaded into their workspace. In this case, the global execution semantics can result in significant computational overhead.

To mitigate the outlined shortcoming, in this paper we propose an extension of the RETE approach that enables local, yet fully incremental execution of graph queries, while still guaranteeing completeness of results with respect to the relevant subgraph.

We empirically evaluate the presented approach via experiments inspired by a scenario from software development and an independent social network benchmark. The experimental results indicate that the proposed technique can significantly improve performance regarding memory consumption and execution time in favorable cases, but may incur a noticeable overhead in unfavorable cases.

1 Introduction

In model-driven engineering, models constitute important development artifacts [30]. With complex development projects involving large, interconnected models, performance of automated model operations becomes a primary concern.

Incremental graph query execution based on the RETE algorithm [20] has been demonstrated to be an adequate solution in scenarios where an evolving model is repeatedly queried for the same information [34]. In this context, the RETE algorithm essentially tackles the problem of querying a usually graph-based model representation via operators from relational algebra [13]. With these operators not designed to exploit the locality found in graph-based encodings, current incremental querying techniques require processing of the entire model to guarantee complete results. However, in certain situations, global query execution is not required and may be undesirable due to performance considerations.

© The Author(s), under exclusive license to Springer Nature Switzerland AG 2024
R. Harmer and J. Kosiol (Eds.): ICGT 2024, LNCS 14774, pp. 118–137, 2024.
https://doi.org/10.1007/978-3-031-64285-2_7

For instance, a developer may be working on only a part of a model loaded into their workspace, with a large portion of the full model still stored on disk. A concrete example of this would be a developer using a graphical editor to modify an individual block diagram from a collection of interconnected block diagrams, which together effectively form one large model. As the developer modifies the model part in their workspace, they may want to continuously monitor how their changes impact the satisfaction of some consistency constraints via incremental model queries. In this scenario, existing techniques for incremental query execution require loading and querying the entire collection of models, even though the user is ultimately only interested in the often local effect of their own changes.

The global RETE execution semantics then results in at least three problems: First, the computation of query results for the full model, of which only a portion is relevant to the user, may incur a substantial overhead on initial query execution time. Second, incremental querying techniques are known to create and store a large number of intermediate results [34], many of which can in this scenario be superfluous, causing an overhead in memory consumption. Third, query execution requires loading the entire model into memory, potentially causing an overhead in loading time and increasing the overhead in memory consumption.

These problems can be mitigated to some extent by employing local search [6,14,23], which better exploits the locality of the problem and lazy loading capabilities of model persistence layers [1,15], instead of a RETE-based technique. However, resorting to local search can result in expensive redundant search operations that are only avoided by fully incremental solutions [34].

In this paper, we instead propose to tackle the outlined shortcoming of incremental querying techniques via an extension of the RETE approach. This is supported by a relaxed notion of completeness for query results that accounts for situations where the computation for the full model is unnecessary. Essentially, this enables a distinction between the full model, for which query results do not necessarily have to be complete, and a relevant model part, for which complete results are required. The extended RETE approach then anchors query execution to the relevant model part and lazily fetches additional model elements required to compute query results that meet the relaxed completeness requirement. Our approach thereby avoids potentially expensive global query execution and allows an effective integration of incremental queries with model persistence layers.

The remainder of this paper is structured as follows: Sect. 2 provides a summary of our notion of graphs, graph queries, and the RETE mechanism for query execution. Section 3 discusses our contribution in the form of a relaxed notion of completeness for query results and an extension to the RETE querying mechanism that enables local, yet fully incremental execution of model queries. A prototypical implementation of our approach is evaluated regarding execution time and memory consumption in Sect. 4. Section 5 discusses related work, before Sect. 6 concludes the paper and provides an outlook on future work.

2   Preliminaries

In the following, we briefly reiterate the definitions of graphs and graph queries and summarize the RETE approach to incremental graph querying.

2.1   Graphs and Graph Queries

As defined in [18], a *graph* is a tuple $G = (V^G, E^G, s^G, t^G)$, with V^G the set of vertices, E^G the set of edges, and $s^G : E^G \rightarrow V^G$ and $t^G : E^G \rightarrow V^G$ functions mapping edges to their source respectively target vertices.

A mapping from a graph Q into another graph H is defined via a *graph morphism* $m : Q \rightarrow H$, which is characterized by two functions $m^V : V^Q \rightarrow V^H$ and $m^E : E^Q \rightarrow E^H$ such that $s^H \circ m^E = m^V \circ s^Q$ and $t^H \circ m^E = m^V \circ t^Q$.

A graph G can be typed over a type graph TG via a graph morphism $type^G : G \rightarrow TG$, forming the *typed graph* $G^T = (G, type^G)$. A *typed graph morphism* from $G^T = (G, type^G)$ into another typed graph $H^T = (H, type^H)$ with type graph TG is a graph morphism $m^T : G \rightarrow H$ such that $type^H \circ m^T = type^G$.

We say that G is *edge-dominated* if $\forall e_{TG} \in E^{TG} : |\{e \in E^G | type(e) = e_{TG}\}| \geq max(|\{v \in V^G | type(v) = s^{TG}(e_{TG})\}|, |\{v \in V^G | type(v) = t^{TG}(e_{TG})\}|)$.

A *model* is then simply given by a (typed) graph. In the remainder of the paper, we therefore use the terms graph and model interchangably.

A *graph query* as considered in this paper is characterized by a (typed) query graph Q and can be executed over a (typed) host graph H by finding (typed) graph morphisms $m : Q \rightarrow H$. We also call these morphisms *matches* and denote the set of all matches from Q into H by $\mathcal{M}_H^Q$. Typically, a set of matches is considered a complete query result if it contains all matches in $\mathcal{M}_H^Q$:

Definition 1. (Completeness of Query Results) *We say that a set of matches M from query graph Q into host graph H is complete iff $M = \mathcal{M}_H^Q$.*

2.2   Incremental Graph Queries with RETE

The RETE algorithm [20] forms the basis of mature incremental graph querying techniques [35]. Therefore, the query graph is recursively decomposed into simpler subqueries, which are arranged in a second graph called RETE net. In the following, we will refer to the vertices of RETE nets as *(RETE) nodes*.

Each RETE node n computes matches for a subgraph $query(n) \subseteq Q$ of the full query graph Q. This computation may depend on matches computed by other RETE nodes. Such dependencies are represented by edges from the dependent node to the dependency node. For each RETE net, one of its nodes is designated as the *production node*, which computes the net's overall result. A RETE net is thus given by a tuple (N, p), where the *RETE graph* N is a graph of RETE nodes and dependency edges and $p \in V^N$ is the production node.

We describe the *configuration* of a RETE net (N, p) during execution by a function $\mathcal{C} : V^N \rightarrow \mathcal{P}(\mathcal{M}_\Omega)$, with $\mathcal{P}$ denoting the power set and $\mathcal{M}_\Omega$ the set of all graph morphisms. $\mathcal{C}$ then assigns each node in V^N a *current result set*.

For a starting configuration $\mathcal{C}$ and host graph H, executing a RETE node $n \in V^N$ yields an updated configuration $\mathcal{C}' = execute(n, N, H, \mathcal{C})$, with

$$\mathcal{C}'(n') = \begin{cases} \mathcal{R}(n, N, H, \mathcal{C}) & \text{if } n' = n \\ \mathcal{C}(n') & \text{otherwise,} \end{cases}$$

where $\mathcal{R}$ is a function defining the *target result set* of a RETE node n in the RETE graph N for H and $\mathcal{C}$ such that $\mathcal{R}(n, N, H, \mathcal{C}) \subseteq \mathcal{M}_H^Q$, with $Q = query(n)$.

We say that $\mathcal{C}$ is *consistent* for a RETE node $n \in V^N$ and host graph H iff $\mathcal{C}(n) = \mathcal{R}(n, N, H, \mathcal{C})$. If $\mathcal{C}$ is consistent for all $n \in V^N$, we say that $\mathcal{C}$ is consistent for H. If H is clear from the context, we simply say that $\mathcal{C}$ is consistent.

Given a host graph H and a starting configuration $\mathcal{C}_0$, a RETE net (N, p) is executed via the execution of a sequence of nodes $O = n_1, n_2, ..., n_x$ with $n_i \in V^N$. This yields the trace $\mathcal{C}_0, \mathcal{C}_1, \mathcal{C}_2, ..., \mathcal{C}_x$, with $\mathcal{C}_y = execute(n_y, N, H, \mathcal{C}_{y-1})$, and the result configuration $execute(O, N, H, \mathcal{C}_0) = \mathcal{C}_x$.

(N, p) can initially be executed over H via a sequence O that yields a consistent configuration $\mathcal{C}_x = execute(O, N, H, \mathcal{C}_0)$, where $\mathcal{C}_0$ is an empty starting configuration with $\forall n \in V^N : \mathcal{C}_0(n) = \emptyset$. Incremental execution of (N, p) can be achieved by retaining a previous result configuration and using it as the starting configuration for execution over an updated host graph. This requires incremental implementations of RETE node execution procedures that update previously computed result sets for changed inputs instead of computing them from scratch.

In the most basic form, a RETE net consists of two kinds of nodes, *edge input nodes* and *join nodes*. An edge input node $[v \rightarrow w]$ has no dependencies, is associated with the query subgraph $query([v \rightarrow w]) = (\{v, w\}, \{e\}, \{(e, v)\}, \{(e, w)\})$, and directly extracts the corresponding (trivial) matches from a host graph. A join node $[\bowtie]$ has two dependencies n_l and n_r with $Q_l = query(n_l)$ and $Q_r = query(n_r)$ such that $V^{Q_l \cap Q_r} \neq \emptyset$ and is associated with $query([\bowtie]) = Q_l \cup Q_r$. $[\bowtie]$ then computes matches for this union subgraph by combining the matches from its dependencies along the overlap graph $Q_\cap = Q_l \cap Q_r$.

In the following, we assume query graphs to be weakly connected and contain at least one edge. A RETE net that computes matches for a query graph Q can then be constructed as a tree of join nodes over edge input nodes. The join nodes thus gradually compose the trivial matches at the bottom into matches for more complex query subgraphs. We call such tree-like RETE nets *well-formed*. An execution sequence that always produces a consistent configuration is given by a reverse topological sorting of the net. The root node of the tree then computes the set of all matches for Q and is designated as the net's production node.

Connected graphs without edges consist of only a single vertex, making query execution via *vertex input nodes*, which function analogously to edge input nodes, trivial. Disconnected query graphs can be handled via separate RETE nets for all query graph components and the computation of a cartesian product.

Fig. 1. Example graph query (left) and corresponding RETE net (right)

Figure 1 shows an example graph query from the software domain and an associated RETE net. The query searches for paths of a package, class, and field. The RETE net constructs matches for the query by combining edges from a package to a class with edges from a class to a field via a natural join.

3   Incremental Queries over Subgraphs

As outlined in Sect. 1, users of model querying mechanisms may only be interested in query results related to some part of a model that is relevant to them rather than the complete model. However, simply executing a query only over the relevant subgraph and ignoring context elements is often insufficient, for instance if the full effect of modifications to the relevant subgraph should be observed, since such modifications may affect matches that involve elements outside the relevant subgraph. Instead, completeness in such scenarios essentially requires the computation of all matches that somehow *touch* the relevant subgraph.

In order to capture this need for local completeness, but avoid the requirement for global execution inherent to the characterization in Definition 1, we define completeness under a relevant subgraph of some host graph as follows:

Definition 2. (Completeness of Query Results under Subgraphs) *We say that a set of matches M from query graph Q into host graph H is complete under a subgraph $H_p \subseteq H$ iff $\forall m \in \mathcal{M}_H^Q : (\exists v \in V^Q : m(v) \in V^{H_p}) \Rightarrow m \in M$.*

3.1   Marking-Sensitive RETE

Due to its reliance on edge and vertex input nodes and their global execution semantics, incremental query execution via the standard RETE approach is unable to exploit the relaxed notion of completeness from Definition 2 for query optimization and does not integrate well with mechanisms relying on operation locality, such as model persistence layers based on lazy loading [1,15].

While query execution could be localized to a relevant subgraph by executing edge and vertex inputs nodes only over the subgraph, execution would then only yield matches where *all* involved elements are in the relevant subgraph. This approach would hence fail to meet the completeness criterion of Definition 2.

To enable incremental queries with complete results under the relevant subgraph, we instead propose to anchor RETE net execution to subgraph elements while allowing the search to retrieve elements outside the subgraph that are required to produce complete results from the full model.

Therefore, we extend the standard RETE mechanism by a natural number marking for matches. Intuitively, we will use this marking to encode up to which height in a RETE join tree derived matches have to be propagated to guarantee completeness with respect to Definition 2. In our extension, an intermediate result is therefore characterized by a tuple (m, ϕ) of a match m and a marking $\phi \in \overline{\mathbb{N}}$, where we define $\overline{\mathbb{N}} := \mathbb{N} \cup \{\infty\}$. A marking-sensitive configuration is then given by a function $\mathcal{C}^{\Phi} : V^N \rightarrow \mathcal{P}(\mathcal{M}_{\Omega} \times \overline{\mathbb{N}})$.

Furthermore, in our extension, result computation distinguishes between the full host graph H and the relevant subgraph $H_p \subseteq H$. For marking-sensitive RETE nodes, we hence extend the function for target result sets by a parameter for H_p. The target result set of a marking-sensitive RETE node $n^{\Phi} \in V^{N^{\Phi}}$ with $Q = query(n^{\Phi})$ for H, H_p, and a marking-sensitive configuration $\mathcal{C}^{\Phi}$ is then given by $\mathcal{R}^{\Phi}(n^{\Phi}, N^{\Phi}, H, H_p, \mathcal{C}^{\Phi})$, with $\mathcal{R}^{\Phi}(n^{\Phi}, N^{\Phi}, H, H_p, \mathcal{C}^{\Phi}) \subseteq \mathcal{M}_H^Q \times \overline{\mathbb{N}}$.

We adapt the *join node*, *union node*, and *projection node* to marking-sensitive variants that assign result matches the maximum marking of related dependency matches and otherwise work as expected from relational algebra [13]. We also adapt the *vertex input node* to only consider vertices in the relevant subgraph H_p and assign matches the marking ∞. Finally, we introduce *marking assignment nodes*, which assign matches a fixed marking value, *marking filter nodes*, which filter marked matches by a minimum marking value, and *forward and backward navigation nodes*, which work similarly to edge input nodes but only extract edges that are adjacent to host graph vertices included in the current result set of a designated dependency. Note that an efficient implementation of the backward navigation node requires reverse navigability of host graph edges.

More specifically, for a host graph H, a relevant subgraph $H_p \subseteq H$, a containing marking-sensitive RETE net (N^{Φ}, p^{Φ}), and a configuration $\mathcal{C}^{\Phi}$ for (N^{Φ}, p^{Φ}), we define the target result sets of marking-sensitive RETE nodes as follows:

- The target result set of a *marking-sensitive join node* $[\bowtie]^{\Phi}$ with marking-sensitive dependencies n_l^{Φ} and n_r^{Φ} with $Q_l = query(n_l^{\Phi})$ and $Q_r = query(n_r^{\Phi})$ such that $V^{Q_{\cap}} = V^{Q_l \cap Q_r} \neq \emptyset$ is given by $\mathcal{R}^{\Phi}([\bowtie]^{\Phi}, N^{\Phi}, H, H_p, \mathcal{C}^{\Phi}) = \{(m_l \cup m_r, max(\phi_l, \phi_r)) | (m_l, \phi_l) \in \mathcal{C}^{\Phi}(n_l^{\Phi}) \wedge (m_r, \phi_r) \in \mathcal{C}^{\Phi}(n_r^{\Phi}) \wedge m_l|_{Q_{\cap}} = m_r|_{Q_{\cap}}\}$.
- The target result set of a *marking-sensitive union node* $[\cup]^{\Phi}$ with a set of marking-sensitive dependencies N_{α}^{Φ} such that $Q = query(n_{\alpha}^{\Phi})$ for all $n_{\alpha}^{\Phi} \in N_{\alpha}^{\Phi}$ is given by $\mathcal{R}^{\Phi}([\cup]^{\Phi}, N^{\Phi}, H, H_p, \mathcal{C}^{\Phi}) = \{(m, \phi_{max})) | (m, \phi_{max}) \in \bigcup_{n_{\alpha} \in N_{\alpha}^{\Phi}} \mathcal{C}^{\Phi}(n_{\alpha}^{\Phi}) \wedge \phi_{max} = max(\{\phi' | (m, \phi') \in \bigcup_{n_{\alpha} \in N_{\alpha}^{\Phi}} \mathcal{C}^{\Phi}(n_{\alpha}^{\Phi})\})\}$.
- The target result set of a *marking-sensitive projection node* $[\pi_Q]^{\Phi}$ with a single marking-sensitive dependency n_{α}^{Φ} and $Q = query([\pi_Q]^{\Phi})$ is given by $\mathcal{R}^{\Phi}([\pi_Q]^{\Phi}, N^{\Phi}, H, H_p, \mathcal{C}^{\Phi}) = \{(m|_Q, \phi_{max})) | (m, \phi_{max}) \in \mathcal{C}^{\Phi}(n_{\alpha}^{\Phi}) \wedge \phi_{max} = max(\{\phi' | (m', \phi') \in \mathcal{C}^{\Phi}(n_{\alpha}^{\Phi}) \wedge m'|_Q = m|_Q\})\}$.
- The target result set of a *marking assignment node* $[\phi := i]^{\Phi}$ with a single marking-sensitive dependency n_{α}^{Φ} is given by $\mathcal{R}^{\Phi}([\phi := i]^{\Phi}, N^{\Phi}, H, H_p, \mathcal{C}^{\Phi}) = \{(m, i) | (m, \phi_{\alpha}) \in \mathcal{C}^{\Phi}(n_{\alpha}^{\Phi})\}$.
- The target result set of a *marking filter node* $[\phi > i_{max}]^{\Phi}$ with marking-sensitive dependency n_{α}^{Φ} is given by $\mathcal{R}^{\Phi}([\phi > i_{max}]^{\Phi}, N^{\Phi}, H, H_p, \mathcal{C}^{\Phi}) = \{(m, \phi_{\alpha}) | (m, \phi_{\alpha}) \in \mathcal{C}^{\Phi}(n_{\alpha}^{\Phi}) \wedge \phi_{\alpha} > i_{max}\}$.

- The target result set of a *forward navigation node* $[v \rightarrow_n w]^{\Phi}$ with $Q = query([v \rightarrow_n w]^{\Phi}) = (\{v, w\}, \{e\}, \{(e, v)\}, \{(e, w)\})$ and marking-sensitive dependency n_v^{Φ} with $Q_v = query(n_v^{\Phi}) = (\{v\}, \emptyset, \emptyset, \emptyset)$ is given by $\mathcal{R}^{\Phi}([v \rightarrow_n w]^{\Phi}, N^{\Phi}, H, H_p, \mathcal{C}^{\Phi}) = \{(m, \phi_{max}) | m \in \mathcal{M}_H^Q \wedge \exists (m_v, \phi_{max}) \in \mathcal{C}^{\Phi}(n_v^{\Phi}) : m_v = m|_{Q_v} \wedge \phi_{max} = max(\{\phi' | (m_v', \phi') \in \mathcal{C}^{\Phi}(n_v^{\Phi}) \wedge m_v' = m|_{Q_v}\})\}$.
- The target result set of a *backward navigation node* $[w \leftarrow_n v]^{\Phi}$ with $Q = query([w \leftarrow_n v]^{\Phi}) = (\{v, w\}, \{e\}, \{(e, v)\}, \{(e, w)\})$ and marking-sensitive dependency n_w^{Φ} with $Q_w = query(n_w^{\Phi}) = (\{w\}, \emptyset, \emptyset, \emptyset)$ is given by $\mathcal{R}^{\Phi}([w \leftarrow_n v]^{\Phi}, N^{\Phi}, H, H_p, \mathcal{C}^{\Phi}) = \{(m, \phi_{max}) | m \in \mathcal{M}_H^Q \wedge \exists (m_w, \phi_{max}) \in \mathcal{C}^{\Phi}(n_w^{\Phi}) : m_w = m|_{Q_w} \wedge \phi_{max} = max(\{\phi' | (m_w', \phi') \in \mathcal{C}^{\Phi}(n_w^{\Phi}) \wedge m_w' = m|_{Q_w}\})\}$.
- The target result set of a *marking-sensitive vertex input node* $[v]^{\Phi}$ with $Q = query([v]^{\Phi}) = (\{v\}, \emptyset, \emptyset, \emptyset)$ is given by $\mathcal{R}^{\Phi}([v]^{\Phi}, N^{\Phi}, H, H_p, \mathcal{C}^{\Phi}) = \{(m, \infty) | m \in \mathcal{M}_{H_p}^Q\}$.

To obtain query results in the format of the standard RETE approach, we define the *stripped result set* of a marking-sensitive RETE node n^{Φ} for a marking-sensitive configuration $\mathcal{C}^{\Phi}$ as $\mathcal{R}_X^{\Phi}(n^{\Phi}, \mathcal{C}^{\Phi}) = \{m | (m, \phi) \in \mathcal{C}^{\Phi}(n^{\Phi})\}$, that is, the set of matches that appear in tuples in the node's current result set in $\mathcal{C}^{\Phi}$.

3.2   Localized Search with Marking-Sensitive RETE

Based on these adaptations, we introduce a recursive *localize* procedure, which takes a regular, well-formed[1] RETE net (N, p) as input and outputs a marking-sensitive RETE net, which performs a localized search that does not require searching the full model to produce complete results according to Definition 2.

If $p = [v \rightarrow w]$ is an edge input node, the result of localization for (N, p) is given by $localize((N, p)) = (LNS(p), [\cup]^{\Phi})$. The *local navigation structure* $LNS(p)$ consists of seven RETE nodes as shown in Fig. 2 (left): (1, 2) Two marking-sensitive vertex input nodes $[v]^{\Phi}$ and $[w]^{\Phi}$, (3, 4) two marking-sensitive union nodes $[\cup]_v^{\Phi}$ and $[\cup]_w^{\Phi}$ with $[v]^{\Phi}$ respectively $[w]^{\Phi}$ as a dependency, (5) a forward navigation node $[v \rightarrow_n w]^{\Phi}$ with $[\cup]_v^{\Phi}$ as a dependency, (6) a backward navigation node $[w \leftarrow_n v]^{\Phi}$ with $[\cup]_w^{\Phi}$ as a dependency, and (7) a marking-sensitive union node $[\cup]^{\Phi}$ with dependencies $[v \rightarrow_n w]^{\Phi}$ and $[w \leftarrow_n v]^{\Phi}$.

Importantly, the marking-sensitive vertex input nodes of the local navigation structure are executed over the relevant subgraph, whereas the forward and backward navigation nodes are executed over the full model. Intuitively, the local navigation structure thus takes the role of the edge input node, but initially only extracts edges that are adjacent to a vertex in the relevant subgraph.

If p is a join node, it has two dependencies p_l and p_r with $Q_l = query(p_l)$ and $Q_r = query(p_r)$, which are the roots of two RETE subtrees N_l and N_r. In this case, (N, p) is localized as $localize((N, p)) = (N^{\Phi}, p^{\Phi}) = (N_{\bowtie}^{\Phi} \cup N_l^{\Phi} \cup N_r^{\Phi} \cup RPS_l \cup$

---

[1] The restriction to well-formed RETE nets prevents an optimization where redundant computation of matches for isomorphic query subgraphs is avoided via a non-tree-like structure. However, for queries as in [19], performance is often primarily determined by the computation of matches for a large query subgraph that cannot be reused.

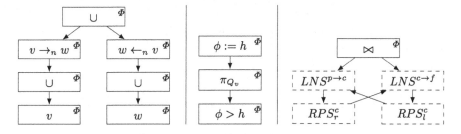

Fig. 2. LNS (left), RPS (center), and localized RETE net (right)

$RPS_r, [\bowtie]^\Phi)$, where $(N_l^\Phi, p_l^\Phi) = localize((N_l, p_l))$, $(N_r^\Phi, p_r^\Phi) = localize((N_r, p_r))$, $N_\bowtie^\Phi$ consists of the marking-sensitive join $[\bowtie]^\Phi$ with dependencies p_l^Φ and p_r^Φ, $RPS_l = RPS(p_l^\Phi, N_r^\Phi)$, and $RPS_r = RPS(p_r^\Phi, N_l^\Phi)$.

The *request projection structure* $RPS_l = RPS(p_l^\Phi, N_r^\Phi)$ contains three RETE nodes as displayed in Fig. 2 (center): (1) A marking filter node $[\phi > h]^\Phi$ with p_l^Φ as a dependency, (2) a marking-sensitive projection node $[\pi_{Q_v}]^\Phi$ with $[\phi > h]^\Phi$ as a dependency, and (3) a marking assignment node $[\phi := h]^\Phi$ with $[\pi_{Q_v}]^\Phi$ as a dependency. The value of h is given by the height of the join tree of which p is the root and Q_v is a graph consisting of a single vertex $v \in V^{Q_l \cap Q_r}$. The request projection structure is then connected to an arbitrary local navigation structure in N_r^Φ that has a marking-sensitive vertex input $[v]^\Phi$ with $Q_v = query([v]^\Phi)$. Therefore, it also adds a dependency from the marking-sensitive union node $[\cup]_v^\Phi$ that already depends on $[v]^\Phi$ to the marking assignment node $[\phi := h]^\Phi$. The mirrored structure $RPS_r = RPS(p_r^\Phi, N_l^\Phi)$ is constructed analogously.

Via the request projection structures, partial matches from one join dependency can be propagated to the subnet under the other dependency. Intuitively, the inserted request projection structures thereby allow the join's dependencies to request the RETE subnet under the other dependency to fetch and process the model parts required to complement the first dependency's results. The marking of a match then signals up to which height in the join tree complementarity for that match is required. Notably, matches involving elements in the relevant subgraph are marked ∞, as complementarity for them is required at the very top to guarantee completeness of the overall result.

The result of applying *localize* to the example RETE net in Fig. 1 is displayed in Fig. 2 (right). It consists of a marking sensitive join and two local navigation structures connected via request projection structures.

A consistent configuration for a localized RETE net then indeed guarantees query results that are complete according to Definition 2:

Theorem 1. *Let H be a graph, $H_p \subseteq H$, (N, p) a well-formed RETE net, and $Q = query(p)$. Furthermore, let $\mathcal{C}^\Phi$ be a consistent configuration for the localized RETE net $(N^\Phi, p^\Phi) = localize((N, p))$. The set of matches given by the stripped result set $\mathcal{R}_X^\Phi(p^\Phi, \mathcal{C}^\Phi)$ is then complete under H_p.*

Proof. (Idea)[2] It can be shown via induction over the height of N that request projection structures ensure the construction of all intermediate results required to guarantee completeness of the overall result under H_p.

Notably, the insertion of request projection structures creates cycles in the localized RETE net, which prevents execution via a reverse topological sorting. However, the marking filter and marking assignment nodes in the request projection structures effectively prevent cyclic execution at the level of intermediate results: Because matches in the result set of a dependency of some join at height h that are only computed on request from the other dependency are marked h, these matches are filtered out in the dependent request projection structure. An execution order for the localized RETE net $(N^\Phi, p^\Phi) = localize((N, p))$ can thus be constructed recursively via an *order* procedure as follows:

If p is an edge input node, the RETE graph given by $N^\Phi = LNS(p)$ is a tree that can be executed via a reverse topological sorting of the nodes in $LNS(p)$, that is, $order((N^\Phi, p^\Phi)) = toposort(LNS(p))^{-1}$.

If p is a join, according to the construction, the localized RETE graph is given by $N^\Phi = N^\Phi_\bowtie \cup N^\Phi_l \cup N^\Phi_r \cup RPS_l \cup RPS_r$. In this case, an execution order for (N^Φ, p^Φ) can be derived via the concatenation $order((N^\Phi, p^\Phi)) = order(RPS_r) \circ order(N^\Phi_l) \circ order(RPS_l) \circ order(N^\Phi_r) \circ order(RPS_r) \circ order(N^\Phi_l) \circ order(N^\Phi_\bowtie)$, where $order(RPS_l) = toposort(RPS_l)^{-1}$, $order(RPS_r) = toposort(RPS_r)^{-1}$, and $order(N^\Phi_\bowtie) = [p^\Phi]$, that is, the sequence containing only p^Φ.

Executing a localized RETE net (N^Φ, p^Φ) via $order((N^\Phi, p^\Phi))$ then guarantees a consistent result configuration for any starting configuration:

Theorem 2. *Let H be a graph, $H_p \subseteq H$, (N, p) a well-formed RETE net, and $\mathcal{C}^\Phi_0$ an arbitrary starting configuration. Executing the marking-sensitive RETE net $(N^\Phi, p^\Phi) = localize((N, p))$ via $O = order((N^\Phi, p^\Phi))$ then yields a consistent configuration $\mathcal{C}^\Phi = execute(O, N^\Phi, H, H_p, \mathcal{C}^\Phi_0)$.*

Proof. (Idea) Follows because the inserted marking filter nodes prevent cyclic execution behavior at the level of intermediate results.

Combined with the result from Theorem 1, this means that a localized RETE net can be used to compute complete query results for a relevant subgraph in the sense of Definition 2, as outlined in the following corollary:

Corollary 1. *Let H be a graph, $H_p \subseteq H$, (N, p) a well-formed RETE net, and $Q = query(p)$. Furthermore, let $\mathcal{C}^\Phi_0$ be an arbitrary starting configuration for the marking-sensitive RETE net $(N^\Phi, p^\Phi) = localize((N, p))$ and $\mathcal{C}^\Phi = execute(order((N^\Phi, p^\Phi)), N^\Phi, H, H_p, \mathcal{C}^\Phi_0)$. The set of matches from Q into H given by $\mathcal{R}^\Phi_X(p^\Phi, \mathcal{C}^\Phi)$ is then complete under H_p.*

Proof. Follows directly from Theorem 1 and Theorem 2.

---

[2] Detailed proofs for theorems in this paper are given in [9].

3.3   Performance of Localized RETE Nets

Performance of a RETE net (N, p) with respect to both execution time and memory consumption is largely determined by the *effective size* of a consistent configuration $\mathcal{C}$ for (N, p), which we define as $|\mathcal{C}|_e := \sum_{n \in V^N} \sum_{m \in \mathcal{C}(n)} |m|$. In this context, we define the size of a match $m : Q \to H$ as $|m| := |m^V| + |m^E| = |V^Q| + |E^Q|$. It can then be shown that localization incurs only a constant factor overhead on the effective size of $\mathcal{C}$ for any edge-dominated host graph:[3]

Theorem 3. *Let H be an edge-dominated graph, $H_p \subseteq H$, (N, p) a well-formed RETE net with $Q = query(p)$, $\mathcal{C}$ a consistent configuration for (N, p) for host graph H, and $\mathcal{C}^{\Phi}$ a consistent configuration for the marking-sensitive RETE net $(N^{\Phi}, p^{\Phi}) = localize((N, p))$ for host graph H and relevant subgraph H_p. It then holds that $\sum_{n^{\Phi} \in V^{N^{\Phi}}} \sum_{(m, \phi) \in \mathcal{C}(n^{\Phi})} |m| \leq 7 \cdot |\mathcal{C}|_e$.*

Proof. (Idea) Follows because localization only increases the number of RETE nodes by factor 7 and marking-sensitive result sets contain no duplicate matches.

By Theorem 3, it then follows that localization of a RETE net incurs only a constant factor overhead on memory consumption even in the worst case where the relevant subgraph is equal to the full model:

Corollary 2. *Let H be an edge-dominated graph, $H_p \subseteq H$, (N, p) a well-formed RETE net, $\mathcal{C}$ a consistent configuration for (N, p) for host graph H, and $\mathcal{C}^{\Phi}$ a consistent configuration for the localized RETE net $(N^{\Phi}, p^{\Phi}) = localize((N, p))$ for host graph H and relevant subgraph H_p. Assuming that storing a match m requires an amount of memory in $O(|m|)$ and storing an element from $\overline{\mathbb{N}}$ requires an amount of memory in $O(1)$, storing $\mathcal{C}^{\Phi}$ requires memory in $O(|\mathcal{C}|_e)$.*

Proof. (Idea) Follows from Theorem 3.

In the worst case, a localized RETE net would still be required to compute a complete result. In this scenario, the execution of the localized RETE net may essentially require superfluous recomputation of match markings, causing computational overhead. When starting with an empty configuration, the number of such recomputations per match is however limited by the size of the query graph, only resulting in a small increase in computational complexity:

Theorem 4. *Let H be an edge-dominated graph, $H_p \subseteq H$, (N, p) a well-formed RETE net for query graph Q, $\mathcal{C}$ a consistent configuration for (N, p), and $\mathcal{C}_0^{\Phi}$ the empty configuration for $(N^{\Phi}, p^{\Phi}) = localize((N, p))$. Executing (N^{Φ}, p^{Φ}) via $execute(order((N^{\Phi}, p^{\Phi})), N^{\Phi}, H, H_p, \mathcal{C}_0^{\Phi})$ then takes $O(T \cdot (|Q_a| + |Q|))$ steps, with $|Q_a|$ the average size of matches in $\mathcal{C}$ and $T = \sum_{n \in V^N} |\mathcal{C}(n)|$.*

---

[3] For non-edge-dominated host graphs, the number of matches for (marking-sensitive) vertex input nodes may exceed the number of matches for related edge-input nodes. If matches for marking-sensitive vertex input nodes make up the bulk of intermediate results, localization then introduces an overhead on effective configuration size that cannot be characterized by a constant factor. However, we expect this situation to be rare in practice, since it requires very sparse host graphs.

Proof. (Idea) Follows since the effort for initial construction of matches by the marking-sensitive RETE net is linear in the total size of the constructed matches and the marking of a match changes at most $|Q|$ times.

Assuming an empty starting configuration, a regular well-formed RETE net (N, p) can be executed in $O(|\mathcal{C}|_e)$ steps, which can also be expressed as $O(T \cdot |Q_a|)$. The overhead of a localized RETE net compared to the original net can thus be characterized by the factor $\frac{|Q|}{|Q_a|}$. Assuming that matches for the larger query subgraphs constitute the bulk of intermediate results, which seems reasonable in many scenarios, $\frac{|Q|}{|Q_a|}$ may be approximated by a constant factor.

For non-empty starting configurations and incremental changes, no sensible guarantees can be made. On the one hand, in a localized RETE net, a host graph modification may trigger the computation of a large number of intermediate results that were previously omitted due to localization. On the other hand, in a standard RETE net, a modification may result in substantial effort for constructing superfluous intermediate results that can be avoided by localization. Depending on the exact host graph structure and starting configuration, execution may thus essentially require a full recomputation for either the localized or standard RETE net but cause almost no effort for the other variant.

4 Evaluation

We aim to investigate whether RETE net localization can improve performance of query execution in scenarios where the relevant subgraph constitutes only a fraction of the full model, considering initial query execution time, execution time for incrementally processing model updates, and memory consumption as performance indicators. We experiment[4] with the following querying techniques:

- **STANDARD**: Our own implementation of a regular RETE net with global execution semantics [8].
- **LOCALIZED**: Our own implementation of the RETE net used for STANDARD, localized according to the description in Sect. 3.2.
- **VIATRA**: The external RETE-based VIATRA tool [35].
- **SDM***: Our own local-search-based Story Diagram Interpreter tool [23]. Note that we only consider searching for new query results. We thus underapproximate the time and memory required for a full solution with this strategy, which would also require maintaining previously found results.

---

[4] Experiments were executed on a Linux SMP Debian 4.19.67-2 computer with Intel Xeon E5-2630 CPU (2.3 GHz clock rate) and 386 GB system memory running OpenJDK 11.0.6. Reported measurements correspond to the arithmetic mean of measurements for 10 runs. Memory measurements were obtained using the Java Runtime class. To represent graph data, all experiments use EMF [2] object graphs that enable reverse navigation of edges. Our implementation is available under [3]. More details and query visualizations can be found in [9].

4.1 Synthetic Abstract Syntax Graphs

We first attempt a systematic evaluation via a synthetic experiment, which emulates a developer loading part of a large model into their workspace and monitoring some well-formedness constraints as they modify the loaded part, that is, relevant subgraph, without simultaneous changes to other model parts.

We therefore generate Java abstract syntax graphs with 1, 10, 100, 1000, and 10000 packages, with each package containing 10 classes with 10 fields referencing other classes. As relevant subgraph, we consider a single package and its contents. We then execute a query searching for paths consisting of a package and four classes connected via fields. After initial query execution, we modify the relevant subgraph by creating a class with 10 fields in the considered package and perform an incremental update of query results. This step is repeated 10 times.

Figure 3 (left) displays the execution times for the initial execution of the query. The execution time of LOCALIZED remains around 120 ms for all model sizes. The execution time for SDM* slowly grows from around 350 ms to 1025 ms due to indexing effort that is necessary for observing model changes. In contrast, the execution time for the other RETE-based strategies clearly scales with model size, with the execution time for STANDARD growing from around 13 ms for the smallest model to more than 184 000 ms for the largest model. On the one hand, localization thus incurs a noticeable overhead in initial execution time for the smallest model, where even localized query execution is essentially global. On the other hand, it significantly improves execution time for the larger models and even achieves better scalability than the local-search-based tool in this scenario.

The average times for processing a model update are displayed in Fig. 3 (center). Here, all strategies achieve execution times mostly independent of model size. While the measurements for STANDARD fluctuate, likely due to the slightly unpredictable behavior of hash-based indexing structures, average execution times remain low overall and below 10 ms for LOCALIZED. Still, localization incurs a noticeable overhead up to factor 6 compared to STANDARD and VIATRA. This overhead is expected, since in this scenario, all considered updates affect the relevant subgraph and thus impact the results of the localized RETE net similarly to the results of the standard RETE nets. Consequently, localization does not reduce computational effort, but causes overhead instead.

Finally, Fig. 3 (right) shows the memory measurements for all strategies and models after the final update. Here, LOCALIZED again achieves a substantial improvement in scalability compared to the other RETE-based strategies, with a slightly higher memory consumption for the smallest model and an improvement by factor 120 over STANDARD for the largest model. This is a result of the localized RETE net producing the same number of intermediate results for all model sizes, with the slight growth in memory consumption likely a product of the growing size of the model itself. SDM*, not storing any matches, performs better for all but the largest model, where memory consumption surpasses the measurement for LOCALIZED. This surprising result can probably be explained by the fact that SDM* has to index the full model to observe modifications, causing an overhead in memory consumption.

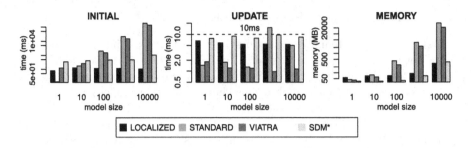

Fig. 3. Measurements for the synthetic abstract syntax graph scenario (log scale)

In addition to these experiments, where the full model is always stored in main memory, we also experimented with a model initially stored on disk via the persistence layer CDO [1]. Measurements mostly mirror those for the main-memory-based experiment. Notably though, in conjunction with CDO the LOCALIZED strategy achieves almost ideal scalability regarding memory consumption for this scenario, with measurements around 70 MB for all model sizes.[5]

4.2 Real Abstract Syntax Graphs

To evaluate our approach in the context of a more realistic application scenario, we perform a similar experiment using real data and queries inspired by object-oriented design patterns [21]. In contrast to the synthetic scenario, this experiment emulates a situation where modifications may concern not only the relevant subgraph but the entire model, for instance when multiple developers are simulatenously working on different model parts.

We therefore extract a history of real Java abstract syntax graphs with about 16 000 vertices and 45 000 edges from a software repository using the MoDisco tool [12]. After executing the queries over the initial commit, we replay the history and perform incremental updates of query results after each commit. As relevant subgraph, we again consider a single package and its contents.

Figure 4 displays the aggregate execution time for processing the commits one after another for the queries where LOCALIZED performed best and worst compared to STANDARD, with the measurement at $x = 0$ indicating the initial execution time for the starting model. Initial execution times are similarly low due to a small starting model and in fact slightly higher for LOCALIZED. However, on aggregate LOCALIZED outperforms STANDARD with an improvement between factor 5 and 18 due to significantly lower update times, which are summarized in Fig. 5 (left).

In this case, the improvement mostly stems from the more precise monitoring of the model for modifications: The RETE nets of both STANDARD and LOCALIZED remain small due to strong filtering effects in the query graphs. However, while STANDARD spends significant effort on processing

---

[5] See [9] for measurement results.

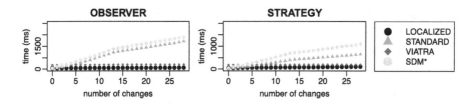

Fig. 4. Execution times for the real abstract syntax graph scenario

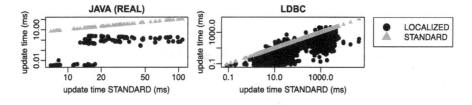

Fig. 5. Summary of update processing times for different scenarios (log scale)

model change notifications due to observing all appropriately typed model elements, this effort is substantially reduced for LOCALIZED, which only monitors elements relevant to query results required for completeness under the relevant subgraph. The execution times of SDM* can be explained by the same effect. Interestingly, VIATRA seems to implement an improved handling of such notifications, achieving improved execution times for particularly small updates even compared to LOCALIZED, but requiring more time if an update triggers changes to the RETE net. Combined with a higher RETE net initialization time, this results in LOCALIZED also outperforming VIATRA for all considered queries.

Regarding memory consumption, all strategies perform very similarly, which is mostly a result of the size of the model itself dominating the measurement and hiding the memory impact of the rather small RETE nets.

4.3 LDBC Social Network Benchmark

Finally, we also experiment with the independent LDBC Social Network Benchmark [5,19], simulating a case where a user of a social network wants to incrementally track query results relating to them personally.

We therefore generate a synthetic social network consisting of around 850 000 vertices, including about 1700 persons, and 5 500 000 edges using the benchmark's data generator and the predefined scale factor 0.1. We subsequently transform this dataset into a sequence of element creations and deletions based on the timestamps included in the data. We then create a starting graph by replaying the first half of the sequence and perform an initial execution of adapted versions of the benchmark queries consisting of plain graph patterns, with a person with a close-to-average number of contacts in the final social network designated as relevant subgraph. After the initial query execution, we replay the remaining changes, incrementally updating the query results after each change.

The resulting execution times for the benchmark queries where LOCALIZED performed best and worst compared to STANDARD are displayed in Fig. 6. A summary of all update time measurements for LOCALIZED in comparison with STANDARD is also displayed in Fig. 5 (right). For all queries, LOCALIZED ultimately outperforms the other approaches by a substantial margin, as the localized RETE version forgoes the computation of a large number of irrelevant intermediate results due to the small relevant subgraph on the one hand and avoids redundant computations on the other hand.

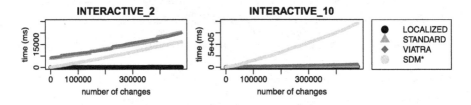

Fig. 6. Execution times for the LDBC scenario

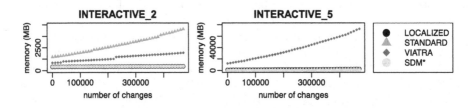

Fig. 7. Memory measurements for the LDBC scenario

The memory measurements in Fig. 7 mostly mirror execution times for RETE-based approaches, with the memory consumption for LOCALIZED always lower than for STANDARD and VIATRA except for a period at the beginning of the execution of the query INTERACTIVE_5, where STANDARD outperforms LOCALIZED. The weaker performance for INTERACTIVE_5 and INTERACTIVE_10 likely stems from the fact that these queries contain cycles, which act as strong filters for subsequent (intermediate) results and achieve a somewhat similar effect as localization. The weaker performance of VIATRA for INTERACTIVE_5 is a product of the usage of a suboptimal RETE net. As expected, memory consumption is lowest for SDM* for all queries.

4.4   Discussion

On the one hand, our experimental results demonstrate that in situations where the relevant subgraph constitutes only a fraction of the full model, RETE net localization can improve the performance of incremental query execution compared to both the standard RETE approach and a solution based on local search. In such scenarios, localization can improve scalability with respect to initial query execution time and memory consumption, as demonstrated in Sects. 4.1 and 4.3 and, if changes are not restricted to the relevant subgraph, also update processing time, as shown in Sects. 4.2 and 4.3.

On the other hand, as demonstrated in Sect. 4.1, localization incurs an overhead on update processing time if changes are only made to the relevant subgraph and on initial execution time and memory consumption if the relevant subgraph contains most of the elements in the full model. While this overhead will essentially be limited to a constant factor in many scenarios, as analyzed in Sect. 3.3, the standard RETE approach remains preferable for query execution with global semantics or if modifications are restricted to the relevant subgraph and initial query execution time and memory consumption are irrelevant.

To mitigate internal threats to the validity of our results resulting from unexpected JVM behavior, we have performed 10 runs of all experiments. However, with reliable memory measurements a known pain point of Java-based experiments, the reported memory consumption values are still not necessarily accurate and can only serve as an indicator. To minimize the impact of the implementation on conceptual observations, we compare the prototypical implementation of our approach to a regular RETE implementation [8], which shares a large portion of the involved code, and to two existing tools [23,35].

We have attempted to address external threats to validity via experiments accounting for different application domains and a combination of synthetic and real-world queries and data, including a setting from an established, independent benchmark. Still, our results cannot be generalized and do not support quantitative claims, but serve to demonstrate conceptual advantages and disadvantages of the presented approach.

5   Related Work

With graph query execution forming the foundation of many applications, there already exists an extensive body of research regarding its optimization.

Techniques based on local search [6,11,14,22,23,25,29] constitute one family of graph querying approaches. While they are designed to exploit locality in the host graph to improve execution time, repeated query execution leads to redundant computations that are only avoided by fully incremental techniques.

In [17], Egyed proposes a scoping mechanism for local search to support incremental query execution, only recomputing query results when a graph element touched during query execution changes. While this approach offers some degree of incrementality, it is limited to queries with designated root elements that serve as an anchor in the host graph and may still result in redundant computations, since query reevaluation is only controlled at the granularity of root elements.

A second family of solutions is based on discrimination networks [10, 26, 35, 36], the most prominent example of which are RETE nets.

VIATRA [35] is a mature tool for incremental graph query execution based on the RETE algorithm [20], which supports advanced concepts for query specification and optimization not considered in this paper. Notably, VIATRA allows reuse of matches for isomorphic query subgraphs within a single RETE net. This is achieved via RETE structures not covered by the rather restrictive definition of well-formedness used in this paper, which points to a possible direction for future work. However, while VIATRA also has a local search option for query execution, it does not integrate local search with the incremental query engine but rather offers it as an alternative.

Beyhl [10] presents an incremental querying technique based on a generalized version of RETE nets, called Generalized Discrimination Networks (GDNs) [26]. The main difference compared to the RETE algorithm is the lack of join nodes. Instead, more complex nodes that directly compute complex matches using local search are employed. The approach however represents more of a means of controlling the trade-off between local search and RETE rather than an integration and still requires a global computation of matches for the entire host graph.

In previous work [7], we have made a first step in the direction of localizing RETE-based query execution. While this earlier technique already allowed anchoring the execution of a RETE net to certain host graph vertices, this anchoring was based on typing information and its results did not meet the definition of completeness introduced in this paper.

Model repositories such as CDO [1] and NeoEMF [15] provide support for query execution over partial models via lazy loading. As persistence layers, these solutions however focus on implementing an interface of atomic model access operations in order to be agnostic regarding the employed query mechanism.

The Mogwaï tool [16] aims to improve query execution over persistence layers like CDO and NeoEMF by mapping model queries to native queries for the underlying database system instead of using the atomic model access operations provided by the layer's API, avoiding loading the entire model into main memory. The tool however does not consider incremental query execution.

Jahanbin et al. propose an approach for querying partially loaded models stored via persistence layers [28] or as XMI files [27]. In contrast to the solution presented in this paper, their approach still aims to always provide complete query results for the full model and is thus based on static analysis and typing information rather than dynamic exploitation of locality.

Query optimization for relational databases is a research topic that has been under intense study for decades [31–33]. Generally, many of the techniques from this field are applicable to RETE nets, which are ultimately based on relational algebra and related to materialized views in relational databases [24]. However, relational databases lack the notion of locality inherent to graph-based encodings and are hence not tailored to exploit local navigation.

This shortcoming has given rise to a number of graph databases [4], which employ a graph-based data representation instead of a relational encoding and form the basis of some model persistence layers like NeoEMF [15]. While these database systems are designed to accommodate local navigation, to the best of our knowledge, support for incremental query execution is still lacking.

6 Conclusion

In this paper, we have presented a relaxed notion of completeness for query results that lifts the requirement of strict completeness of results for graph queries and thereby the need for necessarily global query execution. Based on this relaxed notion of completeness, we have developed an extension of the RETE approach that allows local, yet fully incremental execution of graph queries. An initial evaluation demonstrates that the approach can improve scalability in scenarios with small relevant subgraphs, but causes an overhead in unfavorable cases.

In future work, we plan to extend our technique to accommodate advanced concepts for query specification, most importantly nested graph conditions. Furthermore, we want to investigate whether the proposed solution can be adapted to support bulk loading of partial models in order to reduce overhead caused by lazy loading strategies employed by model persistence layers.

References

1. Eclipse CDO Model Repository. https://projects.eclipse.org/projects/modeling. emf.cdo. Accessed 5 Jan 2024
2. EMF: Eclipse Modeling Framework. https://www.eclipse.org/modeling/emf/. Accessed 27 Jan 2024
3. Localized RETE for Incremental Graph Queries Evaluation Artifacts. https:// github.com/hpi-sam/Localized-RETE-for-Incremental-Graph-Queries. Accessed 5 Feb 2024
4. Angles, R.: A comparison of current graph database models. In: 2012 IEEE 28th International Conference on Data Engineering Workshops, pp. 171–177. IEEE (2012). https://doi.org/10.1109/ICDEW.2012.31
5. Angles, R., et al.: The LDBC Social Network Benchmark (2024). https://doi.org/ 10.48550/arXiv.2001.02299
6. Arendt, T., Biermann, E., Jurack, S., Krause, C., Taentzer, G.: Henshin: advanced concepts and tools for in-place EMF model transformations. In: Petriu, D.C., Rouquette, N., Haugen, Ø. (eds.) MODELS 2010. LNCS, vol. 6394, pp. 121–135. Springer, Heidelberg (2010). https://doi.org/10.1007/978-3-642-16145-2_9
7. Barkowsky, M., Brand, T., Giese, H.: Improving adaptive monitoring with incremental runtime model queries. In: SEAMS 2021, pp. 71–77. IEEE (2021). https:// doi.org/10.1109/SEAMS51251.2021.00019
8. Barkowsky, M., Giese, H.: Host-graph-sensitive RETE nets for incremental graph pattern matching with nested graph conditions. J. Log. Algebr. Methods Program. **131**, 100841 (2023). https://doi.org/10.1016/j.jlamp.2022.100841

9. Barkowsky, M., Giese, H.: Localized RETE for Incremental Graph Queries. arXiv preprint (2024). https://doi.org/10.48550/arXiv.2405.01145

10. Beyhl, T.: A framework for incremental view graph maintenance. Ph.D. thesis, Hasso Plattner Institute at the University of Potsdam (2018)

11. Bi, F., Chang, L., Lin, X., Qin, L., Zhang, W.: Efficient subgraph matching by postponing cartesian products. In: Proceedings of the 2016 International Conference on Management of Data, pp. 1199–1214. ACM (2016). https://doi.org/10.1145/2882903.2915236

12. Bruneliere, H., Cabot, J., Jouault, F., Madiot, F.: MoDisco: a generic and extensible framework for model driven reverse engineering. In: Proceedings of the IEEE/ACM International Conference on Automated Software Engineering (2010). https://doi.org/10.1145/1858996.1859032

13. Codd, E.F.: A relational model of data for large shared data banks. Commun. ACM **13**(6), 377–387 (1970). https://doi.org/10.1145/362384.362685

14. Cordella, L.P., Foggia, P., Sansone, C., Vento, M.: A (sub) graph isomorphism algorithm for matching large graphs. IEEE Trans. Pattern Anal. Mach. Intell. **26**(10), 1367–1372 (2004). https://doi.org/10.1109/TPAMI.2004.75

15. Daniel, G., et al.: NeoEMF: a multi-database model persistence framework for very large models. Sci. Comput. Program. **149**, 9–14 (2017). https://doi.org/10.1016/j.scico.2017.08.002

16. Daniel, G., Sunyé, G., Cabot, J.: Scalable queries and model transformations with the Mogwaï tool. In: Rensink, A., Sánchez Cuadrado, J. (eds.) ICMT 2018. LNCS, vol. 10888, pp. 175–183. Springer, Cham (2018). https://doi.org/10.1007/978-3-319-93317-7_9

17. Egyed, A.: Instant consistency checking for the UML. In: Proceedings of the 28th International Conference on Software Engineering, pp. 381–390 (2006). https://doi.org/10.1145/1134285.1134339

18. Ehrig, H., Ehrig, K., Prange, U., Taentzer, G.: Fundamentals of Algebraic Graph Transformation. Springer, Heidelberg (2006). https://doi.org/10.1007/3-540-31188-2

19. Erling, O., et al.: The LDBC social network benchmark: interactive workload. In: Proceedings of the 2015 ACM SIGMOD International Conference on Management of Data, pp. 619–630. ACM (2015). https://doi.org/10.1145/2723372.2742786

20. Forgy, C.L.: Rete: a fast algorithm for the many pattern/many object pattern match problem. In: Readings in Artificial Intelligence and Databases, pp. 547–559. Elsevier (1989). https://doi.org/10.1016/0004-3702(82)90020-0

21. Gamma, E., Helm, R., Johnson, R., Vlissides, J.: Design patterns: abstraction and reuse of object-oriented design. In: Nierstrasz, O.M. (ed.) ECOOP 1993. LNCS, vol. 707, pp. 406–431. Springer, Heidelberg (1993). https://doi.org/10.1007/3-540-47910-4_21

22. Geiß, R., Batz, G.V., Grund, D., Hack, S., Szalkowski, A.: GrGen: a fast SPO-based graph rewriting tool. In: Corradini, A., Ehrig, H., Montanari, U., Ribeiro, L., Rozenberg, G. (eds.) ICGT 2006. LNCS, vol. 4178, pp. 383–397. Springer, Heidelberg (2006). https://doi.org/10.1007/11841883_27

23. Giese, H., Hildebrandt, S., Seibel, A.: Improved flexibility and scalability by interpreting story diagrams. Electron. Commun. EASST **18** (2009). https://doi.org/10.14279/tuj.eceasst.18.268

24. Gupta, A., Mumick, I.S., et al.: Maintenance of materialized views: problems, techniques, and applications. IEEE Data Eng. Bull. **18**(2), 3–18 (1995)

25. Han, W.S., Lee, J., Lee, J.H.: Turboiso: towards ultrafast and robust subgraph isomorphism search in large graph databases. In: Proceedings of the 2013 ACM SIGMOD International Conference on Management of Data, pp. 337–348 (2013). https://doi.org/10.1145/2463676.2465300

26. Hanson, E.N., Bodagala, S., Chadaga, U.: Trigger condition testing and view maintenance using optimized discrimination networks. IEEE Trans. Knowl. Data Eng. **14**(2), 261–280 (2002). https://doi.org/10.1109/69.991716

27. Jahanbin, S., Kolovos, D., Gerasimou, S.: Towards memory-efficient validation of large XMI models. In: 2023 ACM/IEEE International Conference on Model Driven Engineering Languages and Systems Companion (MODELS-C), pp. 241–250. IEEE (2023). https://doi.org/10.1109/MODELS-C59198.2023.00053

28. Jahanbin, S., Kolovos, D., Gerasimou, S., Sunyé, G.: Partial loading of repository-based models through static analysis. In: Proceedings of the 15th ACM SIGPLAN International Conference on Software Language Engineering, pp. 266–278 (2022). https://doi.org/10.1145/3567512.3567535

29. Jüttner, A., Madarasi, P.: VF2++-an improved subgraph isomorphism algorithm. Discret. Appl. Math. **242**, 69–81 (2018). https://doi.org/10.1016/j.dam.2018.02.018

30. Kent, S.: Model driven engineering. In: Butler, M., Petre, L., Sere, K. (eds.) IFM 2002. LNCS, vol. 2335, pp. 286–298. Springer, Heidelberg (2002). https://doi.org/10.1007/3-540-47884-1_16

31. Krishnamurthy, R., Boral, H., Zaniolo, C.: Optimization of nonrecursive queries. In: VLDB, vol. 86, pp. 128–137 (1986)

32. Lee, C., Shih, C.S., Chen, Y.H.: Optimizing large join queries using a graph-based approach. IEEE Trans. Knowl. Data Eng. **13**(2), 298–315 (2001). https://doi.org/10.1109/69.917567

33. Leis, V., Gubichev, A., Mirchev, A., Boncz, P., Kemper, A., Neumann, T.: How good are query optimizers, really? Proc. VLDB Endow. **9**(3), 204–215 (2015). https://doi.org/10.14778/2850583.2850594

34. Szárnyas, G., Izsó, B., Ráth, I., Varró, D.: The train benchmark: cross-technology performance evaluation of continuous model queries. Softw. Syst. Model. **17**, 1365–1393 (2018). https://doi.org/10.1007/s10270-016-0571-8

35. Varró, D., Bergmann, G., Hegedüs, Á., Horváth, Á., Ráth, I., Ujhelyi, Z.: Road to a reactive and incremental model transformation platform: three generations of the VIATRA framework. Softw. Syst. Model. **15**(3), 609–629 (2016). https://doi.org/10.1007/s10270-016-0530-4

36. Varró, G., Deckwerth, F.: A rete network construction algorithm for incremental pattern matching. In: Duddy, K., Kappel, G. (eds.) ICMT 2013. LNCS, vol. 7909, pp. 125–140. Springer, Heidelberg (2013). https://doi.org/10.1007/978-3-642-38883-5_13

Using Application Conditions to Rank Graph Transformations for Graph Repair

Lars Fritsche[1]([⊠])[iD], Alexander Lauer[2][iD], Andy Schürr[1][iD],
and Gabriele Taentzer[2]

[1] Technical University Darmstadt, Darmstadt, Germany
{lars.fritsche,andy.schuerr}@es.tu-darmstadt.de
[2] Philipps-Universität Marburg, Marburg, Germany
alexander.lauer@uni-marburg.de, taentzer@mathematik.uni-marburg.de

Abstract. When using graphs and graph transformations to model systems, consistency is an important concern. While consistency has primarily been viewed as a binary property, i.e., a graph is consistent or inconsistent with respect to a set of constraints, recent work has presented an approach to consistency as a graduated property. This allows living with inconsistencies for a while and repairing them when necessary. When repairing inconsistencies in a graph, we use graph transformation rules with so-called *impairment- and repair-indicating application conditions* to understand how much repair gain certain rule applications would bring. Both types of conditions can be derived from given graph constraints. Our main theorem shows that the difference between the number of actual constraint violations before and after a graph transformation step can be characterized by the difference between the numbers of violated impairment-indicating and repair-indicating application conditions. This theory forms the basis for algorithms with look-ahead that rank graph transformations according to their potential for graph repair. An initial evaluation shows that graph repair can be well supported by rules with these new types of application conditions.

Keywords: Graph Consistency · Graph Transformation · Graph Repair

1 Introduction

Graph transformation has proven to be a versatile approach for specifying and validating software engineering problems [9]. This is true because graphs are an appropriate means for representing complex structures of interest, the constant change of structures can be specified by graph transformations, and there is a strong theory of graph transformation [4] that has been used to validate software engineering problems. When applying graph transformation, it is typically

This work was partially funded by the German Research Foundation (DFG), project "Triple Graph Grammars (TGG) 3.0".

important that processed graphs are consistent with respect to a given set of constraints. Ensuring graph consistency involves two tasks. First, to specify what a consistent graph is and to check whether a graph is indeed consistent, and second, to ensure that graph transformations preserve or even improve consistency.

Throughout this paper, we consider the Class Responsibility Assignment (CRA) problem as running example [2]. It is concerned with an optimal assignment of features (i.e., methods and attributes) to classes. The constraints enforcing that each feature belongs to one and only one class, are invariants for all class model modifying operations. To validate the quality of a class model, often coupling and cohesion metrics are used. Related design guidelines like "minimize dependencies across class boundaries" may be formulated as constraints, too. This example shows that constraints may serve different purposes; some are considered so essential that they must always be satisfied, while others are used for optimization; they may be violated to a certain extent, but the number of violations should be kept as small as possible.

Nested graph constraints [8] provide a viable means of formulating graph properties. The related notion of constraint consistency introduced in [8] is binary: a graph is either consistent or inconsistent. Since graph repair is often only gradual, it is also interesting to consider graph consistency as graduated property, as was done in [12]. To support gradual repair, graph transformations are analyzed in [12] w.r.t. their potential to improve (sustain) the consistency level of a processed graph, i.e., to strictly decrease (preserve) the number of constraint violations in a graph. A *static analysis approach* is presented in [12] for checking whether or not a graph transformation rule is always consistency-sustaining or -improving. Unfortunately, the approach does not yet support graph constraints with different priorities or propositional logic operators. In addition, scenarios in which a rule either increases or decreases graph consistency depending on the context of the matched and rewritten subgraph are not supported.

To mitigate these issues, we introduce a *new dynamic analysis approach that ranks matches of rules and related rule applications according to their effects on improving graph consistency* as follows:

1. Graph transformation rules are equipped with *impairment-indicating and repair-indicating application conditions.* These conditions do no longer block rule applications, but count the number of constraint violations introduced or removed by a given graph transformation step. They are derived from nested graph constraints based on the constructions presented in [8].
2. Our main theorem shows that *the number of additional constraint violations caused by a rule application can be characterized by the difference between the numbers of violations of associated impairment-indicating and repair-indicating application conditions.*
3. This theory forms the basis for *graph repair algorithms with a look-ahead* for the graph-consistency-improving potential of selectable rule applications. Based on a prototypical implementation of a greedy algorithm, we show in

an initial evaluation that our approach scales well with the available number of rule applications for each consistency-improving transformation step.

2   Running Example

To illustrate the problem addressed, we consider a version of the Class Responsibility Assignment (CRA) problem [2,6] for our running example. The CRA aims to provide a high-quality design for object-oriented class structures. Given a set of features, i.e., methods and attributes, with dependencies but no classes, a solution to the CRA involves a class design with high cohesion within classes and low coupling between classes. We slightly adapt the CRA problem by starting with a predefined class diagram that is then refactored by moving features between classes. The goal of the refactoring steps is to reduce method-attribute dependencies across different classes and group methods with similar attribute dependencies in the same class.

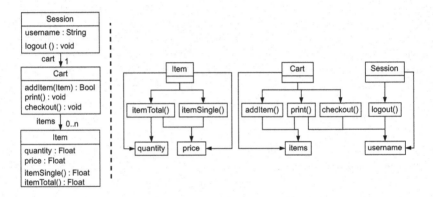

Fig. 1. Class diagram (left) and feature dependencies (right)

Figure 1 shows our running example with a class diagram consisting of three classes modeling an online shopping session on the left. On the right, we see an alternative, graph-like representation of the class diagram, in which classes, methods and attributes are represented as graph nodes. In addition to the class diagram on the left, there are edges between methods and attributes that model access dependencies of methods on attributes[1].

Next, we define two simple refactoring rules and constraints that these rules have to respect in Fig. 2. The rule moveAttribute moves an attribute from one class to another one, while the rule moveMethod does the same for methods. From the set of language constraints on class structures, we select the two constraints h_1 and h_2, which impose two basic properties on class models, and consider them as *hard constraints*. They state that methods and attributes must

---

[1] The original CRA problem also considers call dependencies between methods, which are ignored in this paper to keep the running example as small as possible.

not be contained in more than one class. There are also *weak constraints*, which are concerned with an optimal class design. They specify (quantifiable) quality requirements for the class design. We consider two weak constraints: Constraint w_1 states that two methods within the same class c' should have at least one common attribute dependency on an attribute within the same class c'. Constraint w_2 says that methods should have no dependencies on attributes of other classes. These constraints are weak, which means that violating them is acceptable (at least for a certain time). In fact, the two weak constraints may even contradict each other, since w_2 may cause us to move a method to a class with other methods with which it has no other attribute dependencies in common.

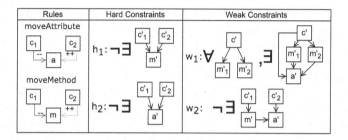

Fig. 2. Rules and constraints

Obviously, neither rule can violate the hard constraints, but moving features between classes can remove or add weak constraint violations, i.e., repair or impair the regarded graph's consistency. For example, moving the print() method from Cart to Session would repair a violation of w_2 because the method depends on the username attribute. However, it would introduce a new violation of the same constraint due to the dependency on the items attribute in the Cart class. In the following, we will study this example using the previously introduced graph (refactoring) rules and nested graph constraints. We will present a new methodology that derives application conditions for each refactoring rule, based on which we can calculate a look-ahead for each refactoring transformation (i.e., rule application) telling us how many repairs and impairments it will perform. Based on this rule ranking information, different CRA optimization algorithms may be implemented including a greedy algorithm that always selects refactoring rule applications with a maximum consistency gain.

3   Preliminaries

In this section, we briefly recall key notions that are used throughout this paper. Our theory for the construction of model repair algorithms is based on graphs or, more precisely, typed graphs, as introduced for graph transformations in [3,4]. In our running example, Classes and Features are represented as nodes. Object references such as the attribute dependencies between a Method and an Attribute are represented as edges.

A *graph* $G = (G_V, G_E, s_G, t_G)$ consists of a set G_V of nodes, a set G_E of edges and two mappings $s_G : G_E \to G_V$ and $t_G : G_E \to G_V$ that assign the source and target nodes for each edge of G. If a tuple as above is not given explicitly, the set of nodes (edges) is denoted by G_V (G_E) and the source (target) mapping is denoted by s_G (t_G). Throughout the paper, we assume that a graph is always finite, i.e. the set of nodes and the set of edges are finite.

A *graph morphism* $f : G \to H$ between two graphs G and H consists of two mappings $f_V : G_V \to H_V$ and $f_E : G_E \to H_E$ that preserve the source and target mappings, i.e., $f_V \circ s_G = s_H \circ f_E$ and $f_C \circ t_G = t_H \circ f_E$. A graph morphism is called *injective(surjective)* if both mappings f_V and f_E are injective (surjective). An injective morphism is denoted by $f : G \hookrightarrow G$. An injective morphism $f : G \hookrightarrow H$ is called *inclusion* if $f_E(e) = e$ for all $e \in G_E$ and $f_V(v) = v$ for all $v \in G_V$.

Given a graph morphism $p : G \hookrightarrow H$ and an inclusion $i : G' \hookrightarrow G$, the *restriction of p to G'*, denoted by $p_{|G'}$, is defined as $p_{|G'} = p \circ i$.

Given a graph TG, called the *type graph*, a *typed graph* over TG is a tuple $(G, type)$ consisting of a graph G and an graph morphism $type : G \to TG$. Given two typed graphs $G = (G', type_G)$ and $H = (H', type_H)$, a *typed graph morphism* $f : G \to H$ is a graph morphism $f : G' \to H'$ such that $type_H \circ f = type_G$.

To formulate hard and weak constraints, we use *nested graph conditions* introduced by Habel and Pennemann [8]. It has been shown that the class of nested graph conditions is equivalent to first-order logic [16] and that almost all OCL formulae can be translated into nested graph constraints [15]. Nested graph constraints, or constraints for short, are nested graph conditions that can be evaluated directly on a given graph, whereas, in general, graph conditions must be evaluated with respect to graph morphisms, which usually represent rule matches.

A *nested graph condition* over a graph P is of the form true, or $\exists(e : P \hookrightarrow Q, d)$ where d is a condition over Q, or $d_1 \vee d_2$ or $\neg d_1$ where d_1 and d_2 are conditions over P.

A condition over the empty graph $\emptyset$ is called *constraint*. We use the abbreviations false $:= \neg$true, $d_1 \wedge d_2 := \neg(\neg d_1 \vee \neg d_2)$, $d_1 \implies d_2 := \neg d_1 \vee d_2$ and $\forall(e : P \hookrightarrow Q, d) := \neg\exists(e : P \hookrightarrow Q, \neg d)$. When e is of the form $e : \emptyset \hookrightarrow Q$, we use the short notation $\exists(Q, d)$ and $\forall(Q, d)$. For a condition $c = \forall(Q, d)$, we call Q the *premise* and d the *conclusion* of c. Throughout the paper, we assume that each constraint is finite, i.e., the graphs used and the number of nesting levels are finite.

Each graph morphism $p : P \hookrightarrow G$ satisfies true. It satisfies a condition $c = \exists(e : P \hookrightarrow Q, d)$, denoted by $p \models c$ if there is a morphism $q : Q \hookrightarrow G$ with $p = q \circ e$ and $q \models d$. For Boolean operators, satisfaction is defined as usual. A graph G satisfies a constraint c, denoted by $G \models c$ if the unique morphism $p : \emptyset \hookrightarrow G$ satisfies c.

Example 1. The graph shown in Fig. 1 satisfies the hard constraints h_1 and h_2. No attribute and no method is contained in two classes. However, the graph does not satisfy the weak constraint w_1 shown in Fig. 2. The methods addItem() and checkout() are contained in the same class Cart, but Cart does not contain an attribute used by both methods. The graph also does not satisfy the weak constraint w_2. This is because checkout() uses the attribute username, which is contained in the class Session.

Graph transformation rules are used to specify state-changing operations for a system of interest. In the case of our CRA example, rules are used to define the set of all available refactoring operations. Their associated application conditions are unsatisfied whenever a rule application changes the consistency state of a rewritten graph. In the following, we recall the formal definitions of *graph transformation rules* and *graph transformations* [3,4].

A *graph transformation rule*, or *rule* for short, $\rho = L \xleftarrow{l} K \xhookrightarrow{r} R$ consists of graph L, called the *left-hand side (LHS)*, K, called the *context*, R, called the *right-hand side (RHS)* and injective morphisms $l \colon K \hookrightarrow L$ and $r \colon K \hookrightarrow R$. The *inverse rule* of ρ, denoted by ρ^{-1} is defined as $\rho^{-1} := R \xleftarrow{r} K \xhookrightarrow{l} L$. A *left (right) application condition* for a rule is a nested condition over its LHS (RHS).

For simplicity, we present the more constructive, set-theoretic definition of graph transformation, which has been shown to be equivalent to the commonly used double-pushout approach, based on category theory [4]. Given a graph G, a rule $\rho = L \xleftarrow{l} K \xhookrightarrow{r} R$ and an injective morphism $m \colon L \hookrightarrow G$, a *graph transformation* t, denoted by $t \colon G \Longrightarrow_{\rho,m} H$, via ρ at m can be constructed by (a) deleting all nodes and edges of L that do not have a preimage in K, i.e., construct the graph $D = G \setminus m(L \setminus (l(K))$ and (b) adding all nodes and edges of R that do not have a preimage in K, i.e., construct the graph $H = D \dot\cup R \setminus r(K)$, where $\dot\cup$ denotes the disjoint union.

The rule ρ is applicable at m if and only if D is a graph, i.e., if it does not contain any dangling edges. In this case, m is called *match* and the newly created morphism $n \colon R \hookrightarrow H$ is called *comatch*. We call G the *original graph*, D the *interface* and H the *result graph* of the transformation t.

The *derived span of* t, denoted by $\mathrm{der}(t)$, is defined as $\mathrm{der}(t) := G \xleftarrow{g} D \xrightarrow{h} H$, where D is the interface, G is the original graph, H is the result graph and g and h are the transformation morphisms of t [4]. The *track morphism of* t [14], denoted by $\mathrm{tr}_t \colon G \dashrightarrow H$, is a partial morphism, defined as

$$\mathrm{tr}_t(e) := \begin{cases} h(g^{-1}(e)) & \text{if } e \in g(D) \\ \text{undefined} & \text{otherwise.} \end{cases}$$

4   Counting Constraint Violations

For ranking rule applications, a method is needed to evaluate the increase or decrease in consistency of weak constraints Here, we rely on the *number of violations* introduced by Kosiol et al. [12]. This notion allows to detect an increase in consistency even if it is not fully recovered. For example, this notion is able to detect whether an occurrence of the premise that does not satisfy the conclusion is deleted or extended so that it satisfies the conclusion in the result graph of the transformation. While this notion was introduced for constraints in so-called *alternating quantifier normal form*, i.e., the set of all nested constraints that do not use Boolean operators, we extend this notion to support universally bounded nested conditions that may use Boolean operators in the conclusion. In particular, this notion evaluates conditions w.r.t. occurrences q of the premise of the condition.

For existentially bounded constraints, the number of violations acts as a binary property, i.e., it is equal to 1 if the constraint is unsatisfied and equal to 0 if the constraint is satisfied. To use a constraint for ranking, it must be all-quantified. Weak constraints are allquantified and have the form $\forall(P, d)$, where $d = \mathsf{false}$ or d is a Boolean formula with conditions of the form $\exists(e' \colon P \hookrightarrow Q, \mathsf{true})$, i.e., all constraints with a nesting level less than or equal to 2 that are universally bound and do not use any boolean operators at the highest nesting level are allowed. We focus on this form because, in our experience, it is the most commonly used in practice. There are no restrictions for hard constraints.

In addition, for a condition with premise P, we allow to restrict violations to injective morphisms i with codomain P if necessary, i.e. if some parts of the violation do not need to be considered.

Definition 1 (Set and number of violations). *Given a condition $c = \forall(e \colon P \hookrightarrow Q, d)$ and graph morphism $p \colon P \hookrightarrow G$, the* set of violations *of c in p, denoted by $\mathrm{sv}_p(c)$, is defined as*

$$\mathrm{sv}_p(c) := \{q \colon Q \hookrightarrow G \mid p = q \circ e \text{ and } q \not\models d\}.$$

The set of violations of a constraint $c = \forall(e \colon \emptyset \hookrightarrow P, d)$ in a graph G, denoted by $\mathrm{sv}_G(c)$, is defined as $\mathrm{sv}_G(c) := \mathrm{sv}_p(c)$ where $p \colon \emptyset \hookrightarrow G$ is the unique morphism into G.

Given a morphism $i \colon Q' \hookrightarrow Q$, such that there exists a morphism $i' \colon P \hookrightarrow Q'$ with $i \circ i' = e$, the set of violations of c in p restricted to i, denoted by $\mathrm{sv}_{p,i}(c)$, is defined as $\mathrm{sv}_{p,i}(c) := \{q \circ i \mid q \in \mathrm{sv}_p(c)\}$.

The number of violations *of a condition c in p restricted to i, denoted by $\mathrm{nv}_{p,i}(c)$, is defined as $\mathrm{nv}_{p,i}(c) := |\mathrm{sv}_{p,i}(c)|$. The number of violations of a constraint c in a graph G, denoted with $\mathrm{nv}_G(c)$, is defined as $\mathrm{nv}_G(c) := |\mathrm{sv}_G(c)|$.*

Example 2. As discussed in Example 1, the graph G in Fig. 1 does not satisfy w_1 and w_2. For w_1 there are two occurrences of the premise that do not satisfy the conclusion of w_1. The premise of w_1 can be mapped to the violation described in Example 1 in two different ways, so $\mathrm{nv}_G(w_1) = 2$. If we restrict $\mathrm{sv}_G(w_1)$ to a morphism $i \colon Q' \hookrightarrow Q$, where Q' contains only the node c', we get $\mathrm{nv}_{q,i}(w_1) = 1$ with $q \colon \emptyset \hookrightarrow G$. For w_2 there are two occurrences of the premise. Both methods print() and checkout() use the attribute username, which is contained in the class Session, so $\mathrm{nv}_G(w_2) = 2$.

5   Application Conditions for Consistency Monitoring

The main idea of our approach is to predict the change in consistency induced by an application of a rule. This allows us to apply the most consistency-increasing rule without using a trial and error approach. To obtain this prediction, we use application conditions that do not block the application of constraint-violating rules, but instead annotate rule matches with the number of constraint impairments and/or repairs caused by the related transformation; thus, the conditions monitor the consistency change.

In addition, we use the observation that a rule introduces a violation of a constraint c if and only if the inverse rule repairs a violation of c. Thus, repair-indicating application conditions for a rule ρ are impairment-indicating application conditions for ρ^{-1}.

To construct the application conditions, we use the well-known techniques introduced by Habel and Pennemann to construct so-called *consistency-preserving* and *consistency-guaranteeing* application conditions [8].

5.1   Preliminaries

In the following we will briefly introduce the preliminaries that are needed for our construction of application conditions.

An *overlap* $\overline{GH} = (i_G, i_H, GH)$ of graphs G and H consists of jointly surjective morphisms[2] $i_G \colon G \hookrightarrow GH$ and $i_H \colon G \hookrightarrow GH$, called the *overlap morphisms of* $\overline{GH}$, and a graph GH called the *overlap graph*.

The *shift along morphism* operator allows shifting a condition c over a graph P along a morphism $i_P \colon P \hookrightarrow PL$, resulting in an equivalent condition over PL. Given a condition c over a graph P and a morphism $i_P \colon P \hookrightarrow PL$, the *shift along* i_P [8], denoted by $\mathrm{Shift}(i_P, c)$, is defined as follows:

- if $c = \mathsf{true}$, $\mathrm{Shift}(i_P, c) = \mathsf{true}$, and
- if $c = \exists(e \colon P \hookrightarrow Q, d)$, $\mathrm{Shift}(i_P, c) = \bigvee_{(e', i_Q)} \exists(e' \colon PL \hookrightarrow QL, \mathrm{Shift}(i_Q, d))$, where (e', i_Q, QL) is an overlap of PL and Q such that $e' \circ i_P = i_Q \circ e$, i.e., the square on the right is commutative, and
- if $c = c_1 \vee c_2$, $\mathrm{Shift}(i_P, c) = \mathrm{Shift}(i_P, c_1) \vee \mathrm{Shift}(i_P, c_2)$, and
- if $c = \neg c_1$, $\mathrm{Shift}(i_P, c) = \neg \mathrm{Shift}(i_P, c_1)$.

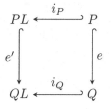

Note that, when shifting a constraint $\forall(P, d)$ over a morphism $p \colon \emptyset \hookrightarrow P'$, the first step is to construct all overlaps $\overline{PP'}$.

The *shift over rule* operator transforms a right-application condition into an equivalent left-application condition and vice versa. Given a rule $\rho = L \xleftarrow{l}$

---

[2] Two morphisms $p \colon P \hookrightarrow G$ and $q \colon Q \hookrightarrow G$ into the same graph G are called *jointly surjective* if for each element $e \hookrightarrow G$ either there is an element $e' \in P$ with $p(e') = e$ or there is an element $e' \in Q$ with $q(e') = e$..

$K \xrightarrow{r} R$ and a condition c over R, the *shift of c over ρ* [8], denoted by $\mathrm{Left}(\rho, c)$ is defined as follows:

- If $c = \mathsf{true}$, $\mathrm{Left}(\rho, c) = \mathsf{true}$, and
- If $c = \exists(e\colon R \hookrightarrow P, d)$, $\mathrm{Left}(\rho, c) := \exists(n\colon L \hookrightarrow P', \mathrm{Left}(\mathrm{der}(t)^{-1}, d))$, where P' is the resulting graph and n the comatch of the transformation $t\colon P \Longrightarrow_{\rho^{-1}, e} P'$. If there is no such transformation, we set $\mathrm{Left}(\rho, c) = \mathsf{false}$.
- If $c = c_1 \vee c_2$, $\mathrm{Left}(\rho, c) := \mathrm{Left}(\rho, c_1) \vee \mathrm{Left}(\rho, c_1)$.
- If $c = \neg c_1$, $\mathrm{Left}(\rho, c) := \neg \mathrm{Left}(\rho, c_1)$.

5.2   Shifting of Overlaps and Conditions

We will now discuss the construction of *impairment- and repair-indicating application conditions*. These are actually *annotated application conditions*, i.e. pairs of application conditions and overlaps, which contain the first morphism of the application condition. This additional information is used to directly obtain occurrences of the premise of a weak constraint to be repaired (or violated) when evaluating the set of violations of the application condition.

Definition 2 (Annotated application condition). *Given a rule $\rho = L \xleftarrow{l} K \xrightarrow{r} R$, an application condition $c = \forall(i_L\colon L \hookrightarrow PL, d)$, and an overlap $\overline{PL} = (i_L, i_P, PL)$. Then, the pair $(c, \overline{PL})$ is called a* left annotated application condition *for ρ. Right annotated application conditions for ρ are defined analogously.*

In order to count violations of application conditions, we require that the match $m\colon L \hookrightarrow G$ of a transformation factors through the inclusion $i_L\colon L \hookrightarrow PL$ of the LHS in the overlap graph and an inclusion $p\colon PL \hookrightarrow G$ of the overlap graph in the original graph of the transformation, i.e., $m = p \circ i_L$. So we only need to consider overlaps of the LHS of a rule and the premise of a constraint, where the rule is applicable at the inclusion of the LHS in the overlap graph.

We decompose the set of overlaps into the sets O_{prem} and O_{con}, where O_{prem} contains each overlap where applying the rule at the inclusion of the LHS destroys the occurrence of the constraint premise, and O_{con} contains each overlap where applying the rule at the inclusion of the LHS does not destroy the occurrence of the premise, so that the impact on the constraint conclusion can be investigated.

In particular, the set O_{prem} is used to construct application conditions that check whether an occurrence of the premise to be deleted satisfies the conclusion before the transformation. If that occurrence does not satisfy the conclusion, then deleting that occurrence would increase the consistency. The set O_{con} is used to construct application conditions that check whether an occurrence of the premise that is neither deleted nor created by the transformation satisfies the conclusion in the original and the resulting graph of the transformation. If it does not satisfy the conclusion in the original graph, the consistency is increased. Otherwise, the consistency is decreased.

Definition 3 (Overlaps of a rule and a graph). *Given a rule $\rho = L \xleftarrow{l}$ $K \xrightarrow{r} R$ and a graph P. The set of* overlaps *of ρ and P, denoted by $\mathrm{O}(\rho, P) :=$ $\mathrm{O}_{\mathrm{prem}}(\rho, P) \cup \mathrm{O}_{\mathrm{con}}(\rho, P)$, contains every* overlap $\overline{PL} = (i_P, i_L, PL)$, *so that ρ is applicable at i_L, i.e., there is a transformation $PL \Longrightarrow_{\rho,i_L} PR$, where*

$$\mathrm{O}_{\mathrm{prem}}(\rho, P) := \{(i_P, i_L, PL) \mid i_P(P) \cap i_L(L \setminus l(K)) \neq \emptyset\} \quad and$$
$$\mathrm{O}_{\mathrm{con}}(\rho, P) := \{(i_P, i_L, PL) \mid i_P(P) \cap i_L(L \setminus l(K)) = \emptyset\}.$$

The *shift of an overlap over a rule* transforms an overlap of the RHS of the rule and a graph P into an overlap of the LHS of the rule and the graph P. The inverse rule is applied to the inclusion of the RHS in the overlap graph. If the newly created elements of the rule (i.e. the elements in $R \setminus r(K)$) and P overlap non-emptily, the occurrence of P is destroyed during the shift. The result is an overlap of the LHS with a subgraph of P called the *remaining graph of P after Shift*.

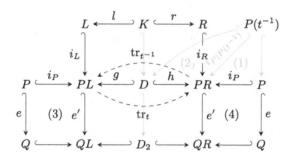

Fig. 3. Construction of impairment-indicating and repair-indicating application conditions and shift of overlaps

Definition 4 (Remaining graph of an overlap after Shift). *Given a rule $\rho = L \xleftarrow{l} K \xrightarrow{r} R$, a graph P, and an overlap $\overline{PR} = (i_P.i_R, PR) \in \mathrm{O}(\rho^{-1}, P)$, the* remaining graph of P after *the transformation $t^{-1}: PR \Longrightarrow_{i_R,\rho^{-1}} PL$, denoted by $P(t^{-1})$, is defined as*

$$P(t^{-1}) := i_P^{-1}(i_P(P) \setminus i_R(R \setminus r(K))),$$

i.e., (1) and (2) in Fig. 3 are commutative.

Note that, by construction, the morphism $\mathrm{tr}_{t^{-1}} \circ i_{P|P(t^{-1})}$ is total and that $P(t^{-1}) = P$ if $\overline{PR} \in \mathrm{O}_{\mathrm{con}}(\rho^{-1}, P)$.

The *shift of an annotated condition over a rule* shifts a right annotated application condition to a left one. Both the shift over a rule and the shift of an overlap over a rule are used.

Definition 5 (Shift of overlap and annotated condition over rule).
Given a rule $\rho = L \xleftarrow{l} K \xrightarrow{r} R$ and a graph P:

1. *Given an overlap $\overline{PR} = (i_R, i_P, PR) \in O(\rho^{-1}, P)$, the shift of $\overline{PR}$ over ρ, denoted by $\mathrm{Left}_{ol}(\rho, \overline{PR})$, is defined as*

$$\mathrm{Left}_{ol}(\rho, \overline{PR}) := (i_L, \mathrm{tr}_{t^{-1}} \circ i_{P|P(t^{-1})}, PL),$$

   where $i_L \colon L \hookrightarrow PL$ is the comatch of the transformation $t^{-1} \colon PR \Longrightarrow_{\rho^{-1}, i_R} PL$.

2. *Given an annotated application condition $ac = (c, \overline{PR})$, where c is a right application condition for ρ, the shift of ac over ρ, denoted by $\mathrm{Left}_{ann}(\rho, ac)$, is defined as $\mathrm{Left}_{ann}(\rho, ac) := (\mathrm{Left}(\rho, c), \mathrm{Left}_{ol}(\rho, \overline{PR}))$.*

The following lemma allows us to simplify the application conditions if we assume that hard constraints are always satisfied. If we consider a hard constraint of the form $c = \neg\exists(P)$ (which is satisfied by a graph G if P is not a subgraph of G) and a condition $\exists(e \colon P' \hookrightarrow Q, d)$, a morphism $p \colon P' \hookrightarrow G$ cannot satisfy $\exists(e \colon P' \hookrightarrow Q, d)$ if $Q \not\models c$, otherwise P is a subgraph of G. Therefore, when simplifying an application condition, we can replace any condition of the form $\exists(e \colon P' \hookrightarrow Q, d)$ with $Q \not\models c$ by false. Since $\forall(e \colon P' \hookrightarrow Q, d) = \neg\exists(e \colon P' \hookrightarrow Q, \neg d)$, conditions of the form $\forall(e \colon P' \hookrightarrow Q, d)$ can be replaced by true if $Q \not\models c$. In our running example, this simplification drastically reduces the number and complexity of derived impairment- and repair-indicating application conditions.

Lemma 1 (Simplification of graph conditions). *Let a constraint $c = \forall(e \colon \emptyset \hookrightarrow P, \mathsf{false})$ and a condition $c' = \exists(e' \colon P' \hookrightarrow Q', d)$ with $Q' \not\models c$ be given, then $G \models c \implies p \not\models c'$, for each morphism $p \colon P' \hookrightarrow G$.*

The proof of this Lemma can be found in Appendix A in [7].

5.3 Construction of Application Conditions

When constructing impairment- and repair-indicating application conditions, all cases in which an impairment can be introduced or a violation can be repaired must be considered. A transformation $t \colon G \Longrightarrow H$ can introduce impairments of a constraint $c = \forall(P, d)$ in the following two ways:

1. *Impairment of the premise*: A new occurrence of P is introduced that does not satisfy d. That is, there is an occurrence $p \colon P \hookrightarrow H$ with $p \not\models d$ so that there is no $p' \colon P \hookrightarrow G$ with $\mathrm{tr}_t \circ p' = p$.
2. *Impairment of the conclusion*: An occurrence of P satisfies d in G and not in H. That is, there is an occurrence $p \colon P \hookrightarrow G$ with $p \models d$ so that $\mathrm{tr}_t \circ p$ is total and $\mathrm{tr}_t \circ p \not\models d$.

Similarly, the transformation t can repair violations in the following ways:

1. *Repair of the premise:* An occurrence of P in G that does not satisfy d is deleted. That is, there is a morphism $p\colon P \hookrightarrow G$ with $p \not\models d$ so that $\mathrm{tr}_t \circ p$ is not total.
2. *Repair of the conclusion:* There is an occurrence of P which satisfies d in H but does not satisfy d in G. That is, there is a morphism $p\colon P \hookrightarrow G$ with $p \not\models d$, $\mathrm{tr}_t \circ p$ is total and $\mathrm{tr}_t \circ p \models d$.

Note that we use the same construction of impairment-indicating and repair-indicating application conditions for the premises and conclusions of constraints. The only difference is that we switch the roles of the LHS and the RHS. Thus, intuitively, the repair-indicating application conditions can be obtained by computing the impairment-indicating application conditions of the inverse rule and shifting them to the LHS.

As a necessary prerequisite, we introduce *overlap-induced pre- and post-conditions.* Intuitively, given an overlap of a constraint with the LHS of a given rule, an overlap-induced pre-condition checks whether an occurrence of the premise of a constraint satisfies the conclusion of the constraint in the original graph of a transformation. The overlap-induced post-condition checks whether an occurrence of the premise satisfies the conclusion in the result graph of a transformation. All graphs and morphisms are visualised in Fig. 3. A detailed example of the construction process can be found in Appendix B in [7].

Definition 6 (Overlap-induced pre- and post-conditions). *Given a rule $\rho = L \xleftarrow{l} K \xrightarrow{r} R$, an overlap $\overline{PL} = (i_P, i_L, PL) \in O(\rho, P)$, and a condition d over P, the overlap-induced pre- and post-conditions of ρ w.r.t. $\overline{PL}$ and d, denoted by $\mathrm{Pre}_\rho(\overline{PL}, d)$ and $\mathrm{Post}_\rho(\overline{PL}, d)$, are defined as*

1. $\mathrm{Pre}_\rho(\overline{PL}, d) := \mathrm{Shift}(i_P, d)$, *and*
2. $\mathrm{Post}_\rho(\overline{PL}, d) := \begin{cases} \mathrm{Left}(\mathrm{der}(t), \mathrm{Shift}(\mathrm{tr}_t \circ i_P, d)) & \text{if } \overline{PL} \in O_{\mathrm{con}}(\rho, P) \\ \textsf{false} & \text{if } \overline{PL} \in O_{\mathrm{prem}}(\rho, P), \end{cases}$

with $t\colon PL \Longrightarrow_{\rho, i_L} PR$.

Note that $\mathrm{Post}_\rho(\overline{PL}, d) = \mathrm{Pre}_{\rho^{-1}}(\overline{PR}, d)$ if $\overline{PL} = \mathrm{Left}_{\mathrm{ol}}(\rho, \overline{PR})$. The correctness of overlap-induced pre- and post-conditions is formalized and proven by Lemma 5, which can be found in Appendix A in [7].

Definition 7 (Impairment-indicating application condition). *Given a constraint $c = \forall(e\colon \emptyset \hookrightarrow P, d)$ and a rule $\rho = L \xleftarrow{l} K \xrightarrow{r} R$. The set of impairment-indicating application conditions for ρ w.r.t. c consists of a set $\mathrm{Imp}_{\mathrm{prem}}(\rho, c)$ of impairment-indicating application conditions for the premise and a set $\mathrm{Imp}_{\mathrm{con}}(\rho, c)$ of impairment-indicating application conditions for the conclusion, where*

$$\mathrm{Imp}_{\mathrm{prem}}(\rho, c) := \{\mathrm{Left}_{\mathrm{ann}}(\rho, (\forall(i_R\colon R \hookrightarrow PR, \mathrm{Pre}_\rho(\overline{PR}, d)), \overline{PR}))$$
$$| \ \overline{PR} \in O_{\mathrm{prem}}(\rho^{-1}, P)\}$$

and

$$\text{Imp}_{\text{con}}(\rho, c) := \{(\forall(i_L \colon L \hookrightarrow PL, \text{Pre}_\rho(\overline{PL}, d) \implies \text{Post}_\rho(\overline{PL}, d)), \overline{PL})$$
$$| \ \overline{PL} \in \text{O}_{\text{con}}(\rho, P)\}.$$

Definition 8 (Repair-indicating application condition). *Given a constraint* $c = \forall(e \colon \emptyset \hookrightarrow P, d)$ *and a rule* $\rho = L \xleftarrow{l} K \xrightarrow{r} R$, *the set of repair-indicating application conditions of* ρ *for* c *consists of a set* $\text{Rep}_{\text{prem}}(\rho, c)$ *of repair-indicating application conditions for the premise and a set* $\text{Rep}_{\text{con}}(\rho, c)$ *of impairment-indicating application conditions for the conclusion, where*

$$\text{Rep}_{\text{prem}}(\rho, c) := \{(\forall(i_L \colon L \hookrightarrow PL, \text{Pre}_{\rho^{-1}}(\overline{PL}, d)), \overline{PL}) \ | \ \overline{PL} \in \text{O}_{\text{prem}}(\rho, P)\}$$

and

$$\text{Rep}_{\text{con}}(\rho, c) := \{\text{Left}_{\text{ann}}(\rho, (\forall(i_R \colon R \hookrightarrow PR, \text{Pre}_{\rho^{-1}}(\overline{PR}, d) \implies \text{Post}_{\rho^{-1}}(\overline{PR}, d)),$$
$$\overline{PR})) \ | \ \overline{PR} \in \text{O}_{\text{con}}(\rho^{-1}, P)\}.$$

Note that for the examples and the implementation we have simplified the constructed application conditions with respect to the hard constraints h_1 and h_2 using Lemma 1.

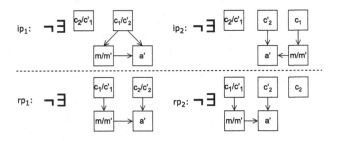

Fig. 4. Impairment-indicating (top) and repair-indicating (bottom) application conditions of moveMethod for w_2

Example 3. Consider the annotated application conditions shown in Figs. 4 and 5. The annotating overlaps are implicitly given by the node names. The impairment-indicating and repair-indicating application conditions of moveMethod for w_2 are shown in Fig. 4. Note that moveMethod can only repair and impair the premise, since the conclusion of w_2 is equal to false.

The impairment-indicating and repair-indicating application conditions of moveAttribute for w_1 are shown in Fig. 5. The premise of w_1 does not contain an edge from a class to an attribute, so moveAttribute can only introduce impairments and repairs of the conclusion. For the impairment-indicating application condition ic_1, the left side of the implication checks whether the occurrence of

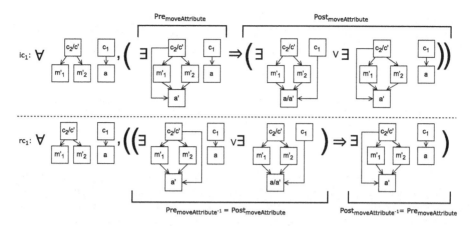

Fig. 5. Impairment-indicating (top) and repair-indicating (bottom) application conditions of moveAttribute for w_1

the premise of w_1 in the annotating overlap satisfies the conclusion of w_1 in the original graph of the transformation. The right side of the implication checks whether this occurrence will satisfy the conclusion of w_1 in the result graph of the transformation.

For the repair-indicating application condition rc_1, this implication is reversed, i.e. the left-hand side of the implication checks whether this occurrence will satisfy the conclusion of w_1 in the result graph of the transformation and the right-hand side checks whether the occurrence satisfies the conclusion of w_1 in the original graph of the transformation. The detailed construction of these application conditions can be found in Appendix B in [7].

When evaluating a repair-indicating application condition $(c', (i_L, i_P, PL))$ w.r.t. to a rule match m, we restrict the set of violations of c' in m to i_P. Then, the set of violations of c' in m restricted to i_P contains occurrences of the premise of the constraint that are repaired by the transformation. We proceed in a similar way for impairment-indicating application conditions.

Our main theorem states that the difference $\mathrm{nv}_H(c) - \mathrm{nv}_G(c)$ in the number of violations of a transformation $t\colon G \Longrightarrow_{\rho,m} H$ can be evaluated by computing the difference in the number of violations of associated impairment-indicating and repair-indicating application conditions. This allows us to evaluate the change in inconsistency before the transformation is performed by simply counting violations of application conditions.

Theorem 1 (Main theorem). *Let* $\overline{PL} = (i_P, i_L, PL)$. *Given a transformation* $t\colon G \Longrightarrow_{\rho,m} H$ *and a constraint* c, *then*

$$\mathrm{nv}_H(C) - \mathrm{nv}_G(C) =$$

$$| \biguplus_{(c',\overline{PL})\in\mathrm{Imp}_{\mathrm{prem}}(\rho,c)} \mathrm{sv}_{m,i_P}(c')| \quad + | \bigcup_{(c',\overline{PL})\in\mathrm{Imp}_{\mathrm{con}}(\rho,c)} \mathrm{sv}_{m,i_P}(c')|$$

$$-| \bigcup_{(c',\overline{PL})\in\mathrm{Rep}_{\mathrm{prem}}(\rho,c)} \mathrm{sv}_{m,i_P}(c')| \quad - | \bigcup_{(c',\overline{PL})\in\mathrm{Rep}_{\mathrm{con}}(\rho,c)} \mathrm{sv}_{m,i_P}(c')|.$$

Proof. Follows by applying Lemma 6 from [7].   □

Note that we use the disjoint union for the impairment-indicating application conditions for the premise, even though we use the union for the other application conditions. As discussed earlier, the overlap of a pair $(c',\overline{PL}) \in \mathrm{Imp}_{\mathrm{prem}}(\rho,c)$ is an overlap of the LHS and a proper subgraph of the premise of the constraint. If c' is violated, the transformation extends an occurrence of this subgraph to at least one occurrence of the premise that does not satisfy the conclusion of the constraint. Using the union, this occurrence would be counted once, even though it could be extended to multiple occurrences of the premise that do not satisfy the conclusion. To deal with this loss of information, we use disjoint union so that an occurrence of this subgraph can be counted multiple times.

Example 4. As discussed in Example 2, we have $\mathrm{nv}_G(\mathsf{w}_2) = 2$ for the graph G shown in Fig. 1. When using the rule moveMethod to move checkout() from the class Cart to the class Session, the application conditions ip_1 and ip_2 are satisfied, so $|\biguplus_{(c',\overline{PL})\in\mathrm{Imp}_{\mathrm{prem}}(\rho,c)}\mathrm{sv}_{m,i_P}(c')| = 0$. There is a violation of rp_1, while rp_2 is satisfied; so $|\bigcup_{(c',\overline{PL})\in\mathrm{Rep}_{\mathrm{prem}}(\rho,c)} \mathrm{sv}_{m,i_P}(c')| = 1$. So this transformation does not introduce an impairment of w_2 but repairs a violation, i.e., $\mathrm{nv}_H(\mathsf{w}_2) = 1$, where H is the result graph of this transformation.

Kosiol et al. [12] introduced the notions of *consistency-sustaining* and *consistency-improving* transformations, i.e., a transformation $t\colon G \Longrightarrow H$ is *consistency-sustaining w.r.t. a constraint* c if $\mathrm{nv}_H(c) - \mathrm{nv}_G(c) \leq 0$; it is called *consistency-improving w.r.t.* c if $\mathrm{nv}_H(c) - \mathrm{nv}_G(c) < 0$. Our main theorem implies that we can predict whether a transformation is *consistency-sustaining (-improving)* by evaluating application conditions that indicate impairment and repair.

6   Evaluation

To evaluate the practical relevance of our approach, we implemented a greedy graph optimization algorithm for the CRA case study (cf. Sect. 2). This was done using the graph transformation tool eMoflon[3], which incrementally computes all matches to a given graph for rules and their application conditions. This is a prerequisite for the efficient implementation of this approach, as calculating all matches from scratch after each rule application can easily become

---

[3] www.emoflon.org.

a serious performance bottleneck. Based on these matches provided by eMoflon, our implementation then counts violations of derived application conditions and ranks rule matches w.r.t. the number of constraint violations that are removed or added by the application of the considered rule at the considered match (see Theorem 1). Then, the rule application with the highest rank is greedily selected and applied, and the ranking is updated. Note that the application conditions are currently not derived automatically, but designed and implemented by hand in eMoflon based on our formal construction. Also, our evaluation was run on a Ryzen 7 3900x and 64GB RAM on Windows 11 23H2. It is available as a VM[4] with a detailed description of how to reproduce our results.

Finding all the matches for rules and application conditions can become very expensive if there are many matches to find. The CRA case study is particularly challenging because a feature can be moved from one class to any other class, which means that the number of refactoring steps, and thus the number of matches, grows rapidly with an increasing number of classes and features. Therefore, we pose the following research questions: *(RQ1) How does our approach scale with respect to the size of a processed class diagram (graph)?* and *(RQ2) Can we reduce the number of violations?*

To answer the first question, we need to examine the two phases of our approach, which are related to how eMoflon works and incrementally provides us with collected matches. First, we measure the time it takes to compute the initial collection of all rule and application condition matches. Then, we use the application conditions matches to rank the rule matches. Depending on the size of the model, this is expected to take longer than applying a rule and incrementally updating eMoflons internal structures as well as the rule rankings. Second, we measure the time taken to perform 10 repair steps, where we have to judge which repair to apply next, based on an actually selected repair step. For this, we use the most promising (highest ranked) rule application, where repairs and impairments are uniformly weighted with a value of 1.

To investigate the scalability of our approach, we created synthetic class diagrams of varying sizes with increasing numbers of classes, where each class has five methods and five attributes. Each method has two dependencies on attributes of the same class and a further three dependencies on attributes of other classes. Having more dependencies means that it is less likely to move features from one class to another, as more features would form a dependency clique within a class.

Figure 6 shows our evaluation results, where the left plot shows the time in seconds for processing models with up to 2,751 elements. Starting with 276 elements, the initialization time is 2.2 s and performing 10 refactoring steps takes 0.5 s. With 2,751 elements, this time increases to 65 s for the initialization and 2.8 s for the 10 steps. Obviously, the initialization takes 30 times longer for a 10 times larger model, while the incremental updates scale better and takes only 5 times longer. The reason for this non-linear increase lies in the structure of our rules, where the number of matches increases rapidly with each new class. For a

---

[4] www.zenodo.org/records/10727438.

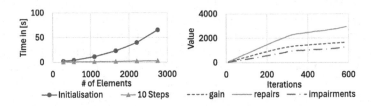

Fig. 6. Performance measurements for increasing model size (left), and consistency improvement of a model with 2201 elements (right)

model with 276 elements, we collected 600 rule matches and 36,000 application condition matches, while for the largest model with 2,751 elements, we found 622,500 rule matches and 3,7 million condition matches. So, while it takes 30 times as long to run a model 10 times the size, we found 118 times as many matches. This shows that even for a challenging scenario like the CRA use case, our approach scales reasonably well given the number of repair steps available (RQ1).

To answer RQ2, we took a model with 2,201 elements and measured the aggregated number of impairments and repairs after n iterations (repair steps) along with their differences (gains). As before, in each iteration we chose the rule application with the highest gain and continued until there was no rule application with a positive gain left, i.e. the application of a rule would have no effect or cause more impairments than repairs at this point. The results are shown in Fig. 6 on the right. After 589 iterations with a total gain of 1,661, the process terminated after finding a (local) optimum. So the resulting model contains 1,661 fewer violations than before, which answers RQ2.

As shown, our approach can incrementally maintain the necessary rule-ranking information (RQ1) and improve consistency in a rather low number of iterations (RQ2). This is particularly interesting considering the fact that the search space of all rule matches can grow very rapidly, which is a particularly challenging scenario for our approach.

Threads to Validity. Currently, we only investigate the CRA case using synthetic class diagrams, which grow only by adding more classes with a fixed number of features and dependencies. To evaluate the general scalability of our approach, more scenarios should be investigated, preferably using real-world data, e.g., extracted from public code repositories. Also, the application conditions are currently constructed by hand (following our formal construction process), which is a source of error. In addition, since we only have a look-ahead of 1, there may not always be a good next step to improve consistency, even though a better overall solution exists. Therefore, future work should investigate how our approach performs when the greedy strategy is replaced by another strategy such as simulated annealing.

7   Related Work

Habel and Pennmann [8] introduced the original process for generating application conditions from nested graph constraints, which are consistency-guaranteeing meaning that applying a rule with such conditions is guaranteed to produce a graph consistent with the given (hard) constraints. In our paper, we extend the binary case of satisfying hard constraints, by ranking rule applications based on how many constraint violations they add or remove w.r.t. a set of weak constraints. Kosiol et al. [12] also count violations but still consider only one type of constraint (hard constraints). While they consider constraints in alternating normal form with arbitrary nesting levels but without Boolean operators, we focus on constraints up to level 2 but allow Boolean operators on level 2. Our experience has shown that this kind of constraints is mostly used in practical applications. Since the resulting set of application conditions can be large and thus expensive to evaluate, Nassar et al. [13] showed that some subconditions can be filtered if they check for cases that cannot occur. We also filter the resulting application conditions, but based on our set of additional hard constraints, e.g., by filtering out conditions that check for features contained in multiple classes simultaneously. In [12], application conditions are constructed to make rule applications consistency-sustaining, but there is no such construction for consistency-improving rule applications. Moreover, the rule applications are consistency-sustaining in the strict sense that no new constraint violations are allowed. In contrast to all existing literature, we also construct application conditions for consistency improvement and use them to rank rule applications, not to block them. This approach can be used in a transformation engine like eMoflon, which supports the incremental computation of rule matches.

In [11], a similar ranking approach is presented for model repair, identifying impairments as negative side effects and repairs as positive side effects of model repair sequences. All constraints have equal priority. Given a model with a set of violations, all possible repairs are computed and ranked according to their side effects. A repair consists of a sequence of repair actions of limited length, each of which repairs only a single model element or a single property of an element. The ranking is not done in advance, but is determined by first executing all computed repair action sequences and then observing their effects on the number of constraints violated. In contrast, we define repair rules that can be arbitrarily complex and perform multiple repair actions in parallel (in one rewrite rule step). Furthermore, we determine the positive and negative effects of all options to apply a given repair rule to all (weak) constraints simultaneously without changing the (model) graph for this purpose.

In [1], constraints are also detected by incremental graph pattern matching. They distinguish between ill-formedness and well-formedness constraints, where occurrences of the former are desirable and the latter are to be avoided. Using genetic algorithms with a set of graph-modifying rules as mutators, they then search for a graph that maximizes the number of occurrences of well-formedness constraints minus the number of occurrences of ill-formedness constraints. Compared to our approach, they have to change the graph to track consistency and detect violations and repairs, whereas we use a look-ahead to plan next steps.

The running example is based on the well-known CRA problem [2], which was the focus of attention at the TTC 2016 [6]. Solutions like [10] have shown that greedy-based solutions like ours achieve quite good results, our approach additionally provides a look-ahead that can be used to find local optima faster.

8   Conclusion

In this paper, we introduce a new dynamic analysis approach that ranks rule matches based on their potential for graph repair. This potential is computed using application conditions that are automatically derived from a set of nested graph constraints. While some of these conditions indicate repair steps, others detect violations. We formally showed that the potential of a rule application can indeed be characterized by the difference between the number of violations of repair-indicating and violation-indicating application conditions. We illustrated and evaluated our approach in the context of the well-known CRA problem, and showed that even for a worst-case scenario, the performance scales reasonably well. For the future, we want to fully automate the ranking of graph transformations based on an automated construction of violation-indicating and repair-indicating application conditions. In addition, we will investigate different strategies besides greedy-based algorithms for specifying and optimizing graphs in different scenarios. To further stengthen our ranking approach with a lookahead of 1, it may be advantageous to combine several repair rules into a larger one by composing concurrent rules [4,5].

References

1. Abdeen, H., et al.: Multi-objective optimization in rule-based design space exploration. In: Crnkovic, I., Chechik, M., Grünbacher, P. (eds.) ACM/IEEE International Conference on Automated Software Engineering, ASE 2014, Vasteras, Sweden - 15–19 September 2014, pp. 289–300. ACM (2014). https://doi.org/10.1145/2642937.2643005
2. Bowman, M., Briand, L.C., Labiche, Y.: Solving the class responsibility assignment problem in object-oriented analysis with multi-objective genetic algorithms. IEEE Trans. Software Eng. **36**(6), 817–837 (2010). https://doi.org/10.1109/TSE.2010.70
3. Ehrig, H.: Introduction to the algebraic theory of graph grammars (a survey). In: Claus, V., Ehrig, H., Rozenberg, G. (eds.) Graph Grammars 1978. LNCS, vol. 73, pp. 1–69. Springer, Heidelberg (1979). https://doi.org/10.1007/BFb0025714
4. Ehrig, H., Ehrig, K., Prange, U., Taentzer, G.: Fundamentals of Algebraic Graph Transformation. MTCSAES, Springer, Heidelberg (2006). https://doi.org/10.1007/3-540-31188-2
5. Ehrig, H., Habel, A.: Concurrent transformations of graphs and relational structures. In: Nagl, M., Perl, J. (eds.) Proceedings of the WG 1983, International Workshop on Graphtheoretic Concepts in Computer Science, pp. 76–88. Universitätsverlag Rudolf Trauner, Linz (1983)

6. Fleck, M., Troya Castilla, J., Wimmer, M.: The class responsibility assignment case. In: TTC 2016: 9th Transformation Tool Contest, co-located with the 2016 Software Technologies: Applications and Foundations (STAF 2016), vol. 1758, pp. 1–8 (2016). https://ceur-ws.org/Vol-1758/paper1.pdf

7. Fritsche, L., Lauer, A., Schürr, A., Taentzer, G.: Using application conditions to rank graph transformations for graph repair (2024). https://arxiv.org/abs/2405.08788

8. Habel, A., Pennemann, K.: Correctness of high-level transformation systems relative to nested conditions. Math. Struct. Comput. Sci. **19**(2), 245–296 (2009). https://doi.org/10.1017/S0960129508007202

9. Heckel, R., Taentzer, G.: Graph Transformation for Software Engineers - With Applications to Model-Based Development and Domain-Specific Language Engineering. Springer, Cham (2020). https://doi.org/10.1007/978-3-030-43916-3

10. Hinkel, G.: An NMF solution to the class responsibility assignment case. In: García-Domínguez, A., Krikava, F., Rose, L.M. (eds.) Proceedings of the 9th Transformation Tool Contest, co-located with the 2016 Software Technologies: Applications and Foundations (STAF 2016), Vienna, Austria, 8 July 2016. CEUR Workshop Proceedings, vol. 1758, pp. 15–20. CEUR-WS.org (2016). https://ceur-ws.org/Vol-1758/paper3.pdf

11. Khelladi, D.E., Kretschmer, R., Egyed, A.: Detecting and exploring side effects when repairing model inconsistencies. In: Proceedings of the 12th ACM SIGPLAN International Conference on Software Language Engineering, pp. 113–126. ACM (2019). https://doi.org/10.1145/3357766.3359546

12. Kosiol, J., Strüber, D., Taentzer, G., Zschaler, S.: Sustaining and improving graduated graph consistency: a static analysis of graph transformations. Sci. Comput. Program. **214**, 102729 (2022). https://doi.org/10.1016/J.SCICO.2021.102729

13. Nassar, N., Kosiol, J., Arendt, T., Taentzer, G.: Constructing optimized constraint-preserving application conditions for model transformation rules. J. Log. Algebr. Methods Program. **114**, 100564 (2020). https://doi.org/10.1016/J.JLAMP.2020.100564

14. Plump, D.: Confluence of graph transformation revisited. In: Middeldorp, A., van Oostrom, V., van Raamsdonk, F., de Vrijer, R. (eds.) Processes, Terms and Cycles: Steps on the Road to Infinity. LNCS, vol. 3838, pp. 280–308. Springer, Heidelberg (2005). https://doi.org/10.1007/11601548_16

15. Radke, H., Arendt, T., Becker, J.S., Habel, A., Taentzer, G.: Translating essential OCL invariants to nested graph constraints for generating instances of metamodels. Sci. Comput. Program. **152**, 38–62 (2018). https://doi.org/10.1016/J.SCICO.2017.08.006

16. Rensink, A.: Representing first-order logic using graphs. In: Ehrig, H., Engels, G., Parisi-Presicce, F., Rozenberg, G. (eds.) ICGT 2004. LNCS, vol. 3256, pp. 319–335. Springer, Heidelberg (2004). https://doi.org/10.1007/978-3-540-30203-2_23

Deriving Delay-Robust Timed Graph Transformation System Models

Mustafa Ghani⑩, Sven Schneider$^{(\boxtimes)}$⑩, Maria Maximova⑩,
and Holger Giese⑩

Hasso Plattner Institute, University of Potsdam, Potsdam, Germany
{mustafa.ghani,sven.schneider,maria.maximova,holger.giese}@hpi.de

Abstract. Distributed Cyber-Physical Systems (DCPSs) are omnipresent and their analysis against provided specifications is a central challenge. Hereby, distribution results in communication delays among agents that have to be adequately taken into account by software models to avoid race conditions. However, engineering DCPSs at a higher level of detail by incorporating communication delays explicitly inflates model size and impedes analysis.

In this paper, we employ Timed Graph Transformation Systems (TGTSs) to model DCPSs and distinguish between local immediate and remote δ-delayed observations, requiring up to δ time units. We then (a) demonstrate potential absence of δ-delay robustness for TGTS models, (b) provide a procedure widening safe behavioral options of a verified 0-delay system model to derive a δ-delay robust TGTS model, and (c) analyze the restrictiveness of the widening and the resulting TGTS model for new unsafe behavior. As a running example, we consider a DCPS in which multiple distributed autonomous shuttles locally coordinate their movement on a track topology to avoid collisions.

Keywords: model driven engineering · formal modeling · quantitative analysis · model checking

1 Introduction

We consider Distributed Cyber-Physical Systems (DCPSs) in which multiple agents perform discrete steps locally and communicate local observations to remote agents via delayed message passing while time may elapse between agents' steps. An abundance of MDE approaches have been applied for designing, understanding, and improving such DCPSs at various levels of abstraction. They ideally support fully-automatic analysis techniques (such as simulation or model checking) allowing to ensure specification satisfaction for safety-critical DCPSs.

As a running example of such a DCPS, we consider the RAILCAB [30, 32, 37] system of autonomous shuttles traversing a track topology while locally coordinating their driving behavior using wireless inter-shuttle communication. Timing constraints are used to capture the time required for shuttles to traverse between track segments. Shuttles approaching a join track segment from different predecessor tracks locally coordinate to determine an ordering to avoid collisions;

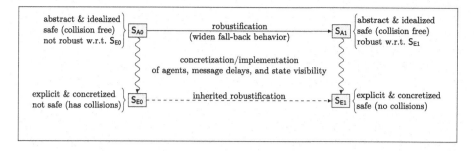

Fig. 1. Overview of approach

in particular, primary shuttle driving behavior is overruled by fall-back shuttle braking behavior to give way to a detected remote shuttle in critical distance.

For the modeling of DCPSs, we employ a Timed Graph Transformation System (TGTS) variation as follows. First, TGTSs subsume Graph Transformation Systems (GTSs) [18] employing a rule-based modeling of local graph modifications assuming that system states can be captured by graphs. Secondly, TGTSs support the modeling of timing constraints as in Timed Automata (TA) [1] and Probabilistic Timed Graph Transformation Systems (PTGTSs) [30] based on clocks (using clock resets, clock guards, and clock invariants) and the values of such clocks are also recorded in system states. Thirdly, TGTSs rely on our novel concept of actor-based rule priorities obtained as a refinement of the standard concept of globally evaluated GT rule priorities to improve descriptive expressiveness of rules for the multi-agent context. Model-checking support of TGTSs w.r.t. timed safety is derived as for PTGTSs.

System models of DCPSs employed in MDE are usually idealized by abstracting from potentially significant real-world aspects as their integration in the model would inflate model size also leading to intractably large state spaces impeding analysis as either the entire state space must be constructed for model checking or rare violations are unlikely to be found using simulation. For DCPSs such real-world aspects may include inter-agent message delays (consisting of delays for transmission, propagation, queuing, and processing [20, 46]), which may result in the storage and reception of outdated information on other agents' states leading to real-world system errors not captured by idealized system models. Information on other agents being outdated may impact or even invalidate quantitative/qualitative assurances in the form of (timed) safety properties. For our running example, outdated information may falsely indicate that all remote shuttles are in a safe distance critically postponing collision avoidance maneuvers. First, we consider the problem of determining for a given abstract system model whether the real-world aspect of inter-agent message delays is insignificant (calling the system model robust in this case) in the sense that qualitative assurances are valid for the relevant concretizations of the system model as well capturing such message delays explicitly. Second, for non-robust system models, instead of manipulating the explicit concretized system model directly, we consider the problem of deriving system model repairs directly at the comparatively more comprehensive abstract system model, which also allows to focus on core behavioral and interaction patterns or zoom in on other previously omitted real-world aspects once the system model has been repaired.

In our approach, as visualized in Fig. 1, we consider a given idealized system model S_{A0} as input for which safety has been established. We then derive a concretization S_{E0} of S_{A0} defining actor identities and making distribution aspects such as inter-actor message exchange, message delays, and state visibility explicit. For our running example, we focus on the peer-to-peer interpretation discussed above but also briefly consider a centralized interpretation as well. We then conclude the need to adapt S_{A0} conditioned on the explicit model S_{E0} exhibiting, as for our running example, safety violations. Given a classification of the rules of S_{A0} into primary and fall-back agent behavior rules, we propose an automatic procedure for deriving additional fall-back rules widening their application to critical contexts reachable based on the concrete message delay bound δ implemented in S_{E1}. These additional rules are added to S_{A0} resulting in S_{A1}. To derive the additional fall-back rules, we apply a timed-shift operation on the preconditions of existing fall-back rules to identify the δ-past of remote agents (relying on a timed version of the proof technique of discrete k-induction) and extend the fall-back rules to apply to states of that δ-past to compensate for the δ-delayed messages. With this approach, we intend to benefit from modeled information on how to avoid safety violations and to avoid unnecessary restrictions of the agents' primary behavior.

Given the complexity and diversity of challenges related to distributed systems and the idea to not only restrict the system behavior but to identify additional steps for which subsequent behavioral rules cannot be automatically derived, our approach can only support MDE but not guarantee a robust resulting system model. In particular, the derived fall-back rules may not prevent the reachability of all safety violations in S_{E1} reachable in S_{E0} and the application of derived fall-back rules may lead to safety violating states in S_{E1} not reachable in S_{E0}. Consequently, analysis of S_{A1}, possibly indicating further missing kinds of fall-back rules, is required to obtain a safe system model.

Nevertheless, our rule generation approach supports the MDE approach to system design by partially automating such manual efforts. In particular, not applying our automatic rule generation approach requires the full manual revision/repair of the given TGTS S_{A0}, which is complicated by several key factors: first, analysis of S_{E0} may not reveal all safety violations when resorting to simulation due to prohibitive model checking costs; second, the applied analysis methods may not result in instructive descriptions of how safety violations could be reached requiring significant efforts to resolve issues automatically resolved by our approach; third, as we intend to retain the idealized view (allowing to focus on varying important aspects separately), we require the revision/repair of S_{A0} based on analysis results for S_{E0} noting that this cross-level issue comprehension-repair is a complex problem in itself. Hence, deriving additional fall-back rules supports manual system revision/repair.

This paper is structured as follows. In Sect. 2, we discuss preliminaries of TGTSs in our notation. In Sect. 3, we introduce TGTS with agent-based priori-

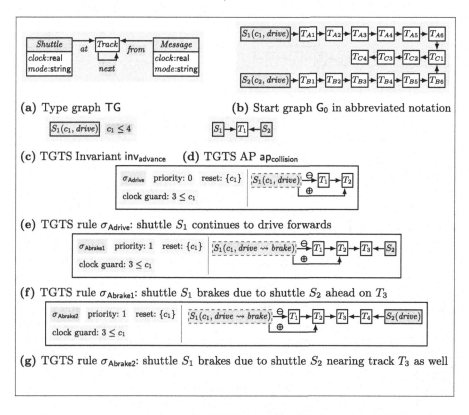

Fig. 2. Type graph, start graph, TGTS AP, and TGTS rules for TGTS S_{A0}

ties and the modeling of our running example. In Sect. 4, we discuss concretizations formalizing interpretations of distribution and exemplify potential non-robustness against such concretizations. In Sect. 5, we present our approach for robustness violation resolution based on necessary widening of fall-back behavior. In Sect. 6, we consider related work. Finally, in Sect. 7, we close the paper with a conclusion and an outlook on future work.

2 Preliminaries and Notation

We employ typed attributed graphs (graphs) [18] and injective graph morphism $m : G_1 \hookrightarrow G_2$ between them. For our running example, we rely on the type graph TG from Fig. 2a and the start graph from Fig. 2b. Here and subsequently, we use an abbreviated notation for graphs in which *(a)* node types are indicated by the node names (S, T, and M indicating shuttles, tracks, and messages), *(b)* edge types are omitted as edge types can be derived uniquely for TG, *(c)* node attributes and their values are, if present, given behind the node name (e.g., $S_1(c_1, drive)$ in G_0 is a shuttle node with a *clock* attribute c_1 of unspecified value and *mode* attribute of value *drive*).

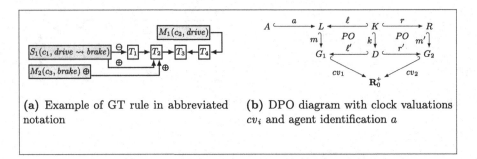

(a) Example of GT rule in abbreviated notation

(b) DPO diagram with clock valuations cv_i and agent identification a

Fig. 3. Example of GT rule and basic DPO diagram

Following the DPO approach in [18], GT steps are derived by applying GT rules $\rho = (\ell : K \hookrightarrow L, r : K \hookrightarrow R)$ (where we assume ℓ and r to be inclusions) for a match $m : L \hookrightarrow G_1$. The morphisms ℓ and r capture elements to be preserved in K, deleted in $L - \ell(K)$, and created in $R - r(K)$. When depicting GT rules as in Fig. 3a, we *(a)* use a standard integrated notation for ℓ and r providing a single graph where $\ominus$ and $\oplus$ indicate the elements to be deleted and created and *(b)* depict the change of an attribute as $v_1 \leadsto v_2$ for values v_1 and v_2. In the DPO approach, GT rules are applied by constructing a DPO diagram (see Fig. 3b) for which we capture the seven morphisms in the set of all GT steps $(G_1, (\rho, m : L \hookrightarrow G_1, k : K \hookrightarrow D, \bar{m} : R \hookrightarrow G_2, \ell' : D \hookrightarrow G_1, r' : D \hookrightarrow G_2), G_2) \in$ GTsteps (where we assume ℓ' and r' to be inclusions).

TA [1] and TGTSs employ clocks and Clock Constraints (CCs) thereon. For a set of clock variables C, CCs $\psi \in \mathsf{CC}(C)$ are finite conjunctions of clock comparisons of the form $c_1 \sim r$ and $c_1 - c_2 \sim r$ where $c_1, c_2 \in C$, $\sim \in \{<, >, \leq, \geq\}$, and $r \in \mathbf{N} \cup \{\infty\}$. Clock valuations $cv : C \to \mathbf{R}_0^+$ satisfy CCs ψ, written $cv \models \psi$, as expected. For a clock valuation cv and a set of clocks C', $cv[C' := 0]$ is the clock valuation mapping the clocks from C' to 0 and all other clocks according to cv. For a clock valuation cv and a duration $\delta \in \mathbf{R}_0^+$, $cv + \delta$ is the clock valuation mapping each clock x to $cv(x) + \delta$.

3   Timed GTS and Running Example

We now discuss the syntax and semantics of TGTSs *with agent-based priorities* and present the TGTS model S_{A0} of our running example.

Each TGTS employs a type graph and contains a single start graph (for our running example, see again Fig. 2). In our running example, the shuttles S_1 and S_2 are initially in mode *drive* and will advance on the track topology eventually traversing the join track T_{C1} unless braking beforehand.

TGTSs identify for each graph G a set of clock attributes $\mathsf{CA}(G)$. For our running example, these are the *clock* attributes of shuttle and message nodes.

TGTS states (G, cv) contain a graph G and, for the clocks $\mathsf{CA}(G)$ of G, a clock valuation $cv : \mathsf{CA}(G) \to \mathbf{R}_0^+$. The start state of a TGTS consists of the start graph and the clock valuation mapping all clocks to 0.

TGTS invariants (I, ψ) contain a graph I and a CC $\psi \in \mathsf{CC}(\mathsf{CA}(I))$. They are used to state clock constraints based on graph patterns that must be satisfied by all traversable TGTS states. A TGTS state (G, cv) satisfies a TGTS invariant (I, ψ), if $cv \circ m \models \psi$ for every match $m : I \hookrightarrow G$. For our running example, the shuttles' clocks are reset to 0 upon each advancement and the TGTS invariant $\mathsf{inv}_{\mathsf{advance}}$ from Fig. 2c then prevents timed steps when a shuttle in mode *drive* has a clock value of 4 time units capturing that such a shuttle must advance.

TGTS Atomic Propositions (APs) are given by graphs; each TGTS state (G, cv) is labeled with the TGTS APs A for which a match $m : A \hookrightarrow G$ exists. In our running example, we employ the TGTS AP $\mathsf{ap}_{\mathsf{collision}}$ from Fig. 2d capturing states where two shuttles are located on a common track.

(TGTS) rules σ describe how discrete steps between TGTS states are to be derived. Rules $\sigma = (\rho, a, \psi, CR, p)$ consist of a GT rule $\rho = (\ell : K \hookrightarrow L, r : K \hookrightarrow R)$, an actor embedding $a : A \hookrightarrow L$, a clock guard $\psi \in \mathsf{CC}(\mathsf{CA}(L))$ over the clocks of L, a reset set $CR \subseteq \mathsf{CA}(R)$ of clocks to be reset, and a priority $p \in \mathbf{N}$.

For our running example, see Fig. 2 for the visualizations of the rules employed where we depict the actor subgraph A (consisting of a shuttle node with its attributes each time) using a red, dashed border: each of the three rules represents the advancement of a shuttle to the next track after at least 3 time units; the rule σ_{Adrive} with priority 0 represents the primary driving behavior where the shuttle remains in mode *drive* and the rule $\sigma_{\mathsf{Abrake1}}$ and $\sigma_{\mathsf{Abrake2}}$ with priority 1 represent fall-back braking behavior where the shuttle changes its mode to *brake* to avoid potential subsequent collision with a nearby shuttle S_2.

We deviate from PTGTSs, by refining the standard (global) GT rule priority concept to an agent-based (local) priority concept. The standard GT rule priority concept is insufficient for TGTSs modeling multi-agent systems because a high priority step s_1 for an actor A_1 prevents a lower priority step s_2 of another actor A_2, which can lead to the modeling error of timelocks when A_2 must perform s_2 before A_1 can perform s_1 (for our running example, a shuttle that needs to brake using $\sigma_{\mathsf{Abrake1}}$ or $\sigma_{\mathsf{Abrake2}}$ prevents a step using σ_{Adrive} to be executed by all other shuttles). To evaluate priorities for agents separately, we rely on the actor embedding $a : A \hookrightarrow L$ in the rules. While local agent-based priorities can be encoded using application conditions they increase the descriptive expressiveness of TGTSs and simplify our presentation later on.

A TGTS $S = (G_0, \mathcal{P}, \mathcal{I}, \mathcal{AP})$ consists of a start graph G_0, a set of rules $\mathcal{P}$, a set of TGTS invariants $\mathcal{I}$, and a set of TGTS APs $\mathcal{AP}$. For our running example, see Fig. 2 for the elements of the TGTS $\mathsf{S}_{\mathsf{A0}} = (\mathsf{G}_0, \{\sigma_{\mathsf{Adrive}}, \sigma_{\mathsf{Abrake1}}, \sigma_{\mathsf{Abrake2}}\}, \{\mathsf{inv}_{\mathsf{advance}}\}, \{\mathsf{ap}_{\mathsf{collision}}\})$. For such a TGTS S, we use $\mathsf{states}(S)$ to capture its TGTS states satisfying all its TGTS invariants $\mathcal{I}$.

The single step relation of a TGTS S on TGTS states (G, cv) defines *(a)* timed steps in which time elapses by adding the same delay to all clocks requiring that the implicitly traversed steps are TGTS states as well and *(b)* discrete steps where a rule is applied by applying its underlying GT rule and resetting the clocks in the reset set to 0 requiring that the clock guard of the rule is satisfied and that no rule of S with an overlapping actor-embedding and higher priority is applicable. Consequently, the passage of time competes with possibly multiple rule applications of possibly multiple actors resulting in non-determinism.

Definition 1 (TGTS Steps). *A TGTS $S = (G_0, \mathcal{P}, \mathcal{I}, \mathcal{AP})$ defines a set of timed and discrete labeled steps* steps(S) *among TGTS states of S.*

- *Timed Step:* $((G_1, cv_1), \delta, (G_1, cv_1 + \delta)) \in$ steps(S) *if* $(G_1, cv_1) \in$ states(S), $\delta \in \mathbf{R}^+$, *and* $(G_1, cv_1 + \delta') \in$ states(S) *for each delay* $\delta' \in [0, \delta]$.
- *Discrete Step:* $((G_1, cv_1), (\sigma, m), (G_2, cv_2)) \in$ steps(S) *if* $(G_1, cv_1), (G_2, cv_2) \in$ steps(S), $\sigma = (\rho, a : A \hookrightarrow L, \psi, CR, p) \in \mathcal{P}$, $\rho = (\ell : K \hookrightarrow L, r : K \hookrightarrow R)$, $m : L \hookrightarrow G_1$ *(match),* $(G_1, (\rho, m, k, \bar{m}, \ell', r'), G_2) \in$ GTsteps *(GT step),* $cv_1 \models \psi$ *(clock guard satisfaction),* $cv_2 = cv_1[\bar{m}(CR) := 0]$ *(preservation/reset of clocks), and* $\nexists \sigma' = (\rho', a' : A' \hookrightarrow L', \psi', CR', p') \in \mathcal{P}$. $((G_1, cv_1), (\sigma', m'), \_) \in$ steps$(S) \wedge (m \circ a)(A)$ *and* $(m' \circ a')(A')$ *overlap* $\wedge\, p' > p$ *(priorities).*

Model-checking of TA [1], PTGTSs [30–33], and, here, TGTSs w.r.t. timed reachability properties such as "is an *ap*-labeled state reachable within time t?" relies directly (or indirectly using the PRISM model checker [26,27]) on the construction of a symbolic state space. The states of the (backwards constructed) symbolic state space are of the form (G, ψ) representing TGTS states (G, cv) where $cv \models \psi$. The syntactic restrictions of CCs are sufficient to ensure the finiteness of the symbolic state space when a finite number of discrete state components (locations for TA and graphs for (P)TGTSs) can be reached.

4   Interpretations and Consequences of Distribution

We now discuss how TGTSs can capture distribution aspects of DCPSs, how a TGTS S_{A0} abstracting from such aspects can be concretized to a TGTS S_{E0} making such aspects explicit exemplifying this step for our running example, and how such a concretization can reveal non-robustness of S_{A0} against distribution.

Distribution aspects are not explicitly enforced by TGTSs requiring careful modeling. For our running example, we focus on the distribution aspects of *(a)* actor identities, *(b)* an inter-agent message passing overlay network with an upper message delay bound δ (ruling out synchronous $\delta = 0$ and asynchronous $\delta = \infty$ communication) allowing for message reordering faults, *(c)* actor knowledge of static global information (which does not need to be communicated), *(d)* limited visibility of the TGTS state (G, cv) by agents, and *(e)* (non)atomic rule applications by one/multiple agents. Further important aspects of distribution not explicitly considered in our running example are, e.g., agent faults preventing further control decisions and their detection, message faults such as loss, duplication, or byzantine, and clock drifts among agents.

In contrast, TGTS models employing (e.g., mutual exclusion) coordination mechanisms between agents using monitors/semaphores based on atomic test-and-set operations implicitly assume that such agents are not spatially distributed and are executed on a single processor (with shared memory) ensuring an interleaving semantics. Nevertheless, the TGTS interleaving semantics is adequate for DCPS modeling since, as for system models with true-concurrency semantics, whether spatially distributed agents perform steps at the same global time is not preventable/exploitable in the context of messages with non-zero delays. Hence, for our running example, the mutual exclusion problem of ensuring that two shuttles are never on a common track requires distributed mutual exclusion mechanisms when making (additional distribution aspects such as) message delays explicit.

To exemplify the possibility of diverse (subjective) interpretations of distribution aspects for a given TGTS, we now consider a Peer-to-Peer (P2P) and a Region-Coordinator (RC) interpretation of distribution for the TGTS S_{A0} of our running example. Above all, these interpretations differ in the identification of actors, their memory, and by which agents the rules of the TGTS are applied and agree in that the track topology is static and known by all agents.

Peer-to-Peer (P2P) Interpretation: Shuttles (the agents) send messages to remote shuttles upon traversing (driving/braking) to a new track segment. Such messages are stored upon reception by attaching the message data to the track segment from which that message was send in the local view on the track topology. Upon traversal to a new track segment, shuttles query this local view to apply a (high-priority) fall-back rule (σ_{Abrake1} or σ_{Abrake2}) or to apply a primary rule (σ_{Adrive}). The TGTS state of S_{A0} thereby captures the different views of all shuttles but abstracts from message transmission.

Region-Coordinator (RC) Interpretation: In addition to the shuttles being agents again, an RC is an agent as well. This RC is not captured explicitly in the TGTS state as this TGTS state represents the memory stored by the RC itself. Based on this state the RC derives driving/braking instructions that are sent to the shuttles. The shuttles are shallow agents only executing these instructions when traversing to the next track and informing the RC about their new location upon traversal. Compared to the P2P interpretation, rules not only include implicit message sending but are also interpreted to be executed non-atomically and by different agents (deriving and executing instructions by RC and shuttles). Information on shuttle traversals are relayed by the RC resulting in two message delays between shuttles.

For the P2P interpretation, we now discuss an operation conc for concretizing in particular the abstract idealized S_{A0} into an explicit TGTS $S_{E0} = \text{conc}(S_{A0})$. The resulting TGTS $S_{E0} = (G_0, \{\sigma_{\text{Edrive}}, \sigma_{\text{Ebrake1a}}, \sigma_{\text{Ebrake1b}}, \sigma_{\text{Ebrake2a}}, \sigma_{\text{Ebrake2b}}\}, \{\text{inv}_{\text{advance}}\}, \{\text{ap}_{\text{collision}}\})$ is equipped with different rules given in Fig. 4 to implement the spatial and temporal decoupling of agents making, above all, message exchange explicit. As stated above, in each of these rules, the traversing shuttle creates a message at the reached track segment to allow other shuttles to be informed as early as possibly about this traversal also communicating its current mode. In the fall-back rules σ_{Ebrake1a}, σ_{Ebrake1b}, σ_{Ebrake2a}, and σ_{Ebrake2b}, we use the clock of such messages to determine whether *(a)* the message may have been received yet in σ_{Ebrake1a} and σ_{Ebrake2a}, making, due to the priority 0, braking an alternative to driving or *(b)* the message must have been received in σ_{Ebrake1b} and σ_{Ebrake2b} without the sending shuttle having performed another driving step already, making, due to the priority 1, braking the only option.

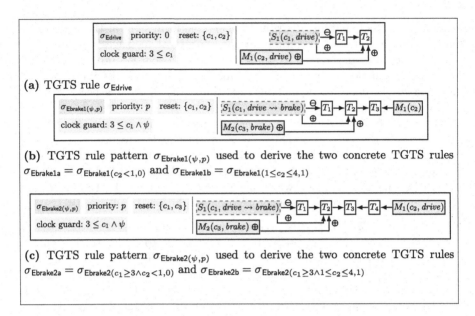

(a) TGTS rule σ_{Edrive}

(b) TGTS rule pattern $\sigma_{\text{Ebrake1}(\psi,p)}$ used to derive the two concrete TGTS rules $\sigma_{\text{Ebrake1a}} = \sigma_{\text{Ebrake1}(c_2<1,0)}$ and $\sigma_{\text{Ebrake1b}} = \sigma_{\text{Ebrake1}(1\leq c_2 \leq 4,1)}$

(c) TGTS rule pattern $\sigma_{\text{Ebrake2}(\psi,p)}$ used to derive the two concrete TGTS rules $\sigma_{\text{Ebrake2a}} = \sigma_{\text{Ebrake2}(c_1 \geq 3 \wedge c_2 <1,0)}$ and $\sigma_{\text{Ebrake2b}} = \sigma_{\text{Ebrake2}(c_1 \geq 3 \wedge 1 \leq c_2 \leq 4,1)}$

Fig. 4. The TGTS rules of S_{E0}

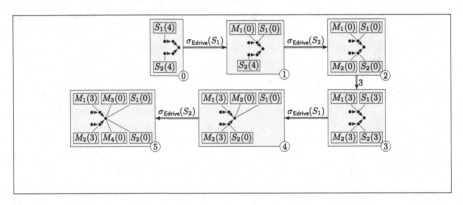

Fig. 5. A critical path of S_{E0} leading to a collision. Both shuttles have a *mode* attribute of *drive* throughout the path, which is omitted for brevity.

While collisions are not possible in S_{A0} due to successful fall-back braking behavior, we present a critical path in Fig. 5 (where track segments are given by black dots for brevity) for S_{E0} demonstrating that distribution with non-zero message delays threatens system robustness. The collision can be reached (in contrast to S_{A0}) because the message M_1 is not received in time allowing shuttle S_2 to drive from state 1 to state 2 (the fall-back rule σ_{Ebrake2a} is applicable but does not prevent σ_{Edrive} as both have priority 0). This path can be prevented

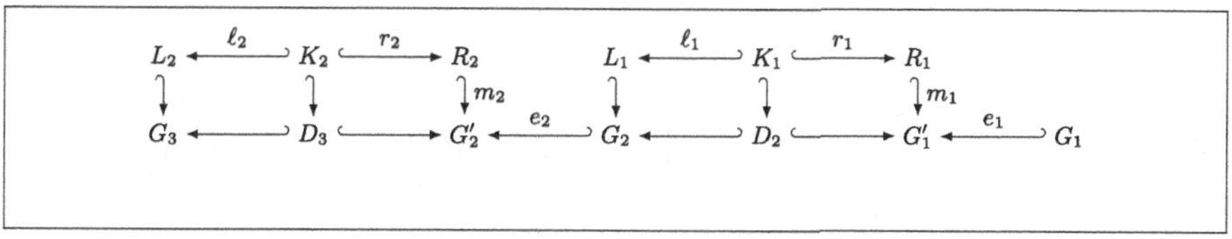

Fig. 6. Reverse DPO steps from graph G_1 where e_i morphisms are used to extend the current graph minimally to enable reverse GT rule application (i.e., (m_i, e_i) are jointly epimorphic)

using an additional fall-back braking rule (preventing the application of σ_{Edrive} in state 1) that checks for a remote shuttle one track farther from the join track.

The problem of collision reachability could also be resolved by adapting S_{E1} directly, but we intend to retain the view on the idealized, simpler modeling level of S_{A0}. Clearly, the required additional braking behavior must be derived from S_{A0}, the concretization operation leading to $S_{E0} = \text{conc}(S_{A0})$, and the set of already provided fall-back rules, which are assumed to be applied more often.

5   Fall-Back Widening Approach for Increased Robustness

In this section, we introduce the technique of δ-induction to derive for critical states their δ-past of states preceding such states by up to δ time units. We then present our approach using a δ-shift operation based on δ-induction generating (see Fig. 1 for the overview in the introduction) for the given TGTS S_{A0} additional fall-back rules (leading to the TGTS S_{A1}) to prevent safety violations occurring in the concretization $S_{E0} = \text{conc}(S_{A0})$. Lastly, we discuss our post-generation analysis evaluating the necessity of generated rules and the robustness of the resulting TGTS S_{A1} w.r.t. its concretization $S_{E1} = \text{conc}(S_{A1})$.

Recall that, for our running example (see Fig. 2), S_{A0} is equipped with the primary rule σ_{Adrive} and the fall-back rules σ_{Abrake1} and σ_{Abrake2}, safety violations are given by shuttle collisions (see $\text{ap}_{\text{collision}}$), which are avoided in S_{A0} by braking shuttles early enough, and we need to generate a further rule for braking shuttles earlier as indicated by the critical S_{E0}-path from Fig. 5 with a collision.

Focusing here on the distribution aspect of message delays (implying also agent-local state visibility), robustness critical state patterns in S_{A0} are those where an agent A_1 implicitly uses information from another agent A_2, which is δ-delayed in S_{E0}. To compensate for the δ-delay ensuring that agent A_1 has the information in time (i.e., when he would have had that information in S_{A0}), we explore the δ-past of that state (where A_1 uses the information) rolling back remote agents to create then rules allowing A_2 to send the required information δ-time earlier. The derivation of suitable additional fall-back rules from this δ-past allows then to eclipse (using higher priority fall-back rules) the primary agent behavior to avoid the critical state patterns in the resulting TGTS S_{A1}.

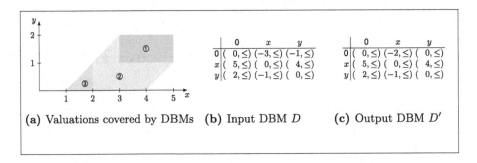

(a) Valuations covered by DBMs   **(b)** Input DBM D **(c)** Output DBM D'

Fig. 7. An application of the adapted timed predecessor operation.

5.1   δ-Induction for Deriving δ-Past of System States

We now introduce the verification technique of δ-induction to be used later on.

The notion of δ-induction is a clock-based continuous time adaptation of the static invariant verification technique of k-induction [11,25,42]. The idea of k-induction is to apply steps backwards from unsafe states passing only through safe states capturing why and how unsafe states can be reached; then, if no such paths of some length k can be constructed, safety violations are unreachable unless a start state of the system has been reached at some point. For the GT context, see Fig. 6 for an example path of length 2 from a given unsafe state G_1 where graphs must (in general) be extended (minimally) first (here using extension morphisms e_i to allow for comatches (here m_i) for the reverse GT rule application). Since the extension morphisms only accumulate context backwards, potential violations of priorities (or application conditions when considered) require that this context is propagated forwards through the path as well to enable the check for such violations. The existing GT-based approaches for k-induction [2–4,12–17,38–40] differ in their support for arbitrary values of k or just $k = 1$, the expressiveness of safety properties (e.g., relying on nested graph conditions), (nested) application condition of rules[1], timing constraints (e.g., time abstract, discrete time, or continuous time with fixed durations for steps), and single vs. multi-actor systems.

Assumed invariants are used to prevent backward steps in k-induction that are unreachable from the start state. For our running example, we use assumed invariants stating that each node has at most one attribute of each attribute type, that there are no parallel edges, that the track topology of derived graphs is contained in the start graph, and that there are at most two shuttles.

For δ-induction, we vary from k-induction by deriving backward sequences that have a duration of at least δ instead of stopping after k discrete steps. We now discuss our adaptation of the symbolic state space generation procedure for

---

[1] For the case with application conditions, the L and shift operations are used to translate application conditions across backward GT steps (see [12,23]).

TGTSs (mentioned at the end of Sect. 3) to determine when a backward path has a duration of at least δ allowing to stop further backward extension.[2]

We use the operations dPre and tPre for deriving discrete and timed predecessors of symbolic states (G, ψ, I) where I is an interval such as $[a, \infty)$ or (a, ∞) stating a lower bound on the duration of the path constructed so far, which is initially $[0, \infty)$. For $\mathsf{dPre}((G, \psi), \sigma, e, m)$, the GT rule of σ is applied backwards on G for a comatch m and extension morphism e to obtain a new source graph G', the modification of the CC ψ is defined (as for TA model checking) based on the clock guards and clock resets of applied rules and the invariants derived for G'. For $\mathsf{tPre}((G, \psi), \delta)$, we determine the least lower bound llb as either $[a, \infty)$ or (a, ∞) as stated using $a \leq c$ or $a < c$ for any clock, record this additional delay by updating I to $I + llb$ to record the duration of the path constructed so far, and do not apply tPre on states where this duration has reached δ.

Finite Difference Bound Matrices (DBMs) [5,10] developed initially for TA straightforwardly supporting tPre and dPre are used to represent the CCs contained in extended symbolic states. Given a CC ψ and a fixed ordering $(c_i)_{1 \leq i \leq n}$ of the clocks occurring in ψ, there is a a unique normalized DBM $\mathsf{nDBM}(\psi) = (D_{i,j})_{0 \leq i,j \leq n+1}$ storing pairs $(\mathbf{Z} \cup \{\infty\}) \times \{\leq, <\}$ where $D_{i,j} = (k, \lesssim)$ represents the CC $c_i - c_j \lesssim k$ (assuming that $c_0 = 0$).

For an example of an application of tPre and our adaptation of it, consider Fig. 7a. The (blue) area ① is given as DBM D in Fig. 7b stating the lower bounds $3 \leq x$ (in $D_{0,1}$) and $1 \leq y$ (in $D_{0,2}$). The (green) area ② extends D to DBM D' in Fig. 7c by decreasing the lower bounds of x and y to 2 and 0. The (red) area ③ would be included by the regular timed predecessor operation (used in TA model checking) also dropping the lower bound on x to 0.

For our running example, we apply δ-induction for $\delta = 1$ due to the assumed maximal message delay of 1. In Fig. 8a, we apply δ-induction to the state 0 representing the precondition of $\sigma_{\mathsf{Abrake1}}$ (given by the LHS graph of its GT rule, its clock guard, and the CC derived from the TGTS invariant $\mathsf{inv}_{\mathsf{advance}}$ being instantiated for S_1 (for S_2 the lack of attributes prevents an instantiation of $\mathsf{inv}_{\mathsf{advance}}$)). Note that we discuss the step labeled unclock from state 1 to 0 and the (intentional) omission of certain discrete steps later and that we omit steps $\mathsf{tPre}(\cdot, 0)$ and do not apply tPre twice in a row. Applying σ_{Adrive} backwards from state 1 for shuttle S_2 leads to two different graphs (depending on whether S_2 stays on the already given track topology fragment or whether S_2 was on a join track implying that it may have come from a different predecessor track segment) where we then also know that the shuttle S_2 was in driving mode before (as the shuttles' mode is checked by the rule) and its clock c_2 must have satisfied the clock guard of σ_{Adrive} and the TGTS invariant $\mathsf{inv}_{\mathsf{advance}}$. Applying σ_{Adrive} backwards again is not possible because the clock of shuttle S_2 would need to be 0 because its clock gets reset to 0 in every traversal step. However, from states 2 and 3 we can apply tPre backwards decreasing the least lower bound on

---

[2] Since we may traverse unreachable states, even a non-zeno system S_{A0} may lead to infinite symbolic paths resulting in non-terminating path generation.

the clock c_2 of S_2 from 3 to 2 by a value of 1 as recorded in the interval given in the last compartment.

5.2 δ-Shift Operation Based on δ-Induction

We now discuss, based on the technique of δ-induction, our δ-shift operation to discover the remote δ-past of the given TGTS system w.r.t. a given selected actor to then derive appropriate additional fall-back rules.

The δ-shift operation first identifies the minimal root states $\mathsf{minRoot}(\sigma_{in}) = (G, \psi, [0, \infty))$ satisfying the preconditions of fall-back (input) rules σ_{in} already present in the given TGTS $\mathsf{S_{A0}}$. Again, this means that the graph G is the LHS graph of the GT rule of σ_{in} and the CC ψ is the conjunction of the clock guard of σ_{in} and the CCs derived by evaluating the TGTS invariants of the TGTS against graph G.

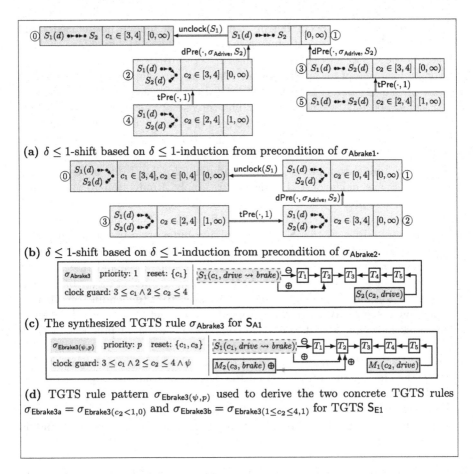

(a) $\delta \leq 1$-shift based on $\delta \leq 1$-induction from precondition of $\sigma_{\mathsf{Abrake1}}$.

(b) $\delta \leq 1$-shift based on $\delta \leq 1$-induction from precondition of $\sigma_{\mathsf{Abrake2}}$.

(c) The synthesized TGTS rule $\sigma_{\mathsf{Abrake3}}$ for $\mathsf{S_{A1}}$

(d) TGTS rule pattern $\sigma_{\mathsf{Ebrake3}(\psi, p)}$ used to derive the two concrete TGTS rules $\sigma_{\mathsf{Ebrake3a}} = \sigma_{\mathsf{Ebrake3}(c_2 < 1, 0)}$ and $\sigma_{\mathsf{Ebrake3b}} = \sigma_{\mathsf{Ebrake3}(1 \leq c_2 \leq 4, 1)}$ for TGTS $\mathsf{S_{E1}}$

Fig. 8. Reverse structural-temporal reachability analysis using $\delta \leq 1$-shift and TGTS rules for $\mathsf{S_{A1}}$ and $\mathsf{S_{E1}}$ derived from this analysis

For our running example, for the two fall-back rules σ_{Abrake1} and σ_{Abrake2}, these states are marked 0 in Fig. 8a and Fig. 8b. For each such minimal root state, the δ-shift operation then consists of the following three steps.

δ-**Shift Step S1:** To discover only the *remote* δ-past of a selected minimal root state $\text{minRoot}(\sigma_{in})$, backward steps of the local actor must be excised from the backward state space to retain him in his position of the intended action as described by the selected input rule σ_{in} at hand. To achieve this, the clocks of the local actor (and all CCs mentioning them) are pruned from the root state using a backward step unclock.

For our running example, this means that the clock c_1 of actor shuttle S_1 is removed with its constraints in Fig. 8a and Fig. 8b leading in both cases to states labeled 1; here, assuming the P2P interpretation to distribution, the local actor is identical to the actor identified by the actor morphism in σ_{in}.

δ-**Shift Step S2:** To derive the δ-past of the remote actor, we apply the technique of δ-induction as discussed in the previous subsection. However, we excise the generation of dPre steps for the selected actor and note that the removal of the local clocks in the previous step was required to allow here for unrestricted applications of tPre.

For our running example, we generate the remainder of Fig. 8a and Fig. 8b not generating backward steps of (the selected actor) shuttle S_1. In both symbolic backward state spaces, we reach (green) leaf states where time has rolled back by precisely 1 time unit.

δ-**Shift Step S3:** We generate the additional fall-back rules based on the generated symbolic backwards state space. For this purpose, we consider all states except *(a)* the minimal root state 0 and the state 1 reached using unclock (as rules generated from these states would not be different) and *(b)* the target states of timed steps as the CCs in the source states of such steps are wider leading to generated rules being applicable more often.

For our running example, we thereby only consider the three (green) leaf states in the symbolic backward state spaces Fig. 8a and Fig. 8b.

These states, when reinserting the clocks and CCs removed in the unclock step before, constitute the modified preconditions for the fall-back rules σ_{fb} to be derived. Additionally, σ_{fb} then applies the same graph, attribute, and clock modifications as the original rule σ_{in}. Technically, following Fig. 9, assume that for a considered state k *(a)* σ_{in} contains the GT rule span (ℓ_{in}, r_{in}) where ℓ_{in} has the graph of state 0 as codomain, *(b)* the step from state 1 to 0 labeled with unclock is captured by a GT span (ℓ_{uc}, r_{uc}) where r_{uc} is an identity morphism, and *(c)* the step sequence from state k to 1 is captured by the GT span sequence $\pi = (\ell_{k-1}, r_{k-1}) \ldots (\ell_1, r_1)$ (assuming spans of identities for timed steps). The step (ℓ_{uc}, r_{uc}) is sequentially independent (due to local vs. remote steps) from the steps in π and π can therefore be moved across (ℓ_{uc}, r_{uc}) (see [19] for technical details on sequential independence) effectively reversing the unclock step resulting in a GT span sequence π_{ac} (π with actor clocks) ending in state 0. Also, the GT span (ℓ_{in}, r_{in}) is sequentially independent from the steps in π_{ac}

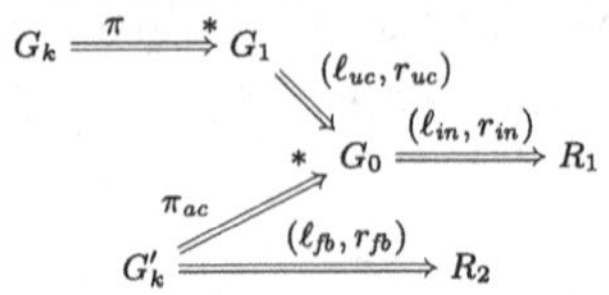

Fig. 9. Visualization for technical construction of additional fall-back TGTS rules.

and (ℓ_{in}, r_{in}) can therefore be moved across π_{ac} effectively propagating the local change of σ_{in} for the selected actor across the remote steps in π_{ac} resulting in a span (ℓ_{fb}, r_{fb}), which is the GT rule span of σ_{fb}. The clock guard, clock reset, and priority of σ_{fb} are preserved from σ_{in} (where clock identities are preserved alongside when moving (ℓ_{in}, r_{in}) over π_{ac} into (ℓ_{fb}, r_{fb})).

For our running example, we generate three fall-back rules. First, from state 4 in Fig. 8a we generate a rule, which is identical to the GT part of $\sigma_{Abrake2}$ only imposing an additional clock constraint $c_2 \in [2, 4]$; we omit this rule subsequently as $\sigma_{Abrake2}$ is applicable whenever this rule would be; this observation also indicates that $\sigma_{Abrake2}$ may have been derived from $\sigma_{Abrake1}$ informally applying our approach. Second, from state 5 in Fig. 8a we generate a rule, which considers a situation in which a collision cannot be avoided; we omit this rule subsequently as it cannot help to obtain a robust system. Third, from state 3 in Fig. 8b we generate the rule $\sigma_{Abrake3}$, which is similar to $\sigma_{Abrake2}$ but where a message that is sent by the remote shuttle from an earlier track is used to initiate braking. The resulting TGTS S_{A1} then contains, in comparison to S_{A0}, the additional rule $\sigma_{Abrake3}$ from Fig. 8c and its concretization $\text{conc}(S_{A1}) = S_{E1}$ contains, in comparison to S_{E0}, the two additional rules $\sigma_{Ebrake3a}$ and $\sigma_{Ebrake3b}$ from Fig. 8d.

The discussion above is tailored to the P2P interpretation of distribution from Sect. 4; for the RC interpretation, we note that also the step performed by shuttle S_1 will be delayed due to a secondary message delay between RC and shuttle S_1. To accommodate for this interpretation in which also shuttle S_1 is part of the remote part of the TGTS (w.r.t. the location of the RC), we do *(a)* assume an 2-hop message delay of 0–2 time units (i.e., $\delta = 2$ instead of $\delta = 1$), *(b)* unclock in step S1 does not remove any clocks as the RC has no explicitly represented clocks in the TGTS state, and *(c)* do not omit discrete backward steps of shuttle S_1 in step S2. For the RC interpretation, we therefore obtain an additional fall-back rule in which the shuttle S_1 brakes earlier (itself being a track farther from the join track).

5.3   Post-extension Robustness Analysis

The resulting TGTS S_{A1} must be analyzed for robustness w.r.t. its concretization S_{E1} because our approach does not guarantee the absence of safety violations in S_{E1} for two reasons. Firstly, applications of generated fall-back rules can lead to states (previously not reachable in S_{A0}) from which safety violations may be reachable in a way not previously considered. Secondly, the existing fall-back rules (especially when no such fall-back rules are provided) may be fundamentally insufficient potentially leading also to insufficient generated fall-back rules.

For our running example, we report that post-extension robustness analysis via model checking as described at the end of Sect. 3 confirmed the success of our approach in the sense that *(a)* S_{E0} suffers from collisions and that S_{A0} is therefore not robust w.r.t. its concretization S_{E0} and *(b)* S_{E1} has no reachable collisions and that S_{A1} is therefore robust w.r.t. its concretization S_{E1}. The cost of analysis is higher for S_{E0} and S_{E1} compared to S_{A0} and S_{A1} because the additional clocks of messages increase the cost of model checking of the resulting TA.

We may reduce the cost of model checking based analysis in this post-extension robustness analysis. While a full model checking analysis can be performed for S_{E1} from its start state, we may exploit the similarity of S_{E0} and S_{E1} by reusing model checking results of S_{E0} when model checking S_{E1}. Hence, the symbolic state space of S_{E1} only needs to be adapted to the rule set change between S_{E0} and S_{E1}. In particular, we may verify that *(a)* every collision in S_{E0} is avoided by a prior step of a new fall-back rule in S_{E1}, *(b)* each new fall-back rule in S_{E1} is used to prevent some collision reachable in S_{E0} (to prevent some unnecessary behavior restrictions), and *(c)* every state reached by applying such a step cannot lead to a collision later on.

5.4 Iterative Application

Our explanations for the running example revealed the potential importance of its iterative application. In particular, as stated above, our approach generates the rule $\sigma_{Abrake2}$ (almost precisely) from the fall-back rule $\sigma_{Abrake1}$ and generates the rule $\sigma_{Abrake3}$ from the fall-back rule $\sigma_{Abrake2}$. Applying our rule generation approach iteratively for the rules generated so far may yield further rules required for robustness. For our running example, this is due to the limited distance the local shuttle S_1 monitors the current state in terms of connected track segments; for $\sigma_{Abrake1}$, $\sigma_{Abrake2}$, and $\sigma_{Abrake3}$, we observe a number of 3, 4, and 5 track segments between the two shuttles. Thereby, an iterative application of our approach does not only shift the remote global time backwards applying discrete steps as required but also gradually enlarges the locally considered context (within one application of our approach but also across multiple such applications) allowing to derive local rules that match a larger region of the local state.

6 Related Work

We introduced our TGTS formalism as a non-probabilistic restriction of PTGTSs [30] extended with agent-based priorities. Regarding the modeling of timing constraints, GTSs have been extended to variations of timed GTSs in [4, 21, 22, 34, 41] without relying on clocks as in TA [1] and to variations based on clocks in [29, 30, 33]. Regarding the modeling of agent-based priorities, we are only aware of the usage of application conditions (e.g., in [32]) to encode such agent-based local priorities leading to sometimes large application conditions for which maintenance across system evolutions is error-prone.

Distribution aspects complicating design solutions as considered for our context in Sect. 4 have been surveyed in [47] listing further aspects of general relevance. Clock-related effects such as clock drifts and imprecise clock measurements (jittering) have been investigated thoroughly in the fundamental setting of TA [1] as surveyed in [6–8] but also for Hybrid Automata [9,44,45].

The running example based on the RailCab [37] system has been considered before in, e.g., [30,32] analyzing system correctness at the abstract idealized level. In particular, in [30,32] shuttle convoy establishment was analyzed where message delays have been implicitly considered (by limiting the number communications per time) in the context of probabilistic communication failures. In [28,43], similar examples on self-driving cars employing delayed unreliable wireless communication is considered where cars communicate information on current or future locations and are manually designed at a concrete, explicit level to adjust information from received messages based on delay estimates. Lastly, in a discrete time setting, delay-robustness of timed distributed system has been investigated in [49] in the context of supervisory control theory.

Trace refinement [24], considers the refinement of a model A into a model B such that the trace set T_A of A contains the trace set T_B of B. The absence of specification violating paths in T_B can be derived from their absence in T_A. Examples of refinements include narrowing of guards of steps or the prevention of some non-deterministic alternatives. In comparison, we do not only restrict the set of traces from S_{A0} to S_{A1} as constructed rules are applied instead of previously existing rules leading to both systems having incomparable trace sets. Similarly, from S_{A0} to S_{E0} (and accordingly from S_{A1} to S_{E1}), we employ different rules on different graphs but (deviating from the trace refinement idea) may also have safety violations in S_{E0} and S_{E1} even though S_{A0} and S_{A1} are safe. However, as discussed, the symbolic state spaces of the considered systems may overlap in large parts making points where they deviate from each other most relevant.

Timed action refinement was considered in [35,36] relying on the idea that abstract atomic steps are refined to multiple steps in a refined model. While strong results can be obtained in certain cases, this approach appears to be not suitable for the refinement w.r.t. aspects of distribution considered here.

Lastly, in [48] design-time and run-time analysis are integrated whereby symbolic paths to violations are derived (backwards as in k-induction) at design-time to increase the look-ahead towards such violations at run-time. However, this approach considers untimed systems and attempts to disable steps leading to unsafe states while we benefit here from the system model capturing safe fallback behaviors to generate alternative steps systematically at design time.

7 Conclusion and Future Work

MDE for DCPSs focusing on core behavioral and interaction patterns may, at a high level of abstraction, assume inter-agent message delays of 0. Zooming in on this aspect assuming a maximal message delay of $\delta > 0$ in concretizations may reveal the insufficiency of provided fall-back rules to avoid safety violations. In this paper, we proposed an approach to derive additional fall-back rules from provided fall-back rules widening their application to further critical situations. Already for our running example, we conclude that our automatic rule derivation approach can ease MDE for DCPSs.

In the future, we intend to consider coordination patterns such as convoy formation using consensus algorithms subject to probabilistic message loss as previously considered at an abstract level in [30, 32] for PTGTS. Moreover, we intend to develop the technique of δ-induction formally for TGTSs including complications such as (nested) application conditions. Lastly, we are designing a tool-based mechanization for the proposed rule derivation approach.

Acknowledgments. This research was partially funded by the HPI research school on Service-Oriented Systems Engineering.

References

1. Alur, R., Dill, D.L.: A Theory of Timed Automata. Theor. Comput. Sci. **126**(2), 183–235 (1994). https://doi.org/10.1016/0304-3975(94)90010-8
2. Becker, B., Giese,H.: Cyber-Physical Systems with Dynamic Structure: Towards Modeling and Verification of Inductive Invariants. Technical report. 64. Hasso Plattner Institute, University of Potsdam (2012). https://nbn-resolving.org/urn:nbn:de:kobv:517-opus-62437
3. Becker, B., Giese, H.: Incremental verification of inductive invariants for the run-time evolution of self-adaptive software-intensive systems. In: 23rd IEEE /ACM International Conference on Automated Software Engineering - Workshop Proceedings (ASE Workshops 2008), L'Aquila, Italy, 15–16 September 2008., pp. 33–40. IEEE (2008). isbn: 978-1-4244-2776-5. https://doi.org/10.1109/ASEW.2008.4686291. https://ieeexplore.ieee.org/xpl/mostRecentIssue.jsp?punumber=4674379
4. Becker, B., Giese, H.: On safe service-oriented real-time coordination for autonomous vehicles. In: 11th IEEE International Symposium on Object- Oriented Real-Time Distributed Computing (ISORC 2008), Orlando, Florida, USA, 5–7 May 2008, pp. 203–210. IEEE Computer Society (2008). isbn: 978-0-7695-3132-8. https://doi.org/10.1109/ISORC.2008.13. https://ieeexplore.ieee.org/xpl/mostRecentIssue.jsp?punumber=4519543
5. Bengtsson, J., Yi, W.: Timed automata: semantics, algorithms and tools. In: Desel, J., Reisig, W., Rozenberg, G. (eds.) ACPN 2003. LNCS, vol. 3098, pp. 87–124. Springer, Heidelberg (2004). https://doi.org/10.1007/978-3-540-27755-2_3
6. Bouyer, P.: Timed automata. In: Pin, J. (ed.) Handbook of Automata Theory, pp. 1261–1294. European Mathematical Society Publishing House, Zürich, Switzerland (2021). https://doi.org/10.4171/Automata-2/12

7. Bouyer, P., Gastin, P., Herbreteau, F., Sankur, O., Srivathsan, B.: Zone-based verification of timed automata: extrapolations, simulations and what next? In: Bogomolov, S., Parker, D. (eds.) Formal Modeling and Analysis of Timed Systems - 20th International Conference, FORMATS 2022, Warsaw, Poland, 13–15 September 2022, Proceedings, vol. 13465, pp. 16–42. Springer, Heidelberg (2022). https://doi.org/10.1007/978-3-031-15839-1_2

8. Bouyer, P., Kupferman, O., Markey, N., Maubert, B., Murano, A., Perelli, G.: Reasoning about quality and fuzziness of strategic behaviours. In: Kraus, S. (ed.) Proceedings of the Twenty-Eighth International Joint Conference on Artificial Intelligence, IJCAI 2019, Macao, China, 10–16 August 2019, pp. 1588–1594. ijcai.org (2019). https://doi.org/10.24963/ijcai.2019/220

9. Denis, B., Lesage, J., Juárez-Orozco, Z.: Performance verification of discrete event systems using hybrid model-checking. In: Cassandras, C.G., Giua, A., Seatzu, C., Zaytoon, J. (eds.) 2nd IFAC Conference on Analysis and Design of Hybrid Systems, ADHS 2006, Alghero, Italy, 7–9 June 2006, vol. 39. IFAC Proceedings, vol. 5, pp. 365–370. Elsevier (2006). https://doi.org/10.3182/20060607-3-IT-3902.00067

10. Dill, D.L.: Timing assumptions and verification of finite-state concurrent systems. In: Sifakis, J. (ed.) CAV 1989. LNCS, vol. 407, pp. 197–212. Springer, Heidelberg (1990). https://doi.org/10.1007/3-540-52148-8_17

11. Donaldson, A.F., Haller, L., Kroening, D., Rümmer, P.: Software verification using k-induction. In: Yahav, E. (ed.) SAS 2011. LNCS, vol. 6887, pp. 351–368. Springer, Heidelberg (2011). https://doi.org/10.1007/978-3-642-23702-7_26 isbn: 978-3-642-23701-0

12. Dyck, J.: Increasing expressive power of graph rules and conditions and automatic verification with inductive invariants. MA thesis. University of Potsdam, Hasso Plattner Institute, Potsdam, Germany (2012)

13. Dyck, J.: Verification of Graph Transformation Systems with k-Inductive Invariants. PhD thesis. University of Potsdam, Hasso Plattner Institute, Potsdam, Germany (2020). https://doi.org/10.25932/publishup-44274

14. Dyck, J., Giese, H.: Inductive Invariant Checking with Partial Negative Application Conditions. In: Parisi-Presicce, F., Westfechtel, B. (eds.) ICGT 2015. LNCS, vol. 9151, pp. 237–253. Springer, Cham (2015). https://doi.org/10.1007/978-3-319-21145-9_15 isbn: 978-3-319-21144-2

15. Dyck, J., Giese, H.: Inductive invariant checking with partial negative application conditions. Technical report. 98. Potsdam, Germany: Hasso Plattner Institute at the University of Potsdam (2015)

16. Dyck, J., Giese, H.: k-inductive invariant checking for graph transformation systems. In: de Lara, J., Plump, D. (eds.) ICGT 2017. LNCS, vol. 10373, pp. 142–158. Springer, Cham (2017). https://doi.org/10.1007/978-3-319-61470-0_9 isbn: 978-3-319-61469-4

17. Dyck, J., Giese, H.: k-Inductive invariant checking for graph transformation systems. Technical report. 119. Potsdam, Germany: Hasso Plattner Institute at the University of Potsdam (2017)

18. Ehrig, H., Ehrig, K., Prange, U., Taentzer, G.: Fundamentals of Algebraic Graph Transformation. MTCSAES, Springer, Heidelberg (2006). https://doi.org/10.1007/3-540-31188-2. isbn: 978-3-540-31187-4

19. Ehrig, H., Golas, U., Habel, A., Lambers, L., Orejas, F.: M-adhesive transformation systems with nested application conditions. Part 1: parallelism, concurrency and amalgamation. Math. Struct. Comput. Sci. **24**(4), 240406 (2014)

20. Forouzan, B.A.: Data communications and networking. Huga Media (2007)

21. Gyapay, S., Heckel, R., Varró, D.: Graph transformation with time: causality and logical clocks. In: Corradini, A., Ehrig, H., Kreowski, H.-J., Rozenberg, G. (eds.) ICGT 2002. LNCS, vol. 2505, pp. 120–134. Springer, Heidelberg (2002). https://doi.org/10.1007/3-540-45832-8_11

22. Gyapay, S., Varró, D., Heckel, R.: Graph Transformation with Time. Fund. Inf. **58**(1), 1–22 (2003). https://content.iospress.com/articles/fundamenta-informaticae/fi58-1-02

23. Heckel, R., Wagner, A.: Ensuring consistency of conditional graph rewriting - a constructive approach. In: Corradini, A., Montanari, U. (eds.) Joint COMPUGRAPH/SEMAGRAPH Workshop on Graph Rewriting and Computation, SEGRAGRA 1995, Volterra, Italy, 28 August–1 September 1995. Electronic Notes in Theoretical Computer Science, vol. 2, pp. 118–126. Elsevier (1995). https://doi.org/10.1016/S1571-0661(05)80188-4

24. Heizmann, M., Hoenicke, J., Podelski, A.: Refinement of trace abstraction. In: Palsberg, J., Su, Z. (eds.) SAS 2009. LNCS, vol. 5673, pp. 69–85. Springer, Heidelberg (2009). https://doi.org/10.1007/978-3-642-03237-0_7

25. Khasidashvili, Z., Korovin, K., Tsarkov, D.: EPR-based k-induction with Counterexample Guided Abstraction Refinement. In: Gottlob, G., Sutcliffe, G., Voronkov, A. (eds.) Global Conference on Artificial Intelligence, GCAI 2015, Tbilisi, Georgia, 16–19 October 2015. EPiC Series in Computing, vol. 36, pp. 137–150. EasyChair (2015). https://doi.org/10.29007/scv7

26. Kwiatkowska, M., Norman, G., Parker, D.: PRISM 4.0: verification of probabilistic real-time systems. In: Gopalakrishnan, G., Qadeer, S. (eds.) CAV 2011. LNCS, vol. 6806, pp. 585–591. Springer, Heidelberg (2011). https://doi.org/10.1007/978-3-642-22110-1_47 isbn: 978-3-642-22109-5

27. Kwiatkowska, M., Norman, G., Sproston, J., Wang, F.: Symbolic model checking for probabilistic timed automata. In: Lakhnech, Y., Yovine, S. (eds.) FORMATS/FTRTFT -2004. LNCS, vol. 3253, pp. 293–308. Springer, Heidelberg (2004). https://doi.org/10.1007/978-3-540-30206-3_21 isbn: 3-540-23167-6

28. Lee, G., Jung, J.: Decentralized platoon join-in-middle protocol considering communication delay for connected and automated vehicle. Sensors **21**(21), 7126 (2021). https://doi.org/10.3390/s21217126

29. Maximova, M., Giese, H., Krause, C.: Probabilistic timed graph transformation systems. In: de Lara, J., Plump, D. (eds.) ICGT 2017. LNCS, vol. 10373, pp. 159–175. Springer, Cham (2017). https://doi.org/10.1007/978-3-319-61470-0_10 isbn: 978-3-319-61469-4

30. Maximova, M., Giese, H., Krause, C.: Probabilistic timed graph transformation systems. J. Log. Algebr. Meth. Program. **101**, 110–131 (2018). https://doi.org/10.1016/j.jlamp.2018.09.003

31. Maximova, M., Schneider, S., Giese, H.: Compositional analysis of probabilistic timed graph transformation systems. In: FASE 2021. LNCS, vol. 12649, pp. 196–217. Springer, Cham (2021). https://doi.org/10.1007/978-3-030-71500-7_10

32. Maximova, M., Schneider, S., Giese, H.: Compositional analysis of probabilistic timed graph transformation systems. Formal Aspects Comput. **35**(3), 1–79 (2023). https://doi.org/10.1145/3572782. issn: 0934-5043

33. Maximova, M., Schneider, S., Giese, H.: Interval probabilistic timed graph transformation systems. In: Gadducci, F., Kehrer, T. (eds.) ICGT 2021. LNCS, vol. 12741, pp. 221–239. Springer, Cham (2021). https://doi.org/10.1007/978-3-030-78946-6_12

34. Neumann, S.: Modellierung und Verifikation zeitbehafteter Graphtransformationssysteme mittels Groove. MA thesis. University of Paderborn (2007)

35. Normann, H., Debois, S., Slaats, T., Hildebrandt, T.T.: Zoom and enhance: action refinement via subprocesses in timed declarative processes. In: Polyvyanyy, A., Wynn, M.T., Van Looy, A., Reichert, M. (eds.) BPM 2021. LNCS, vol. 12875, pp. 161–178. Springer, Cham (2021). https://doi.org/10.1007/978-3-030-85469-0_12
36. Qin, G., Wu, J.: Action refinement for real-time concurrent processes with urgency. J. Comput. Sci. Technol. **20**(4), 514–525 (2005). https://doi.org/10.1007/s11390-005-0514-2
37. RailCab Team. RailCab Project. https://www.hni.uni-paderborn.de/cim/projekte/railcab
38. Schneider, S., Dyck, J., Giese, H.: Formal verification of invariants for attributed graph transformation systems based on nested attributed graph conditions. In: Gadducci, F., Kehrer, T. (eds.) ICGT 2020. LNCS, vol. 12150, pp. 257–275. Springer, Cham (2020). https://doi.org/10.1007/978-3-030-51372-6_15
39. Schneider, S., Maximova, M., Giese, H.: Invariant analysis for multi-agent graph transformation systems using k-induction. Technical report. 143. Potsdam, Germany: Hasso Plattner Institute at the University of Potsdam (2022). https://doi.org/10.25932/publishup-54585
40. Schneider, S., Maximova, M., Giese, H.: Invariant analysis for multi-agent graph transformation systems using k-induction. In: Behr, N., Strüber, D. (eds.) Graph Transformation - 15th International Conference, ICGT 2022, Held as Part of STAF 2022, Nantes, France, 7–8 July 2022, Proceedings. Lecture Notes in Computer Science, vol. 13349, pp. 173–192. Springer (2022). https://doi.org/10.1007/978-3-031-09843-7_10
41. Schneider, S., Maximova, M., Sakizloglou, L., Giese, H.: Formal testing of timed graph transformation systems using metric temporal graph logic. Int. J. Softw. Tools Technol. Transf. **23**(3), 411–488 (2021). https://doi.org/10.1007/s10009-020-00585-w
42. Sheeran, M., Singh, S., Stålmarck, G.: Checking safety properties using induction and a SAT-solver. In: Hunt, W.A., Johnson, S.D. (eds.) FMCAD 2000. LNCS, vol. 1954, pp. 127–144. Springer, Heidelberg (2000). https://doi.org/10.1007/3-540-40922-X_8 isbn: 3-540-41219-0
43. Shin, D., Yi, K.: Compensation of wireless communication delay for integrated risk management of automated vehicle. In: IEEE Intelligent Vehicles Symposium, IV 2015, Seoul, South Korea, 28 June–1 July 2015, pp. 1355–1360. IEEE (2015). https://doi.org/10.1109/IVS.2015.7225904
44. Silva, B.I., Krogh, B.H.: Modeling and verification of hybrid systems with clocked and unclocked events. In: 40th IEEE Conference on Decision and Control, CDC 2001, Orlando, FL, USA, 4–7 December 2001, pp. 762–767. IEEE (2001). https://doi.org/10.1109/.2001.980198
45. Tsuchie, Y., Ushio, T.: Control-invariance of sampled data hybrid systems with periodically clocked events and jitter. In: Cassandras, C.G., Giua, A., Seatzu, C., Zaytoon, J. (eds.) 2nd IFAC Conference on Analysis and Design of Hybrid Systems, ADHS 2006, Alghero, Italy, 7–9 June 2006, vol. 39. IFAC Proceedings, vol. 5, pp. 417–422. Elsevier (2006). https://doi.org/10.3182/20060607-3-IT-3902.00075
46. Van Steen, M., Tanenbaum, A.S.: A brief introduction to distributed systems. Computing **98**, 967–1009 (2016)
47. Weyns, D., Van Dyke Parunak, H., Michel, F., Holvoet, T., Ferber, J.: Environments for multiagent systems state-of-the-art and research challenges. In: Weyns, D., Van Dyke Parunak, H., Michel, F. (eds.) E4MAS 2004. LNCS (LNAI), vol. 3374, pp. 1–47. Springer, Heidelberg (2005). https://doi.org/10.1007/978-3-540-32259-7_1

48. Xu, H., Schneider, S., Giese, H.: Integrating look-ahead design-time and run-time control-synthesis for graph transformation systems. In: Beyer, D., Cavalcanti, A. (eds.) Fundamental Approaches to Software Engineering, FASE 2024, Held as Part of the Joint European Conferences on Theory and Practice of Software, ETAPS 2024. Proceedings (2024)
49. Zhang, R., Cai, K., Gan, Y., Wonham, W.M.: Delay-robustness in distributed control of timed discrete-event systems based on supervisor localisation. Int. J. Control **89**(10), 2055–2072 (2016). https://doi.org/10.1080/00207179.2016.1147606

Taint Analysis for Graph APIs Focusing on Broken Access Control

Leen Lambers[✉][iD], Lucas Sakizloglou[iD], Osama Al-Wardi[iD], and Taisiya Khakharova[iD]

Brandenburg University of Technology Cottbus-Senftenberg, Cottbus, Germany
leen.lambers@b-tu.de

Abstract. Graph APIs are capable of flexibly retrieving or manipulating graph-structured data over the web. This rather novel type of APIs presents new challenges when it comes to properly securing the APIs against the usual web application security risks, e.g., broken access control. A prominent security testing approach is taint analysis, which traces tainted, i.e., security-relevant, data from sources (where tainted data is inserted) to sinks (where the use of tainted data may lead to a security risk), over the information flow in an application.

We present a first systematic approach to static and dynamic taint analysis for Graph APIs focusing on broken access control. The approach comprises the following. We taint nodes in the Graph API if they represent data requiring specific privileges in order to be retrieved or manipulated, and identify API calls which are related to sources and sinks. Then, we statically analyze whether tainted information flow between API source and sink calls occurs. To this end, we model the API calls using graph transformation rules. We subsequently use critical pair analysis to automatically analyze potential dependencies between rules representing source calls and rules representing sink calls. The static taint analysis (i) identifies flows that need to be further reviewed, since tainted nodes may be created by an API call and used or manipulated by another API call later without having the necessary privileges, and (ii) can be used to systematically design dynamic security tests for broken access control. The dynamic taint analysis checks if potential broken access control risks detected during the static taint analysis really occur. We apply the approach to a part of the GitHub GraphQL API.

Keywords: Graph Transformation · Taint Analysis · Graph APIs · Security Testing

1 Introduction

An increasing number of software applications in the area of service-oriented and cloud computing relies on *Graph APIs* [32,40]. A prominent specification for Graph APIs is GraphQL [13], which specifies a query language for the construction of Graph APIs. GraphQL was introduced by Facebook in 2012 as part of the development of mobile applications. GraphQL has been open source since 2015 and is now used in various web and mobile applications (e.g. GitHub, Pinterest, Paypal, etc.). As Graph APIs emerged to facilitate data management in social networks, they are capable of flexibly retrieving or manipulating graph-structured data over the web.

Security [15] is particularly important in the area of web and mobile applications. Graph APIs are still relatively new, which means that a wide range of methods and tools for security analyses are not yet available for them [29,32]. *Taint analysis* is a typical representative of such security analyses. It is a particular instance of data flow analysis [30] and systematically tracks certain user inputs through the application. *Broken access control* is considered as one of the most prominent security vulnerabilities for web applications [36]. Access control enforces policies such that users cannot act outside their intended permissions. Failures typically lead to unauthorized information disclosure, modification, or destruction of all data or performing a business function outside the user's limits. Typical access control failures occur when an API can be accessed with (i) missing or misunderstood access control, or (ii) wrongly implemented access control. Such failures occur in practice as demonstrated e.g., by the issues 110618, 106598, 85661 in the GitHub community forum [10].

In this paper, we present a *taint analysis for Graph APIs* focusing on *broken access control*. A *security analyst* can use the taint analysis to systematically check for *broken access control* (BAC) vulnerabilities. During *setup* of the analysis, the security analyst can taint nodes in the Graph API that represent sensitive data. From these tainted nodes, we can derive Graph API *sources* (where tainted data is inserted) and *sinks* (where the use of tainted data may lead to a security risk). The taint analysis consists of a static and a dynamic part. The *static taint analysis* supports the security analyst in *validating* the access control policies, i.e. finding missing or misunderstood access control policies. The *dynamic taint analysis* supports the security analyst in *verifying* that the access control policies are implemented correctly. We formalize the taint analysis using concepts from graph transformation [7,16,34]. We demonstrate that the static analysis is *sound* w.r.t. finding BAC vulnerabilities. However, since the static analysis is incomplete. we integrate it with a subsequent *complete* dynamic analysis.

Graph transformation is a formal paradigm with many applications. A *graph transformation system* is a collection of graph transformation rules that, in union, serve a common purpose. In this paper, a *graph transformation system* is used to formally describe a Graph API. For many applications (see [27] for a survey involving 25 papers), it is beneficial to know all conflicts and dependencies that can occur for a given pair of rules. A conflict is a situation in which one rule application renders another rule application inapplicable. A dependency is a situation in which one rule application needs to be performed such that another rule application becomes possible. We will use *dependency analysis* to discover tainted information flow that may lead to broken access control vulnerabilities.

The rest of this paper is structured as follows: Sect. 2 revisits preliminaries. Section 3 presents taint analysis, i.e, the setup, the static, and the dynamic

```
1  type User {                    20  type Project {
2     username: String!           21     name: String!
3     projects: [Project]!        22     id: ID!
4     repos: [Repository]!        23     body: String
5  }                              24  }
6                                 25
7  type Repository {              26  type Query {
8     id: ID!                     27  }
9     name: String!               28
10    description: String!        29  type Mutation {
11    url: String!                30     createRepo(name: String!, description
12    owner: Repository!                    : String!, url: String!): ID
13 }                              31     updateRepo(id: ID!, name: String!,
14                                          description: String!): Boolean
15 type Issue {                   32  }
16    username: String!
17    body: String!
18    repo: Repository!
19 }
```

Fig. 1. The GraphQL schema for the running example

analysis, in further detail, based on a basic running example. Section 4 applies the analysis to a part of the GitHub GraphQL API [11]. Section 5 is dedicated to related work. Section 6 concludes the paper and outlines future work.

2   Preliminaries

First, we recapitulate the basics of Graph APIs and GraphQL, a prominent Graph API discussed further in the paper; in the same section, we introduce the running example used in the remainder. Then we present the basics of access control. Finally, we recall the double-pushout (DPO) approach to graph transformation as presented in [7]. We reconsider dependency notions on the transformation and rule level [25–27] for the DPO approach, including dependency reasons.

2.1   Graph APIs and Running Example

A Graph API models the data managed by a web application based on the graph data model, i.e., data objects are represented by vertices and relationships by edges. Graph APIs emerged to facilitate data management in social networks which similarly relied on graph-structured data. Two familiar examples of Graph APIs are the *Facebook API* [28] and the *GitHub GraphQL API* [11] (henceforth referred to as GitHub API). The GitHub API is based on the *GraphQL* query language specification, which describes the capabilities and requirements of graph data models for web applications [13]. In the remainder, we focus on APIs which are based on GraphQL.

GraphQL uses a *schema*, i.e., a strongly typed contract between the client and the server, to define all exposed types of an API as well as their structure. Schemata of GraphQL APIs are either publicly available or can be discovered using distinguished *introspection* queries. An example of a schema is shown in Fig. 1. The schema captures the GraphQL API of a simplified version of a platform that allows developers to manage their code as well as their development efforts, similar to GitHub. The schema is defined in the GraphQL schema language [14] and contains the following object types: the `User`, which represents a user of the platform; the `Repository`, which represents a code repository hosted on the platform; the `Issue`, which represents a task related to the development work being performed in a repository; and the `Project`, which represents a task-board that can be used for planning and organizing development work. Attributes and relationships are captured by fields within types, where squared brackets refer to a list of objects. The exclamation mark renders a field non-nullable.

GraphQL defines two types of operations that can be performed on the server: *queries* are used to fetch data, whereas *mutations* are used to manipulate, i.e, create, update, or delete data. Queries and mutations are also defined in the schema, within the distinguished types `Query` and `Mutation`, respectively. Figure 1 shows a minimal list of mutations for the platform in our example: the mutations *createRepo*, *updateRepo* create and update an instance of a `Repository`, respectively.

2.2 Access Control Policies

In various types of software systems, e.g., financial and safety-critical, the adequate security of the information and the systems themselves constitutes a fundamental requirement. *Access control* refers to ensuring that users are allowed to perform activities which may access information or resources only when they have the required rights. An access control *policy* is a collection of high-level requirements that manage user access and determine who may access information and resources under which circumstances [18]. Correspondingly, a policy is *broken* when these requirements are not fulfilled.

While there are various methods of realizing policies, we focus on the *role-based* method. In a role-based policy, a security analyst defines roles, i.e., security groups, whose access rights are defined according to roles held by users and groups in organizational functions; the analyst assigns a role to each user who, once authorized, can perform the activities permitted by said role. A role-based access control policy is broken when a user performs an activity that is not permitted by their role.

A requirement of a role-based policy for our running example could state that *a user may only update the information of a repository, if that user is the owner of the repository*. In practice, the Graph API of the platform should ensure that the query *updateRepo* can only be executed by a user who is authorized to update the repository, i.e, who has the role *owner*. An update of the repository information by a user who does not belong to this group would break the policy.

2.3   Graph Transformation and Static Dependency Analysis

Throughout this paper we always consider graphs (and graph morphisms) typed over some fixed type graph TG via a typing morphism $type_G : G \to TG$ as presented in [7]. A graph morphism $m : G \to H$ between two graphs consists of two mappings m_V and m_E between their nodes and edges being both type-preserving and structure-preserving w.r.t. source and target functions. Note that in the main text we denote inclusions by $\hookrightarrow$ and all other morphisms by $\to$.

Graph transformation is the rule-based modification of graphs. A *rule* consists of two graphs: L is the left-hand side (LHS) of the rule representing a pattern that has to be found to apply the rule. After the rule application, a pattern equal to R, the right-hand side (RHS), has been created. The intersection K is the graph part that is not changed; it is equal to $L \cap R$ provided that the result is a graph again. The graph part that is to be deleted is defined by $L \setminus (L \cap R)$, while $R \setminus (L \cap R)$ defines the graph part to be created. We consider a graph transformation system just as a set of rules.

A *direct graph transformation* $G \overset{m,r}{\Longrightarrow} H$ between two graphs G and H is defined by first finding a graph morphism m of the LHS L of rule r into G such that m is injective, and second by constructing H in two passes: (1) build $D := G \setminus m(L \setminus K)$, i.e., erase all graph elements that are to be deleted; (2) construct $H := D \cup m'(R \setminus K)$. The morphism m' has to be chosen such that a new copy of all graph elements that are to be created is added. It has been shown that r is applicable at m iff m fulfills the *dangling condition*. It is satisfied if all adjacent graph edges of a graph node to be deleted are deleted as well, such that D becomes a graph. Injective matches are usually sufficient in applications and w.r.t. our work here, they allow explaining constructions with more ease than for general matches. In categorical terms, a direct transformation step is defined using a so-called double pushout as in the following definition. Thereby step (1) in the previous informal explanation is represented by the first pushout (PO1) and step (2) by the second pushout (PO2) [7].

Definition 1 ((read/write) rule and transformation). *A rule r is defined by $r = (L \overset{le}{\hookleftarrow} K \overset{ri}{\hookrightarrow} R)$ with $L, K,$ and R being graphs connected by two graph inclusions. A rule r is a* read rule *if le and ri are identity morphisms. A rule r is a* write rule *if it is not a read rule.*

A *direct transformation* $G \overset{m,m',r}{\Longrightarrow} H$ applying r to G consists of two pushouts as depicted on the right. Rule r is *applicable* and the injective morphism $m : L \to G$ ($m' : R \to H$) is called a *match* (resp. co-match) if there exists a graph D such that $(PO1)$ is a pushout.

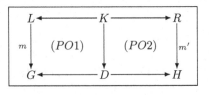

Given a pair of transformations, a *produce-use dependency* [7] occurs if the co-match of the first transformation via rule r_1 produces a match for the second transformation that was not available yet before applying the first transformation. We use an initial pushout construction over the right-hand side morphism

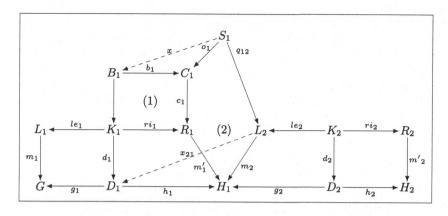

Fig. 2. Illustration of dependency and dependency reason

of rule r_1 to extract its *creation graph* C_1 stripping away as much preserved graph elements from the right-hand side graph R_1 as possible. The remaining graph then contains all elements produced by the rule completed by those preserved nodes that are needed to complete the produced elements to a graph. These additional nodes are called boundary nodes and summarized in the *boundary graph* B_1. The *dependency reason* $C_1 \overset{o_1}{\hookleftarrow} S_1 \overset{q_{12}}{\rightarrow} L_2$ then comprises elements produced by the first and used by the second rule giving rise to the dependency. It is explained in detail in [24] how to obtain using *static dependency analysis* such dependency reasons for a pair of rules for which transformation pairs in produce-use dependency exist. Each transformation pair in produce-use dependency comes with a unique dependency reason (completeness).

Definition 2 (produce-use dependency, dependency graph). *Given a pair of direct transformations* $(t_1, t_2) = (G \overset{m_1, m_1', r_1}{\Longrightarrow} H_1, H_1 \overset{m_2, m_2', r_2}{\Longrightarrow} H_2)$ *applying rules* $r_1 : L_1 \overset{le_1}{\hookleftarrow} K_1 \overset{ri_1}{\hookrightarrow} R_1$ *and* $r_2 : L_2 \overset{le_2}{\hookleftarrow} K_2 \overset{ri_2}{\hookrightarrow} R_2$ *as depicted in Fig. 2. Square (1) in Fig. 2 can be constructed as initial pushout over morphism* ri_1*. It yields the* boundary graph B_1 *and the* creation graph C_1*. The transformation pair* (t_1, t_2) *is in* produce-use dependency *if there does not exist a morphism* $x_{21} : L_2 \to D_1$ *such that* $h_1 \circ x_{21} = m_2$*. The dependency graph* $DG(\mathcal{R})$ *for a set of rules* $\mathcal{R}$ *consists of a node for each rule in* $\mathcal{R}$ *and an edge* (r_1, r_2) *from* r_1 *to* r_2 *if there exists a transformation pair* $(t_1, t_2) = (G \overset{r_1}{\Longrightarrow} H_1, H_1 \overset{r_2}{\Longrightarrow} H_2)$ *in* produce-use dependency.

For a given graph transformation system, a *static dependency analysis* technique is a means to compute a list of all pairwise produce-use dependencies. Inspired by the related concept from term rewriting, *critical pair analysis* (CPA, [31]) has been the established static conflict and dependency analysis technique for three decades. CPA reports each dependency as a critical pair, that is, a minimal example graph together with a pair of rule applications from which a dependency arises. The *multi-granular static dependency analysis* [5,25] supports the computation of dependencies for graph transformation systems on three granularity levels: On binary granularity, it reports if a given rule pair contains a dependency at all. On coarse granularity, it reports *minimal dependency reasons*. Fine granularity is, roughly speaking, the level of granularity provided by critical pairs. Coarse-grained results have been shown to be more usable than fine-grained ones in a diverse set of scenarios and can be used to compute the fine-grained results much faster [27]. We use the coarse granularity also in the following.

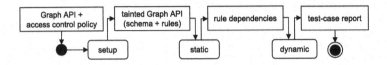

Fig. 3. The segments of the presented approach, illustrated as an activity diagram

3   Taint Analysis for Graph APIs

This section presents our main contribution, the taint analysis for Graph APIs focusing on broken access control. The analysis assumes the following input: a *Graph API schema* (or a Graph API endpoint from which the schema can be introspected—see Sect. 2.1) and a *role-based access control policy* description (henceforth referred to as policy) consisting of high-level access control requirements like the one mentioned in Sect. 2.2. The analysis comprises three segments: the setup, the static analysis, and the dynamic analysis—see Fig. 3. The *setup* segment comprises the derivation of a *tainted Graph API*, i.e, a Graph API *schema with tainted nodes* as well as *source and sink graph transformation rules* derived from the policy; we describe the setup in detail in Sect. 3.1. Once the setup is complete, the *static analysis* can be performed. The output of the static analysis is a set of *rule dependencies* which indicate potential occurrences of a broken policy. We present the formal concepts of the static analysis in Sect. 3.2. The completion of the static analysis enables the execution of the *dynamic analysis*. The dynamic analysis relies on the output of the static analysis and itself generates a *test-case report* which either verifies implementations of policy rules or exposes rules where the policy is broken. We present the formal concepts of the dynamic analysis in Sect. 3.3.

In the following, illustrations of type graphs and graph transformation rules are based on the well-known Henshin project [3], which is also used for tool support by the static analysis.

3.1   Setup

As mentioned in Sect. 1, the analysis relies on a formal description of a Graph API based on a graph transformation system, and focuses on broken access control. Therefore, our approach to taint analysis relies on the security analyst (i) obtaining a representation of the GraphQL schema as a type graph TG and the representation of the API, i.e, all possible API calls, as a set of graph transformation rules $\mathcal{R}$ typed over TG; (ii) having access to a role-based policy specification for the Graph API; (iii) identifying nodes in the Graph API which represent sensitive data to be checked for broken access control by the taint analysis. We formally define these concepts below.

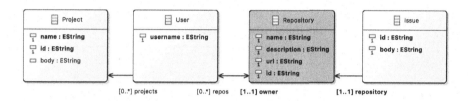

Fig. 4. The tainted type graph, where `Repository` is a tainted node (in gray)

Graph APIs. We formally represent a Graph API by a set of graph transformation rules $\mathcal{R}$ typed over TG. We represent each API call as a direct graph transformation. A sequence of API calls, denoted here as API execution, is a graph transformation of arbitrary length. It starts with a fixed initial graph G_0, representing the initial graph data structure of the API.

Definition 3 (graph API, call, execution). *A Graph API with a schema TG is a set of graph transformation rules $\mathcal{R} = \{r_i | i \in I\}$ typed over TG. A Graph API execution consists of a (possibly infinite) graph transformation sequence $(t : G_0 \overset{r_1}{\Rightarrow} G_1 \overset{r_2}{\Rightarrow} G_2 \overset{r_3}{\Rightarrow} G_3 \dots)$ starting with a fixed initial graph G_0, such that each r_i is an element of $\mathcal{R}$. A Graph API call is a direct transformation within a Graph API execution. We denote with $Sem(\mathcal{R})$ the set of all Graph API executions, and with $Sem_{call}(\mathcal{R})$ the set of all Graph API calls.*

An illustration of a type graph for the platform of the running example is shown in Fig. 4. The type graph corresponds to the entity types in the schema defined in the GraphQL schema language in Fig. 1. Based on the mutations defined within the distinguished type `Mutation` in the GraphQL schema, we also illustrate a minimal API according to Definition 3, i.e, a set of transformation rules typed over the type graph. The API consists of the rules *createRepo* and *updateRepo* in Fig. 5a, corresponding to the eponymous definitions in Fig. 1. In this section, we provide examples based on this minimal API; we extend the API in Sect. 4.

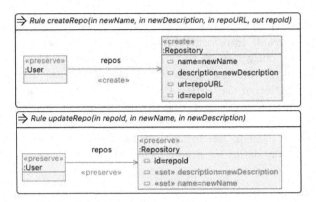

(a) Rules related to **Repository**

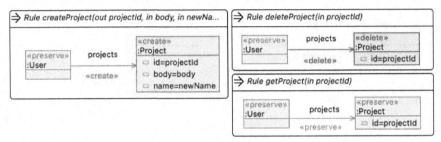

(b) Rules related to **Project**

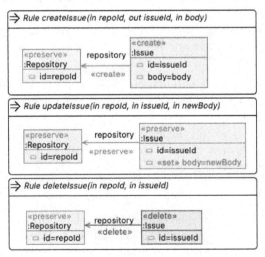

(c) Rules related to **Issue**

Fig. 5. The sources (*create* rules) and sinks (*update, get, delete* rules) used in the running example and Sect. 4

Access Control. In a Graph API, access control, i.e, the fact that certain API calls that may create or manipulate sensitive data can only be executed by users if they have sufficient privileges, can be implemented via a role-based policy. In that case, users of the Graph API can only execute certain calls if they belong to a group with the required privileges. For instance, according to the GitHub access control policy, an API user who has been granted the *scope*, i.e, group of permissions, `repo` has full control over the repository; an API user with the scope `repo_deployment` has access only to the deployment status of the repository and not to the code [12]. We formalize role-based access control policy and API calls whose access control is role-based below. The policy formalization is deliberately kept high-level since we want to support cases where only rather informal policy descriptions are given. Nevertheless, all (formal or informal) policies need to clarify which roles (if any) are allowed to perform a given API call.

Definition 4 (role-based Graph API call, access control policy). *Given a Graph API $\mathcal{R}$ with schema TG. A role specification $(RO, \leq_{RO})$ consists of a finite set of roles RO and a partial order $\leq_{RO} \subseteq RO \times RO$.*

A role-based Graph API execution $(t : G_0 \overset{r_1}{\Rightarrow} G_1 \overset{r_2}{\Rightarrow} G_2 \overset{r_3}{\Rightarrow} G_3 \ldots, role_t)$ is a Graph API execution such that each direct transformation in t is executed with a specific role as given in $role_t : \mathbb{N} \to \mathcal{P}(RO)$, where $role_t(i)$ refers to the subset of roles of the i-th direct transformation $G_i \overset{r_{i+1}}{\Longrightarrow} G_{i+1}$. A role-based Graph API call is a direct transformation within a role-based Graph API execution. We denote with $Sem^r(\mathcal{R})$ the set of all role-based Graph API executions, and with $Sem^r_{call}(\mathcal{R})$ the set of all roll-based Graph API calls.

A role-based access control policy $ACP = ((RO, \leq_{RO}), P)$ consists of a role specification $(RO, \leq_{RO})$ and a policy P, a total mapping $P : Sem^r_{call}(\mathcal{R}) \to \{true, false\}$.

As discussed in Sect. 2.2, a broken access control vulnerability occurs if policy requirements about user access to information and resources may not be fulfilled. In our running example, such a vulnerability occurs if users without the proper privileges get access or are able to manipulate specific elements in the repositories, e.g., issues or projects, or repositories themselves; this would result to a data leak or even the loss of valuable information in the code repository. Formally, a *broken access control vulnerability* is captured by a role-based Graph API execution containing a call that breaks the access control policy.

Definition 5 (broken access control vulnerability). *Given a graph API $\mathcal{R}$ with schema TG, and a role-based access control policy $ACP = ((RO, \leq_{RO}), P)$. A Broken Access Control (BAC) vulnerability is a role-based Graph API execution $(t : G_0 \overset{r_1}{\Rightarrow} G_1 \overset{r_2}{\Rightarrow} G_2 \overset{r_3}{\Rightarrow} G_3 \ldots, roles_t)$ entailing an API call $c : G_i \overset{r_{i+1}}{\Longrightarrow} G_{i+1}$ such that $P(c) = false$.*

Tainting. We taint nodes in the schema representing sensitive data objects. Based on these tainted nodes, we derive a tainted Graph API. A BAC vulnerability is related to tainted nodes if a Graph API call introduces an instance node of a tainted node, and a later call reads or manipulates this node without the necessary privileges. Later, we argue that via static analysis we can find all potential BAC vulnerabilities related to tainted nodes. We formalize these concepts below.

Definition 6 (tainted type graph, tainted nodes). *Given a type graph* $TG = (V_{TG}, E_{TG}, s_{TG}, t_{TG})$, *then* $TG_t = (TG, T)$ *is a* tainted type graph *with* $T \subseteq V_{TG}$ *the set of tainted nodes.*

A BAC vulnerability $(t : G_0 \overset{r_1}{\Rightarrow} G_1 \overset{r_2}{\Rightarrow} G_2 \overset{r_3}{\Rightarrow} G_3 \dots, roles_t)$ *related to a tainted node* s *from* T *entails a call* $c : G_i \overset{(m_{i+1}, r_{i+1})}{\Longrightarrow} G_{i+1}$ *such that a node* n *with* $type_V(n) = s$ *in the image of* m_{i+1} *exists, leading to* $P(c) = false$, *and* n *does not belong already to* G_0. *A* direct BAC vulnerability *related to a tainted node* s *with the call* $c : G_i \overset{(m_{i+1}, r_{i+1})}{\Longrightarrow} G_{i+1}$ *violating the policy creates the node* s *in the call* $G_{i-1} \overset{(m_i, r_i)}{\Longrightarrow} G_i$ *in* t.

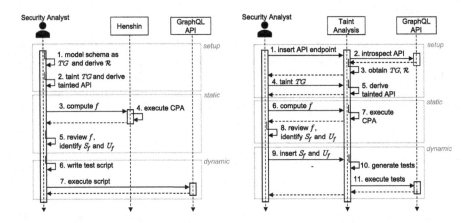

Fig. 6. An overview of the sequence of steps in the presented approach (left), and the vision for the automation of the presented approach (right)

See Fig. 4 for an example of a tainted schema, where we taint the node Repository—tainting is illustrated by the gray background of the node in question. In this case, this node is tainted because we want to further analyze if users without the proper privileges are able to manipulate created repositories.

Finally, to obtain a tainted graph API, we define *source rules*, i.e, rules that create instances of a tainted node in a schema, and *sink rules*, i.e, rules that read or manipulate instances of tainted nodes.

Definition 7 (source rule, sink rule, tainted graph API). *Given a tainted type graph* $TG_t = (TG, T)$ *and a Graph API* $\mathcal{R}$ *with schema* TG. *A rule* $r = (L \overset{le}{\hookleftarrow} K \overset{ri}{\hookrightarrow} R)$ *from* $\mathcal{R}$ *is a* source rule *if there exists a node* n *in* $R \setminus L$

such that $type_{R,V}(n) \in T$. *A rule* $r = (L \overset{le}{\hookleftarrow} K \overset{ri}{\hookrightarrow} R)$ *from* $\mathcal{R}$ *is a* sink rule *if there exists a node* n *in* L *such that* $type_{L,V}(n) \in T$. *A* tainted Graph API $(TG_t, \mathcal{R}, src, sink)$ *consists of a tainted type graph* TG_t, *a Graph API typed over* TG, *and a corresponding set of source and sink rules,* src *and* $sink$.

As an example of a tainted Graph API comprising a set of source and sink rules, see the rules *createRepo* and *updateRepo* in Fig. 5a. The rule *createRepo* creates a tainted node, i.e, an instance of `Repository`, and is therefore a source rule, whereas *updateRepo* refer to instances of `Repository` and is therefore a sink rule.

3.2   Static Analysis

Figure 6 (left) shows an overview of the sequence of steps in the presented approach to taint analysis. The setup segment comprised steps 1 and 2. The next segment is the static analysis, which is based on the output of the setup. The static analysis comprises the following steps: the provision of the schema and tainted Graph API obtained during setup by the security analyst (step 3 in Fig. 6); the execution of the actual static analysis over the input provided by the security analyst, which is performed based on available tool support (step 4); and the review of output of the analysis by the security analyst (step 5) so that potential broken access control vulnerabilities are identified due to an incomplete or incorrect access control policy. We describe this segment in detail below.

Owing to the graph-transformation-based formalization, the static analysis can automatically determine potential dependencies between the rules for a tainted Graph API by using CPA (Sect. 2.3). In particular, we use the coarse granularity level of the *multi-granular static dependency analysis* to determine all minimal *dependency reasons*. The *tainted information flow* for a tainted Graph API is then a symbolic yet complete description of all possibilities to use a tainted node after it has been created. In the running example, a dependency occurs when a `Repository` is created, which is afterward updated. The unique dependency reason between rules *createRepo* and *updateRepo* in Fig. 5a consists of the overlap of the complete right-hand side of the rule *createRepo* with the left-hand side of the rule *updateRepo*.

Definition 8 (tainted information flow). *Let* $(TG_t, \mathcal{R}, src, sink)$ *be a tainted Graph API. We denote with* $src(s)$ *($sink(s)$) the subset of rules in* src *($sink$) creating (using) a node of type* s *from* T, *respectively. Then, the tainted information flow of the API* $f = \cup\{DR(r_{src}, r_{sink}) | r_{src} \in src(s), r_{sink} \in sink(s), s \in T\}$ *with* $DR(r_{src}, r_{sink})$ *the set of dependency reasons for the rule pair* (r_{src}, r_{sink}).

Figure 7 shows the tainted information flow for the tainted Graph API from the running example—see dependency between *createRepo* and *updateRepo*.

A security vulnerability detection technique is *sound* for a category of vulnerabilities if it can correctly conclude that a program has no vulnerabilities of that category. In the following theorem, we argue that a static analysis technique computing the tainted information flow is sound in that it detects all *direct* potential BAC vulnerabilities related to tainted nodes. Devising a detection technique that is sound for potential *indirect* BAC vulnerabilities is part of our planned future work.

Theorem 1 (soundness). *Given a tainted Graph API ($TG_t, \mathcal{R}, src, sink$) with $TG_t = (TG, T)$ and tainted information flow $f = \cup\{DR(r_{src}, r_{sink}) | r_{src} \in src(s), r_{sink} \in sink(s), s \in T\}$. Given a direct BAC vulnerability ($t : G_0 \overset{r_1}{\Rightarrow} G_1 \overset{r_2}{\Rightarrow} G_2 \overset{r_3}{\Rightarrow} G_3 \ldots, roles_t$) related to a tainted node s from T, then $DR(r_{src}, r_{sink})$ in f is non-empty for some r_{src} in $src(s)$ and r_{sink} in $sink(s)$.*

Proof. There exists an API call $c : G_i \overset{(m_{i+1}, r_{i+1})}{\Longrightarrow} G_{i+1}$ in t violating the policy P for a node n in the image of m_{i+1} with $type_V(n) = s$ (Definition 6). We thus have that r_{i+1} equals a sink rule r_{sink} in $sink(s)$, since the preimage of n in the left-hand side of the rule r_{i+1} then also has type s (Definition 7). The call $G_{i-1} \overset{(m_i, r_i)}{\Longrightarrow} G_i$ creates n in t. This means that r_i equals a source rule r_{src} in $src(s)$, since it creates node n of type s being tainted (Definition 7). Because of completeness of dependency reasons [24], then there exists a dependency reason d in $DR(r_{src}, r_{sink})$ such that a node m of type s (instantiated to node n in t) is created by r_{src} and used by r_{sink} as described in d.   □

Effectively, each dependency reason in the tainted information flow constitutes the possibility of a vulnerability that calls for further investigation. But the tainted information flow may still contain *false positives*, i.e. issues that do not turn out to be actual vulnerabilities. Therefore, a manual review of the tainted information flow by the security analyst (step 5 in Fig. 6) as described in the following is necessary. The review helps the analyst identify true from false positives. The analyst reviews each dependency reason between source and sink rules contained in the tainted information flow against the access control policy of the API. The review has one of the following two outcomes. First, if the policy does not properly cover the tainted flow expressed by the dependency reason, it represents *unsecured tainted flow*; the policy thus needs to be adapted such that proper access control for this flow can be implemented; in this case, the dynamic analysis (see Sect. 3.3) can be used to prove that the dependency reason leads to a true positive, i.e., a BAC vulnerability, by demonstrating the occurrence of unsecured access to sensitive data. Alternatively, the dependency reason represents *secured tainted flow*; in this case, the dynamic analysis can be used to verify the correct implementation of the policy.

We formalize these two alternative outcomes below.

Definition 9 (secured and unsecured tainted information flow). *Given a tainted Graph API ($TG_t, \mathcal{R}, src, sink$), with tainted information flow $f = \cup\{DR(r_{src}, r_{sink}) | r_{src} \in src(s), r_{sink} \in sink(s), s \in T\}$ and access control policy $ACP = (RO, \leq_{RO}, P)$, then we can subdivide f in two subsets*

- S_f is the set of secured tainted flows *and consists of all dependency reasons in f for which the security analyst has confirmed the correctness of the ACP based on their manual review.*
- $U_f = f \setminus S_f$ the set of unsecured tainted flows *consisting of all dependency reasons for which the security analyst did not confirm the correctness of the ACP based on their manual review.*

Recall the policy requirement stated in Sect. 2.2 based on the running example: *a user may only update the information of a repository, if that user is the owner of the repository*; additionally, recall the tainted information flow of the running example containing the pair *(createRepo, updateRepo)*. Following a manual review, the security analyst can identify that the pair is covered by the policy. Therefore, in this case, $f = S_f = \{(createRepo, updateRepo)\}$.

3.3 Dynamic Analysis

The static analysis detects all potential BAC vulnerabilities, which may contain false positives, i.e., dependencies which do not cause vulnerabilities. Therefore, the static analysis is not *complete*; to achieve completeness, we complement the static analysis with the dynamic analysis. The dynamic analysis conducts actual API executions verifying BAC vulnerabilities detected by the static analysis; the dynamic analysis is complete, in that, all vulnerabilities it detects are indeed BAC instances. We describe the dynamic analysis segment of our approach below.

The dynamic analysis relies on the secured flows S_f and unsecured flows U_f identified during the static analysis—step 5 in Fig. 6 (left). Drawing on S_f and U_f, the security analyst defines test-cases (step 6 in Fig. 6) which:

- If a dependency reason belongs to S_f, test whether the policy has been implemented correctly for this dependency reason.
- If a dependency reason belongs to U_f, expose the BAC vulnerability stemming from this dependency reason or discover an implicit undocumented implementation of the policy.

We refer to these test-cases as *taint tests*, and formally define them below. We distinguish positive taint tests from negative taint tests. For positive (negative) tests we expect the policy to be fulfilled (violated) by the API executions.

Definition 10 (taint test). *Given a tainted Graph API* $(TG_t, \mathcal{R}, src, sink)$, *with* tainted information flow $f = \cup\{DR(r_{src}, r_{sink}) | r_{src} \in src(s), r_{sink} \in sink(s), s \in T\}$ *and access control policy* $ACP = (RO, \leq_{RO}, P)$, *a taint test* $((t, roles_t), access)$ *consists of a role-based Graph API execution* $(t, roles_t)$ *and an expected outcome represented by the boolean variable* $access \in \{true, false\}$.

The variable access equals false (true) if $(t, roles_t)$ *represents (or does not represent) a BAC vulnerability (see Definition 5). A taint test* $((t, roles_t), access)$ *is* positive *(negative) in case access equals true (false).*

For a systematic method to test-case definition, we define the *flow coverage*. This coverage criterion helps with ascertaining whether taint tests covering the tainted information flow are available. Flow coverage draws on similar approaches for covering rule dependencies [17, 35] in the context of testing.[1]

Definition 11 (flow coverage). *Given a tainted Graph API* $(TG_t, \mathcal{R}, src, sink)$, *with* tainted information flow $f = \cup\{DR(r_{src}, r_{sink})|r_{src} \in src(s), r_{sink} \in sink(s), s \in T\}$ *and access control policy* $ACP = (RO, \leq_{RO}, P)$. *A taint test* $((t, roles_t), access)$ *covers a dependency reason* d *in* $DR(r_{src}, r_{sink})$ *if for* $(t : G_0 \overset{r_1}{\Rightarrow} G_1 \overset{r_2}{\Rightarrow} G_2 \overset{r_3}{\Rightarrow} G_3 \ldots, roles_t)$ *there exists some* $i, j > 0$ *with* $i < j$ *such that* $r_i = r_{src}$ *and* $r_j = r_{sink}$, *and* t *creates and uses elements as described in the dependency reason* d.

A set of taint tests satisfies the (unsecured, or secured) *flow coverage if it entails a positive and negative test covering each dependency reason in* f *(U_f, or S_f, resp.).*

We call a taint test covering a dependency reason *minimal*, when it consists of the source rule application directly followed by the sink rule application.

Based on the flow coverage, a security analyst is in position to generate a set of concrete test-cases for a Graph API implementation. Their execution (step 7 in Fig. 6, left) comprises the execution of API calls, as prescribed by the graph transformation sequence captured by a taint test. We have realized steps 6 and 7 by capturing these concrete test-cases as unit-tests in a script which evaluates tests automatically—see Sect. 4. During evaluation, we observe if an expected exception is included in the API response, e.g., in GitHub such an exception indicates that scopes that have been granted to an access token are insufficient.

Regarding the test-case evaluation results and their correspondence to taint tests, we distinguish the following cases:

- A *positive* taint test $((t, roles_t), true)$ is *successful* if its execution does *not* lead to BAC exception.
- A *positive* taint test $((t, roles_t), true)$ *fails* if its execution leads to a BAC exception.
- A *negative* taint test $((t, roles_t), false)$ is *successful* if its execution leads to a BAC exception.
- A *negative* taint test $((t, roles_t), false)$ *fails* if its execution does *not* lead to a BAC exception.

The dynamic analysis is *complete*, in that, if a positive (or negative) taint test fails, then a BAC vulnerability has been detected. Depending on the context in which the security analysis of the Graph API is performed, if the dynamic analysis focuses on flows from S_f, the security analyst should expect to obtain successful positive and negative tests—a test evaluation that would not meet

---

[1] We focus in this first approach to *covering dependency reasons* for pairs of sink and source rules, instead of covering dependencies between all rule pairs, or even dependency paths in a dependency graph.

these expectations would indicate an incorrect policy implementation; if the dynamic analysis focuses on flows in U_f, the security analyst should expect to obtain failing positive or negative tests.

As an example, recall the (secured) tainted information flow obtained for the running example: $f = \{(createRepo,\ updateRepo)\}$. For covering the flow, a security analyst could design and perform a positive and a negative test. The positive test should entail the execution of the query *createRepo* (see Fig. 1) as a user with an owner role and subsequently the execution of *updateRepo* as the same user. The negative test should entail the execution of *createRepo* as a user with an owner role and subsequently the execution of *updateRepo* as a user that is not the owner. As in this case the flow is secured, the expected outcome is that the positive test should succeed, i.e, the test should lead to no BAC exception and the repository information should be updated, and so should the negative test, i.e, the test should lead to a BAC exception and the repository information should not be updated.

4   Application of the Taint Analysis to GitHub

We applied the presented taint analysis to a basic part of the GitHub API. In particular, we aimed to test part of the access permissions of personal accounts [9].

	createIssue	updateRepo	createProject	updateIssue	getProject	deleteIssue	createRepo	deleteProject
createIssue				CDR		CDDR		
updateRepo								
createProject				CDR				CDDR
updateIssue								
getProject								
deleteIssue								
createRepo	CDR	CDR						
deleteProject								

Fig. 7. The output of the dependency analysis (coarse) in Henshin for all the rules in Fig. 5; CDR stands for *Create Dependency Reason* and CDDR for *Create-delete Dependency Reason*

We completed the setup segment of the analysis (see Fig. 6) as follows. The application of the taint analysis to GitHub builds on the running example: it re-uses the schema in Fig. 4 and the rules *createRepo* and *updateRepo* which are based on the GitHub API GraphQL schema. Besides repositories, the documentation contains information regarding permissions of roles for projects and issues, whose unauthorized manipulation could also compromise the data security of the platform. Therefore, for the application, we also tainted the nodes `Project` and `Issue` in the schema in Fig. 4. According to queries and mutations in the GitHub

API schema, we also defined source and sink rules for the creation, deletion, and updating of instances of these newly tainted nodes. All rules are illustrated in Fig. 5.

We continued with the static analysis segment. As discussed earlier, we use the Henshin project for tool support of the static analysis. Specifically, the schema was manually captured in an Ecore file, whereas the rules were captured in a Henshin diagram. The CPA was then executed using the relevant Henshin feature, presented in [5]. An illustration of the output, i.e., the dependency reasons, of the CPA is shown in Fig. 7. Overall, the CPA returned the following dependencies, i.e, tainted information flow: (*createIssue, updateIssue*), (*createIssue, deleteIssue*), (*createProject, getProject*), (*createRepo, updateRepo*), (*createProject, deleteProject*), (*createRepo, createIssue*). All detected dependencies are addressed by the documentation for repositories, issues, and projects (see [9]), effectively rendering the tainted information flow equal to the set of secured tainted flows, i.e, $f = S_f$—see Definition 9.

For the dynamic analysis segment, we designed a set of positive and negative test-cases for the tainted flows obtained by the static analysis. The set satisfies (secured) flow coverage, as it contains twelve test-cases, i.e, a positive and a negative test case per dependency. Each test case entails a role-based API execution (see Definition 4), i.e, a sequence of API calls realized by GraphQL queries and mutations and executed using a given role. The test cases are *minimal*, i.e, an application of a source rule directly followed by an application of a sink rule. The set of available roles comprises a user who is the repository *owner* and a user who is invited to the repository as a *collaborator*. Additionally, in order to be able to test finer-grained aspects of the documentation, e.g., various possible roles of a collaborator in a project, we added another *collaborator with no permissions*—contrary to the collaborator mentioned above, who has all permissions. Positive tests entail the application of source and sink rules by roles with sufficient permissions, whereas negative tests entail the application of a source rule by a role with sufficient permissions and the application of the sink rule by a role with insufficient permissions.

We wrote a Python script containing the test cases in the form of unit-tests. The script requires three (classic) personal tokens which correspond to the users from above and are granted the appropriate scopes, i.e, groups of permissions: all scopes for the owner and collaborator, and no scopes for the collaborator with no permissions. The unit-tests were executed and evaluated using the Pytest library [21]; the execution of the API calls in the unit-tests was performed using the Python GraphQL client [22]. All positive tests were successful, i.e, they did not lead to a BAC exception (see Sect. 3.3), as were all negative tests, i.e, they led to a BAC exception, verifying that the GitHub access control policy was correctly implemented for the flows in question.

The Henshin artifacts, the Python script, and instructions to replicate the application of the proposed taint analysis in this section are available online [1].

5   Related Work

To the best of our knowledge, there currently does not exist any taint analysis approach for Graph APIs focusing on broken access control. Pagey et al. [29] present a security evaluation framework focusing also on access control vulnerabilities, but it is specifically dedicated to software-as-a-service e-commerce platforms.

Several *static and dynamic taint analysis* approaches for security testing of *mobile or web applications* exist, see e.g. [8,23,37]. These approaches also analyze the data flow of security-relevant or private data, but there is currently no approach dedicated to Graph APIs, nor broken access control in particular.

The following approaches are dedicated to *testing GraphQL APIs*, but do not focus on finding *security vulnerabilities* such as broken access control. Belhadi et al. [4] present a fuzzing approach based on evolutionary search. Karlsson et al. [19] introduce a property-based testing approach. Vargas et al. [38] present a regression testing approach. Zetterlund et al. [39] present an approach to mine GraphQL queries from production with the aim of testing the GraphQL schema.

There exist a number of approaches *formalizing and analyzing graph-based access control policies* [2,20,33]. For example, Koch et al. [20] use such a sophisticated formalization to compare and analyze different policy models. We decided to equip our taint analysis with a first relatively simple access control policy model. It lowers the barrier for applying our analysis and allow cases, where only natural language descriptions for access control policies are given. We plan to integrate more complex formalizations into our taint analysis in future work. Burger et al. [6] present an approach to detect (and correct) *vulnerabilities in evolving design models* of software systems. Graph transformation is used to formalize patterns used to detect security flaws. This approach is not dedicated specifically to Graph APIs, nor broken access control.

6   Conclusion and Future Work

We have presented a *systematic approach supporting security analysts to test for broken access control vulnerabilities in Graph APIs*. The security testing approach is based on taint analysis and consists of a static and dynamic segment. The static taint analysis primarily aims at validating the access control policies. The dynamic taint analysis aims at finding errors in the implementation of the (validated) access control policies.

Future work comprises *increasing the level of automation* of the taint analysis as illustrated in Fig. 6 (right). In particular, it is part of current work to provide automatic translations from Graph APIs such as GraphQL into tainted typed graph transformation systems. Naturally, the validation then still includes a manual review, but the formal specification could at least in parts be mined from the GraphQL implementation. There is also room for improving automation of the dynamic segment of our analysis. Currently, we provided automated test execution scripts, but we are also planning to explore the automation of the test generation step or coverage analysis for existing test suites. Moreover, we want to *evaluate the approach* on further and larger case studies and test its generality, also in the context of other relevant domains such as access control to graph databases. As mentioned in Sect. 3, we plan to devise a sound detection technique for potential *indirect vulnerabilities* related to broken access control. Moreover, as mentioned in Sect. 5, a further line of future work is the integration of more *detailed graph-based formalizations* of (other types of) access control. Finally, although for simplicity we omitted attribute manipulation and concepts like inheritance in the current formalization of Graph APIs, it seems feasible to incorporate such more sophisticated concepts based on the state-of-the-art for graph transformation.

References

1. Al Wardi, O., Khakharova, T., Sakizloglou, L., Lambers, L.: Supplement to Submission, v2 (2024). https://doi.org/10.5281/zenodo.11197370. Accessed 15 May 2024
2. Alves, S., Fernández, M.: A graph-based framework for the analysis of access control policies. Theor. Comput. Sci. **685**, 3–22 (2017). https://doi.org/10.1016/J.TCS.2016.10.018
3. Arendt, T., Biermann, E., Jurack, S., Krause, C., Taentzer, G.: Henshin: advanced concepts and tools for in-place EMF model transformations. In: Petriu, D.C., Rouquette, N., Haugen, Ø. (eds.) MODELS 2010. LNCS, vol. 6394, pp. 121–135. Springer, Heidelberg (2010). https://doi.org/10.1007/978-3-642-16145-2_9
4. Belhadi, A., Zhang, M., Arcuri, A.: Random testing and evolutionary testing for fuzzing GraphQL APIs. ACM Trans. Web **18**(1), 14:1–14:41 (2024). https://doi.org/10.1145/3609427
5. Born, K., Lambers, L., Strüber, D., Taentzer, G.: Granularity of conflicts and dependencies in graph transformation systems. In: de Lara, J., Plump, D. (eds.) ICGT 2017. LNCS, vol. 10373, pp. 125–141. Springer, Cham (2017). https://doi.org/10.1007/978-3-319-61470-0_8
6. Bürger, J., Jürjens, J., Wenzel, S.: Restoring security of evolving software models using graph transformation. Int. J. Softw. Tools Technol. Transf. **17**(3), 267–289 (2015). https://doi.org/10.1007/S10009-014-0364-8
7. Ehrig, H., Ehrig, K., Prange, U., Taentzer, G.: Fundamentals of Algebraic Graph Transformation. MTCSAES, Springer, Heidelberg (2006). https://doi.org/10.1007/3-540-31188-2
8. Enck, W., et al.: Taintdroid: an information flow tracking system for real-time privacy monitoring on smartphones. Commun. ACM **57**(3), 99–106 (2014). https://doi.org/10.1145/2494522
9. GitHub: Access permissions on GitHub. https://docs.github.com/en/get-started/learning-about-github/access-permissions-on-github. Accessed 08 Mar 2024
10. GitHub: GitHub GraphQL API community discussions. https://github.com/orgs/community/discussions/. Accessed 01 Mar 2024
11. GitHub: GitHub GraphQL API documentation. https://docs.github.com/en/graphql. Accessed 01 Mar 2024
12. GitHub: Scopes for OAuth apps. https://docs.github.com/en/apps/oauth-apps/building-oauth-apps/scopes-for-oauth-apps. Accessed 07 Mar 2024

13. The GraphQL Foundation: GraphQL | A query language for your API. https://graphql.org/. Accessed 05 Dec 2023
14. The GraphQL Foundation: Schemas and Types | GraphQL. https://graphql.org/learn/schema/. Accessed 08 Mar 2024
15. Graw, G.M.: Software Security: Building Security in. Addison-Wesley Professional, Boston (2007)
16. Heckel, R., Taentzer, G.: Graph Transformation for Software Engineers - With Applications to Model-Based Development and Domain-Specific Language Engineering. Springer, Cham (2020). https://doi.org/10.1007/978-3-030-43916-3
17. Hildebrandt, S., Lambers, L., Giese, H.: Complete specification coverage in automatically generated conformance test cases for TGG implementations. In: Duddy, K., Kappel, G. (eds.) ICMT 2013. LNCS, vol. 7909, pp. 174–188. Springer, Heidelberg (2013). https://doi.org/10.1007/978-3-642-38883-5_16
18. Hu, V.C., Kuhn, R., Yaga, D.: Verification and test methods for access control policies/models. Technical report, NIST SP 800-192, National Institute of Standards and Technology, Gaithersburg, MD (2017). https://doi.org/10.6028/NIST.SP.800-192
19. Karlsson, S., Causevic, A., Sundmark, D.: Automatic property-based testing of GraphQL APIs. In: 2nd IEEE/ACM International Conference on Automation of Software Test, AST@ICSE 2021, Madrid, Spain, 20–21 May 2021, pp. 1–10. IEEE (2021). https://doi.org/10.1109/AST52587.2021.00009
20. Koch, M., Mancini, L.V., Parisi-Presicce, F.: Graph-based specification of access control policies. J. Comput. Syst. Sci. **71**(1), 1–33 (2005). https://doi.org/10.1016/J.JCSS.2004.11.002
21. Krekel, H., dev team pytest: Pytest: Helps you write better programs — pytest documentation. https://docs.pytest.org/en/8.0.x/. Accessed 06 Mar 2024
22. Krinke, J.: Python-graphql-client: Python GraphQL Client
23. Kuznetsov, K., Gorla, A., Tavecchia, I., Gross, F., Zeller, A.: Mining android apps for anomalies. In: Bird, C., Menzies, T., Zimmermann, T. (eds.) The Art and Science of Analyzing Software Data, pp. 257–283. Morgan Kaufmann/Elsevier (2015). https://doi.org/10.1016/B978-0-12-411519-4.00010-0
24. Lambers, L., Born, K., Kosiol, J., Strüber, D., Taentzer, G.: Granularity of conflicts and dependencies in graph transformation systems: a two-dimensional approach. J. Log. Algebr. Methods Program. **103**, 105–129 (2019). https://doi.org/10.1016/j.jlamp.2018.11.004
25. Lambers, L., Born, K., Orejas, F., Strüber, D., Taentzer, G.: Initial conflicts and dependencies: critical pairs revisited. In: Heckel, R., Taentzer, G. (eds.) Graph Transformation, Specifications, and Nets. LNCS, vol. 10800, pp. 105–123. Springer, Cham (2018). https://doi.org/10.1007/978-3-319-75396-6_6
26. Lambers, L., Ehrig, H., Orejas, F.: Efficient conflict detection in graph transformation systems by essential critical pairs. Electron. Notes Theor. Comput. Sci. **211**, 17–26 (2008). https://doi.org/10.1016/j.entcs.2008.04.026
27. Lambers, L., Strüber, D., Taentzer, G., Born, K., Huebert, J.: Multi-granular conflict and dependency analysis in software engineering based on graph transformation. In: Proceedings of the 40th International Conference on Software Engineering, ICSE 2018, pp. 716–727. Association for Computing Machinery, New York (2018). https://doi.org/10.1145/3180155.3180258
28. Meta: Overview - Graph API - Documentation. https://developers.facebook.com/docs/graph-api/overview/. Accessed 01 Mar 2024

29. Pagey, R., Mannan, M., Youssef, A.: All your shops are belong to us: security weaknesses in e-commerce platforms. In: Proceedings of the ACM Web Conference 2023, WWW 2023, pp. 2144–2154. Association for Computing Machinery, New York (2023). https://doi.org/10.1145/3543507.3583319

30. Pezzè, M., Young, M.: Software Testing and Analysis - Process, Principles and Techniques. Wiley, Hoboken (2007)

31. Plump, D.: Critical pairs in term graph rewriting. In: Prívara, I., Rovan, B., Ruzička, P. (eds.) MFCS 1994. LNCS, vol. 841, pp. 556–566. Springer, Heidelberg (1994). https://doi.org/10.1007/3-540-58338-6_102

32. Quiña-Mera, A., Fernandez, P., García, J.M., Ruiz-Cortés, A.: Graphql: a systematic mapping study. ACM Comput. Surv. **55**(10), 202:1–202:35 (2023). https://doi.org/10.1145/3561818

33. Ray, I., Li, N., France, R.B., Kim, D.: Using UML to visualize role-based access control constraints. In: Jaeger, T., Ferrari, E. (eds.) 9th ACM Symposium on Access Control Models and Technologies, SACMAT 2004, Yorktown Heights, New York, USA, 2–4 June 2004, Proceedings, pp. 115–124. ACM (2004). https://doi.org/10.1145/990036.990054

34. Rozenberg, G. (ed.): Handbook of Graph Grammars and Computing by Graph Transformations, Volume 1: Foundations. World Scientific, Singapore (1997)

35. Runge, O., Khan, T.A., Heckel, R.: Test case generation using visual contracts. Electron. Commun. Eur. Assoc. Softw. Sci. Technol. **58** (2013). https://doi.org/10.14279/TUJ.ECEASST.58.847

36. Team OWASP Top 10: A01 Broken Access Control - OWASP Top 10:2021. https://owasp.org/Top10/A01_2021-Broken_Access_Control/. Accessed 06 Mar 2024

37. Tripp, O., Pistoia, M., Fink, S.J., Sridharan, M., Weisman, O.: TAJ: effective taint analysis of web applications. In: Hind, M., Diwan, A. (eds.) Proceedings of the 2009 ACM SIGPLAN Conference on Programming Language Design and Implementation, PLDI 2009, Dublin, Ireland, 15–21 June 2009, pp. 87–97. ACM (2009). https://doi.org/10.1145/1542476.1542486

38. Vargas, D.M., et al.: Deviation testing: a test case generation technique for GraphQL APIs. In: International Workshop on Smalltalk Technologies (IWST) (2018)

39. Zetterlund, L., Tiwari, D., Monperrus, M., Baudry, B.: Harvesting production graphql queries to detect schema faults. In: 15th IEEE Conference on Software Testing, Verification and Validation, ICST 2022, Valencia, Spain, 4–14 April 2022, pp. 365–376. IEEE (2022). https://doi.org/10.1109/ICST53961.2022.00014

40. Zhang, N., Zou, Y., Xia, X., Huang, Q., Lo, D., Li, S.: Web APIs: features, issues, and expectations - a large-scale empirical study of web APIs from two publicly accessible registries using stack overflow and a user survey. IEEE Trans. Software Eng. **49**(2), 498–528 (2023). https://doi.org/10.1109/TSE.2022.3154769

Tool and Blue Skies Presentations

Checking Transaction Isolation Violations Using Graph Queries

Stefania Dumbrava[1]([✉]), Zhao Jin[1], Burcu Kulahcioglu Ozkan[2],
and Jingxuan Qiu[2]

[1] ENSIIE, Paris, France
`stefania.dumbrava@ensiie.fr`
[2] Delft University of Technology, Delft, Netherlands

Abstract. Distributed databases provide different transaction isolation levels for higher performance and fault tolerance. However, implementing isolation models is challenging, and database systems can produce executions that violate their isolation guarantees. In this work, we propose GRAIL, a new approach that uses graph databases and queries to detect isolation violations expressed as anti-patterns in transactional dependency graphs. We implement the approach on top of the popular ArangoDB and Neo4j graph databases and show its efficiency through an experimental analysis of real executions of ArangoDB as a system under test.

Keywords: Graph Queries · Distributed Databases · Transactions

1 Introduction

Database isolation levels describe the degree to which the updates of a running transaction are isolated from other concurrent transactions. At the strongest level, the transactions are *serializable*, i.e., they produce an execution that is equivalent to running them serially in some order. While serializability provides strong guarantees, implementing it requires strong synchronization in databases with sharding and replication. To achieve higher performance, many databases support isolation levels weaker than serializability. However, it is difficult for them to ensure their claimed guarantees, as witnessed by the many violations discovered in popular distributed databases [24]. Moreover, checking the correctness of the executions for a given isolation level is challenging, e.g., checking serializability or snapshot isolation is NP-complete in general [9,28].

Database isolation violations can be detected by inspecting the *dependency graph* of database executions [7,28], which models read/write transaction dependencies between transactions. In this setting, violations can appear as anti-patterns indicating *isolation anomalies* prohibited in serializable databases. However, existing methods do not leverage their inherent graph structure.

Graph databases have been increasingly used to analyze interconnected data, due to their expressive graph models and custom query languages that allow to

© The Author(s), under exclusive license to Springer Nature Switzerland AG 2024
R. Harmer and J. Kosiol (Eds.): ICGT 2024, LNCS 14774, pp. 203–213, 2024.
https://doi.org/10.1007/978-3-031-64285-2_11

directly extract complex graph patterns [8,10,30]. As transaction dependencies and isolation anti-patterns are modeled by graphs, we investigate the feasibility and performance of graph queries to capture isolation violations. Such a method is easily extensible to analyzing a wider set of execution patterns.

We introduce GRAIL [13,14], a new GRAph-based Isolation Level checking approach that *is the first to use graph queries to detect database isolation violations*. We provide a proof-of-concept implementation of GRAIL through graph queries in AQL (using ArangoDB) and Cypher [16] (using Neo4j [27]). We used it to test ArangoDB [20] database executions, checking these for isolation anomalies. Our evaluation shows that the graph queries are effective and remain scalable with increasing execution lengths and transaction concurrency.

Related Work. Several methods and tools have been designed to check the strong consistency and isolation models, linearizability, and serializability of database transactions using enumerative exploration and/or SMT solvers [2,12,22,23,31, 34]. Recent works [9,19,34,36] focus on checking snapshot isolation (SI), which is weaker than serializability but still provides strong guarantees on transaction conflicts. Different from the existing work, GRAIL uses graph queries for pattern-matching and can check *a spectrum of isolation levels*, including PSI [11], PL-2 [1], and PL-1 [1].

2   Graph-Based Checking of Isolation Violations

Our approach for checking isolation violations using graph queries requires (i) collecting the execution histories of databases, (ii) constructing the dependency graphs from the execution histories, and (iii) checking the graphs for the existence of cycles with violation patterns. The main novelty of our approach lies in (iii), the detection of violations using graph database queries on dependency graphs. First, we generate and collect (i) the test executions using the Jepsen [23] tool. Then, we build the dependency graph (ii), as standard in the literature. A *dependency graph* $\mathcal{G} = \{\mathcal{H}, \mathsf{WR}, \mathsf{WW}, \mathsf{RW}\}$ is an execution history with a finite set of transactions $\mathcal{H} = \{T_1, \ldots, T_n\}$ and *read-dependency* (WR), *write-dependency* (WW), and *anti-dependency* (RW) relations (edges) between its transactions (nodes) [7]. We build the graph by creating a node for each transaction and placing the dependency edges between them, following the read/write operations in the execution history and the database write-ahead logs. Finally, (iii) we run graph database queries on the generated graphs, and we check for anti-patterns to detect violations.

2.1   Database Isolation Levels

Table 1 illustrates the allowed and disallowed serializability anomalies in the SER [28], SI [6], PSI [33], PL-2 [1], and PL-1 [1] transaction isolation levels.

Example 1. Consider an LDBC benchmark [15,35] schema example representing *Persons* who communicate through *Forum* posts. The forums are managed by

Table 1. Anomalies allowed/disallowed by the isolation levels

Level	Write skew	Long fork	Lost update	Dirty read	Anti-Pattern
SER	✗	✗	✗	✗	any cycle
SI	✓	✗	✗	✗	cycle without two consecutive RW edges
PSI	✓	✓	✗	✗	cycle with less than two RW edges
PL-2	✓	✓	✓	✗	cycle without RW edges
PL-1	✓	✓	✓	✓	cycle with only WW edges

moderators who can update their titles. The schema supports *read* and *write* operations to view and update forum titles and the list of moderators. We assume that the initial titles of the two forums *f1* and *f2* are *A1* and *B1*.

In Fig. 1, we present the isolation level anti-patterns as follows. Figure 1a illustrates a *long fork*, as the concurrent transactions T_1 and T_2 write to different objects (the titles of the two forums *f1* and *f2*) and commit, while subsequent transactions T_3 and T_4 only see the effects of one transaction. The WR edges capture that T_1 writes $A2$ to *f1*'s title, read by T_3, and that T_2 writes a value to another object, read by T_4. The RW edges mark that T_3 and T_4 read values $A1$ and $B1$, overwritten by T_1 and T_2. Figure 1b represents a *lost update* anomaly. T_1 and T_2 concurrently write to the forum moderators, adding a new one to the list (we overload the + operation as append). T_1 writes to the variable read by T_2 (the RW edge), and T_2 overwrites the value written by T_1 (the WW edge). The subsequent transaction T_3 only sees the effect of T_2, T_1's update being lost.

We consider the *dirty read* anomaly without restriction to reading from uncommitted transactions, but including *aborted reads*, *intermediate reads*, and *circular information flow* [1]. An aborted read anomaly occurs when a transaction reads a value written by an aborted transaction. An intermediate read anomaly occurs when a transaction reads any intermediate, uncommitted value from another transaction. These two anomalies can be identified by checking whether the values read by the transactions are committed. Circular information flow occurs when transactions have a circular dependency on the values read. Figure 1c shows an example of a *circular information flow*. For concurrent transactions T_1 and T_2, T_2 reads the value $A2$ written by T_1 and, at the same time, T_1 reads the value $B2$ written by T_2 (the WR edges).

The existence of anomalies in transaction executions can be analyzed using their dependency graph. As in early database systems works for conflict serializability [28], isolation level violations can be detected by searching for anti-patterns in the dependency graph. Table 1 lists the anti-patterns that indicate violations to levels SER, SI, PSI, PL-2, and PL-1.

2.2   History Dependency Property Graph Model

Our methodology consists of relying on graph databases for isolation level checking. Graph databases are NoSQL data stores that provide custom support

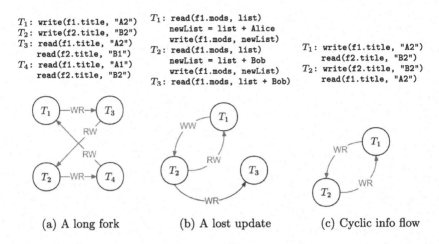

Fig. 1. Isolation anomalies and their graph patterns

for processing interconnected data, leveraging expressive graph models. Among these, the most prominent one is the *property graph* model [3], i.e., a multi-labeled multi-graph, whose nodes and edges can be additionally enriched by sets of key-value properties. We model the history dependency graph as a property graph and depict its corresponding schema [4] in Fig. 2a. Each node represents an event (`WriteEvent` or `ReadEvent`) or a `Transaction`. Event nodes are associated with the transactions that contain them through `BELONG_TO` edges and store their order within the transaction (`evt_order`) and their corresponding object (`var`). Event values are stored on `WriteEvent` nodes using the atomic `val` property, as these are written individually. For `ReadEvent` nodes, we store `val_list` and `val_register` properties, to account for list and register operations. `Event` nodes are interrelated by dependency edges (`EvtDepWW`, `EvtDepWR`, and `EvtDepRW`). We proceed analogously for `Transaction` nodes, explicitly storing, for ease of querying, the identifiers of the events they relate. Identifiers (`id`) for nodes and edges are automatically generated by the graph database.

As a standard graph query language has only just recently been published [18], graph databases have supported the extraction of expressive graph patterns through custom query languages. A leading one [32] is Neo4j's [27] Cypher [16], also used in numerous other databases, e.g., SAP HANA [29], Amazon Neptune [5], Memgraph [26], RedisGraph [25], and AgensGraph [21].

One of our checkers relies on Cypher queries to catch isolation level violations, e.g., the write skew anomaly in Fig. 2b. The reported violation exposes relevant information to the user, e.g., transaction IDs and dependency edge labels, helping with both understandability and explainability.

2.3 Graph Query-Based Anti-pattern Detection

We formulate multiple queries that will serve as checkers for isolation anti-pattern detection to investigate and compare their relative performances.

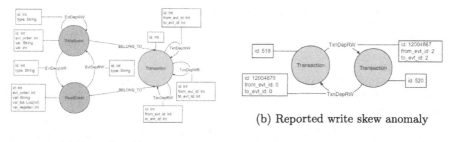

(a) Execution history

(b) Reported write skew anomaly

Fig. 2. Property graph schema of execution histories and example violation

Definition-Based Checkers. In ArangoDB, the `ArangoDB-Cycle` checker detects directed cycles using their definition: non-empty paths where only the first and the last vertices are equal. The corresponding AQL query performs a graph traversal on each transaction vertex. A cycle is detected when the traversal reaches the start vertex again, with a minimum depth of 2. We use ArangoDB's graph traversal, setting the minimum depth to 2 and the maximum one to 4 (see Listing 1.1). The maximum depth limits the number of vertices in a detected cycle and, therefore, becomes an important factor in the effectiveness of the checker. For example, setting the maximum depth to 4 means that the checker can only detect cycles with up to four transactions. The execution histories we collected do not contain any cycles with more than four transactions. As such, we set the maximum depth to 4, and it was sufficient for detecting the isolation anti-patterns (e.g., a cycle of length four in a *long fork* anomaly).

```
FOR start IN txn FOR vertex, edge, path IN 2..4 OUTBOUND start._id GRAPH
    txn_g
FILTER LAST(path.edges[*]._to) == start._id AND NOT REGEX_TEST(
    CONCAT_SEPARATOR(" ",path.edges[*].type),"(^rw.*rw$|rw rw)") LIMIT 1
RETURN path.edges
```

Listing 1.1. Checking SI: ArangoDB-Cycle Checker

Shortest-Path-Based Checkers. In ArangoDB, the `ArangoDB-SP` checker (Listing 1.2) uses the shortest path algorithm to detect cycles. It iterates over dependency edges, using `K_SHORTEST_PATH` to find all the so-called *back paths*, from its end vertex to its starting one. It then detects cycles, by trying to connect each edge to its back paths. Cycles that match the anti-pattern can be filtered using ArangoDB functionalities. In Neo4j, the `Neo4j-APOC` checker runs a Cypher query to detect all the cycles on the set of vertices in the dependency graph. As

ArangoDB-SP, its strategy is based on the shortest path algorithm, finding back paths and forming cycles. After the query returns all cycles, these are filtered to find anti-patterns for a certain isolation level (e.g., SI).

```
LET cycles = ( FOR edge IN dep
FOR p IN OUTBOUND K_SHORTEST_PATHS edge._to TO edge._from GRAPH txn_g
RETURN {edges: UNSHIFT(p.edges, edge),
       vertices: UNSHIFT(p.vertices, p.vertices[LENGTH(p.vertices) - 1])
   })


FOR cycle IN cycles
FILTER NOT REGEX_TEST(CONCAT_SEPARATOR(" ", cycle.edges[*].type), "(^rw.*
   rw$|rw rw)")
LIMIT 1 RETURN cycle
```

Listing 1.2. Checking SI: ArangoDB-SP

SCC-Based Checkers. In ArangoDB, the `ArangoDB-Pregel` checker uses the Pregel SCC algorithm to search for all strongly connected components (SCCs) within a dependency graph and filter those with at least two vertices. This ensures the existence of at least one cycle. In Neo4j, the `Neo4j-GDS-SCC` checker (Listing 1.3) runs a query to find all SCCs with Gabow's path-based SCC algorithm [17] and filters those with at least two vertices. For SER, PL-2, and PL-1, it first filters the graph to remove unwanted edges, as then any SCC will directly lead to an anti-pattern. For SI and PSI, it finds the SCCs and determines whether the subgraph with vertices only from an SCC can form a cycle.

```
CALL gds.alpha.scc.stream('g', {}) YIELD nodeId, componentId WITH
   componentId,
COLLECT(nodeId) AS ns, COUNT(nodeId) AS num WHERE num > 1 RETURN ns

CALL gds.graph.project.cypher('g','MATCH (n:txn) RETURN id(n) AS id',
'MATCH (n:txn)-->(n2:txn) RETURN id(n) AS source, id(n2) AS target')
```

Listing 1.3. Neo4j GDS SCC Checker

Challenges. A major challenge of implementing the checkers in graph databases is that each provides different functionalities. For example, ArangoDB does not support local SCC algorithms, except for the Pregel-based SCC one, while Neo4j supports Gabow's path-based algorithm, which can be directly called from Cypher queries. While this restricts implementations within a graph database, as GRAIL is intended to be system-agnostic, users can choose their framework. Second, database query languages are often not fully-fledged programming languages and their built-in data structures are not always suitable to particular user needs. For example, ArangoDB does not support stacks or hash sets, which makes it difficult to further accelerate the graph queries. As such, we cannot record previous values in a hash set to ensure $O(1)$ access; arrays are the only option. Also, user optimizations are further limited by missing stack structures and by the difficulty of implementing linear-time SCC algorithms, such as Tarjan's. Finally, the query functions of the same algorithm may take different arguments across graph databases and, thus, exhibit performance changes. For example, ArangoDB's shortest path algorithm traverses the whole graph and does not support additional filtering. However, Neo4j can filter a subgraph based on labels provided by users and then execute the traversal on a reduced edge space.

3   Experimental Analysis

We compare the performance of the graph queries presented in Sect. 2.3 for checking anti-patterns and evaluate their performance, effectiveness, and scalability compared to the state-of-the-art isolation checkers. We use execution histories collected on the cluster setting of ArangoDB with an increasing collection time of execution histories (with/without network partitions) and an increasing rate of submitted client transactions. More information about the datasets and their dependency graph characteristics can be found in our repository [14].

Configuration. We ran all experiments on Linux Mint 21 with AMD Ryzen 7 5800H CPU, Radeon Graphics × 8 GPU, and 15.5 GB RAM, using Neo4j Community Ed. 4.4.5, APOC 4.4.0.15, GDS 2.1.13, and ArangoDB 3.9.10.

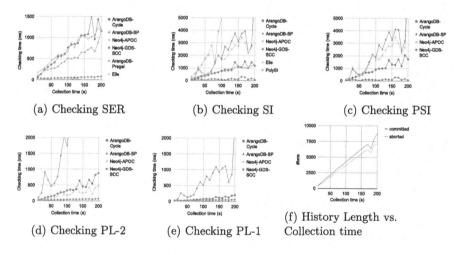

(a) Checking SER   (b) Checking SI   (c) Checking PSI

(d) Checking PL-2   (e) Checking PL-1   (f) History Length vs. Collection time

Fig. 3. Checking anomalies with increasing collection time of list histories

Experiments on Histories with Increasing Collection Time. Figure 3 shows the performance trends of different queries for checking SER, SI, PSI, PL-2, and PL-1 on increasing length of history collection. For `ArangoDB-Cycle`, we used the cycle-length bound $d = 4$, which suffices to detect all violating anti-patterns to all isolation levels in our experiments. Figures 3b–3e omit `ArangoDB-Pregel`, since it can only check for serializability.

Figure 3a shows the evolution of the checking time (**ms**) for SER with increasing collection time. `ArangoDB-Cycle` has the highest increase in analysis time because it traverses the graph to find all cycles with a polynomial complexity in the graph size. `ArangoDB-Pregel` exhibits similar polynomial behavior, as it runs BSP super step computations to search for SCCs with a linear cost for each step. However, `ArangoDB-SP`, `Neo4j-APOC` and `Neo4j-GDS-SCC`, have low analysis times with insignificant history length increases, as they search for the shortest paths and return as soon as they find a cycle.

For SI and PSI (Figs. 3b and 3c), the trends of `ArangoDB-Cycle` and `Neo4j-GDS-SCC` are similar to those for SER. `ArangoDB-SP` and `Neo4j-APOC` take longer to check SI and PSI than SER, as these are more complex and require more graph traversals. For PL-1 and PL-2, `ArangoDB-SP` shows a significant performance degradation, but `Neo4j-APOC` remains efficient and scalable as it combines shortest-path detection with filtering.

We observed that network partitions in the test executions do not result in more anti-patterns. With network faults, history length is reduced without otherwise affecting the dependency graphs. This can be different for other databases. Since ArangoDB does not guarantee serializability in the cluster setting, its executions exhibit isolation anomalies without introducing faults.

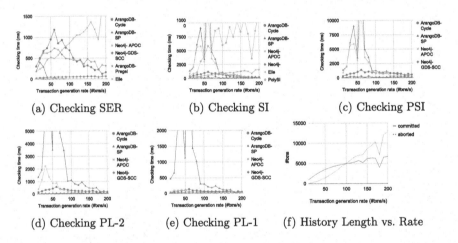

Fig. 4. Runtime for checking anomalies in the increasing rate of transactions

Experiments on Histories with Increasing Transaction Rate. Figure 4 shows the checkers' scalability for SER, SI, PSI, PL-2, and PL-1 when increasing the transaction generation rate. Overall, the relative performances of the checkers are similar to those of increasing history length. That is, when there are violations in the execution history, `ArangoDB-Cycle` requires more analysis time than the shortest-path and SCC-based checkers since it traverses the graph for each vertex. The performances of `ArangoDB-SP` and `Neo4j-APOC` degrade for execution histories without violations since they analyze the shortest paths between all vertices. `Neo4j-GDS-SCC` remains efficient for all isolation levels.

A significant difference from the scalability analysis in Fig. 3 is that increasing the transaction generation rate to 80 **txns/s** or higher *decreases* the analysis time for the graph-based checkers. Our analysis shows that this is caused by the decreasing density of dependency graphs. As we generate a higher number of concurrent transactions, more transactions conflict and abort. The graph density of datasets can be found in the online documentation [14].

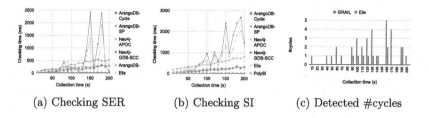

(a) Checking SER   (b) Checking SI   (c) Detected #cycles

Fig. 5. (a–b) Checking register histories (c) Cycle detection

Comparison to the State-of-the-Art Checkers. We compare the graph query-based checkers with state-of-the-art isolation checkers, i.e., Elle [2] for SER, and PolySI [19] for SI, which has been shown to outperform other SI checkers.

Figures 3a, 4a, 5a present the analysis time of checking SER with Elle and, Figs. 3b, 4b 5b presents for checking SI with Elle and PolySI. In the figures, we report the analysis time of the checkers that detect violations. Figure 5 shows the performance of checking SER and SI on register variables with increasing history length. We only present the plots for SER and SI for space reasons, the performance trends of the checkers for other isolation levels are similar to those in Fig. 3. These trends are also similar to those in Fig. 5, except for checking serializability with `ArangoDB-SP` and `Neo4j-APOC`. They need more time to check register histories without violations, since checking them requires a full exploration of the shortest paths. Graph query-based checkers are faster than Elle for histories with lists (Fig. 3a), except for `ArangoDB-Cycle`, due to the cost of its cycle detection by graph traversal on each vertex. For the histories with registers (Fig. 5a), Elle's performance stays more stable, while `ArangoDB-SP` and `Neo4j-APOC` have degraded performance, especially for executions without violations. Similarly, for SI (Figs. 3b and 5b), Elle outperforms `ArangoDB-SP` and `Neo4j-APOC`, has comparable performance with `ArangoDB-Cycle`, and performs worse than `Neo4j-GDS-SCC`. When increasing the transaction generation rate (Fig. 4a), Elle's performance significantly declines. Our results show that `Neo4j-GDS-SCC` significantly outperforms Elle in all cases. Figure 5c compares the number of cycles detected on the execution histories with registers. Elle can detect the isolation anomalies in the histories with list variables using their traceability and recoverability. We also use the database's write-ahead logs to infer dependencies on register variables, and thus can detect more violations.

Figures 3b, 5b, and 4b show that PolySI's performance degrades with increasing history length or transaction generation rate. PolySI introduces and uses the novel polygraph data structure, designed to characterize SI violations. PolySI generates polygraphs, recovering WR edges based on the *unique-writes assumption*, adds the RW edges based on the inferred version order, and then enumerates and prunes possible WW edges to recover the orders among transactions. This however largely expands the number of graphs to analyze.

4 Conclusion and Perspectives

We explore the feasibility and benefits of using graph databases to check isolation violations. As such, we introduce the GRAIL approach, which is *the first, to the best of our knowledge, to propose a generic methodology for isolation violation detection with graph queries*. We implement GRAIL through five checkers, using AQL and Cypher, and evaluate these against the state-of-the-art Elle and PolySI tools. Our results show that the graph query-based checkers provide comparative performance and remain scalable when increasing execution history length and transaction generation rates. The performance of the `ArangoDB-Cycle` definition-based checker highly depends on the characteristics of the analyzed execution, while SCC-based checkers exhibit less performance degradation than shortest-path-based ones, especially for executions without violations. Among these, `Neo4j-GDS-SCC` outperforms all other baselines.

Our empirical analysis also helps distill insights into the challenges of repurposing graph databases and queries as isolation checkers. Future work can analyze a more extensive set of patterns, including other consistency and isolation models. Other perspectives include extending our approach to runtime analyses of database executions and evolving graph processing systems.

References

1. Adya, A., Liskov, B., O'Neil, P.E.: Generalized isolation level definitions. In: ICDE, pp. 67–78. IEEE Computer Society (2000)
2. Alvaro, P., Kingsbury, K.: Elle: inferring isolation anomalies from experimental observations. Proc. VLDB Endow. **14**(3), 268–280 (2020)
3. Angles, R.: The property graph database model. In: AMW. CEUR Workshop Proceedings, vol. 2100. CEUR-WS.org (2018)
4. Angles, R., et al.: PG-schema: schemas for property graphs. Proc. ACM Manag. Data **1**(2), 198:1–198:25 (2023)
5. AWS. Amazon Neptune (2024). https://aws.amazon.com/neptune/
6. Berenson, H., Bernstein, P.A., Gray, J., Melton, J., O'Neil, E.J., O'Neil, P.E.: A critique of ANSI SQL isolation levels. In: SIGMOD Conference, pp. 1–10. ACM Press (1995)
7. Bernstein, P.A., Goodman, N.: Concurrency control in distributed database systems. ACM Comput. Surv. **13**(2), 185–221 (1981)
8. Besta, M., et al.: Demystifying graph databases: analysis and taxonomy of data organization, system designs, and graph queries. ACM Comput. Surv. **56**(2), 31:1–31:40 (2024)
9. Biswas, R., Enea, C.: On the complexity of checking transactional consistency. Proc. ACM Program. Lang. **3**(OOPSLA), 165:1–165:28 (2019)

10. Bonifati, A., Dumbrava, S.: Graph queries: from theory to practice. SIGMOD Rec. **47**(4), 5–16 (2018)
11. Cerone, A., Gotsman, A.: Analysing snapshot isolation. In: PODC, pp. 55–64. ACM (2016)
12. Clark, J.: Verifying serializability protocols with version order recovery (2021). https://doi.org/10.3929/ethz-b-000507577
13. Dumbrava, S., Jin, Z., Kulahcioglu Ozkan, B., Qiu, J.: GRAIL: checking transaction isolation violations with graph queries. In: ICSE Poster Track (2024). https://hal.science/hal-04475697
14. Dumbrava, S., Jin, Z., Ozkan, B.K., Qiu, J.: GRAph-based Isolation Level Checker (GRAIL) Artifact (2023). https://github.com/jasonqiu98/GRAIL-artifact
15. Erling, O., et al.: The LDBC social network benchmark: interactive workload. In: SIGMOD Conference, pp. 619–630. ACM (2015)
16. Francis, N., et al.: Cypher: an evolving query language for property graphs. In: SIGMOD Conference, pp. 1433–1445. ACM (2018)
17. Gabow, H.N.: Path-based depth-first search for strong and biconnected components. Inf. Process. Lett. **74**(3–4), 107–114 (2000)
18. GQL. GQL graph query language (2024). https://www.gqlstandards.org/
19. Huang, K., et al.: Efficient black-box checking of snapshot isolation in databases. Proc. VLDB Endow. **16**(6), 1264–1276 (2023)
20. Inc., A.: ArangoDB (2023). https://www.arangodb.com/
21. Inc., B.G.: AgensGraph (2024). https://bitnine.net/agensgraph/
22. Kingsbury., K.: Gretchen: offline serializability verification. In: clojure (2022). https://github.com/aphyr/gretchen
23. Kingsbury, K.: Jepsen (2022). http://jepsen.io/
24. Kingsbury, K.: Jepsen analyses (2022). https://jepsen.io/analyses
25. Labs, R.: RedisGraph (2017). https://oss.redislabs.com/redisgraph/
26. Memgraph. Memgraph (2024). https://memgraph.com/
27. Neo4j. Neo4j (2023). https://neo4j.com/
28. Papadimitriou, C.H.: The serializability of concurrent database updates. J. ACM **26**(4), 631–653 (1979)
29. Paradies, M.: Graph pattern matching in SAP HANA (2017). https://tinyurl.com/ycxu54pr
30. Sakr, S., et al.: The future is big graphs: a community view on graph processing systems. Commun. ACM **64**(9), 62–71 (2021)
31. Sinha, A., Malik, S., Wang, C., Gupta, A.: Predicting serializability violations: SMT-based search vs. DPOR-based search. In: Eder, K., Lourenço, J., Shehory, O. (eds.) HVC 2011. LNCS, vol. 7261, pp. 95–114. Springer, Heidelberg (2012). https://doi.org/10.1007/978-3-642-34188-5_11
32. solidIT: DB-Engines Ranking (2024). https://db-engines.com/en/ranking_trend/graph+dbms
33. Sovran, Y., Power, R., Aguilera, M.K., Li, J.: Transactional storage for geo-replicated systems. In: SOSP, pp. 385–400. ACM (2011)
34. Tan, C., Zhao, C., Mu, S., Walfish, M.: Cobra: Making transactional key-value stores verifiably serializable. In: OSDI, pp. 63–80. USENIX Association (2020)
35. Waudby, J., Steer, B.A., Karimov, K., Marton, J., Boncz, P., Szárnyas, G.: Towards testing ACID compliance in the LDBC social network benchmark. In: Nambiar, R., Poess, M. (eds.) TPCTC 2020. LNCS, vol. 12752, pp. 1–17. Springer, Cham (2021). https://doi.org/10.1007/978-3-030-84924-5_1
36. Zhang, J., Ji, Y., Mu, S., Tan, C.: Viper: a fast snapshot isolation checker. In: EuroSys, pp. 654–671. ACM (2023)

Can I Teach Graph Rewriting to My Chatbot?

Reiko Heckel[1(✉)] and Issam Al-Azzoni[2(✉)]

[1] University of Leicester, Leicester, UK
rh122@le.ac.uk
[2] Al Ain University, Al Ain, UAE
issam.alazzoni@aau.ac.ae

Abstract. Chatbots based on Large Language Models have been making progress not just in traditional natural language-processing tasks but also in generating from informal requirements formal artefacts such as models and programs. In this short paper we report on experiments in the use of GPT-4 for creating graph rewriting systems. Through two case studies, we guide the bot in the design, extension and analysis of type graphs and rewrite rules in the GXL format of the Groove graph rewriting tool, which we use to visualise our models. While the current experiments are entirely manual, they provide anecdotal evidence for the potential of using large language models in creating and analysing graph rewriting systems and show how a simple tool integration could support a natural language interface for graph rewriting tools.

1 Introduction

Since the rise of ChatGPT and in particular GPT-4 [1], there is considerable interest in how Large Language Models (LLMs) can be used in software development [5,9]. The general consensus is that while chatbots can do well at university-level programming problems, they make too many mistakes to be used for the generation of production code without close expert supervision.

Due to the level of detail, reviewing and correcting automatically generated code is complex and error prone in itself, so an alternative is for domain experts and analysts to use LLMs to generate higher-level models, confirm their correctness and then use them in a model-driven approach to derive implementations using model transformations. Apart from being easier to review, models also raise the level of abstraction, allowing more concise expression of requirements and design solutions. This has the potential to improve the performance of LLMs who impose limits on the number of tokens in input prompts and responses.

Graph rewriting is an expressive high-level modelling language with a precise semantics, tool support for execution and analysis and a visual representation to support expert review of models. Our question is whether chatbots can be instructed to create, modify and analyse graph rewriting systems.

We report on two experiments using GPT-4 based on a simple data structure model of binary trees and a more application-oriented example of a voting system

R. Harmer and J. Kosiol (Eds.): ICGT 2024, LNCS 14774, pp. 214–222, 2024.
https://doi.org/10.1007/978-3-031-64285-2_12

both realised as graph rewriting system in Groove [7]. We introduce our approach in Sect. 2, discuss our experiments in Sects. 3 and 4, and the related work in Sect. 5. Section 6 concludes the paper.

2   Approach

Although LLMs are starting to support image data, they perform better on text-based interactions. We use Groove's textual representation for type graphs, start graphs and rules based on the graph exchange format GXL. GPT-4 allows to provide a knowledge base of background documents, so we uploaded the book [8], the Groove manual and quick reference guide, and a basic binary tree model (see Fig. 1 below) with comments[1].

Prompting the bot with a type graph or a graph rewrite system, the bot is tasked with generating or extending type graph and rules based on short, informal descriptions. The generated artefacts are subject to review and correction. This iterative process relies on the background material and prompting until the results are deemed correct. Groove's utility in this context is primarily for confirming the correct syntax and visualising the rules to ensure they align with the informal requirements provided.

Where needed, we prompted the bot with a condensed version of the generated model, seeking modifications and extensions to the existing structure. This reduced the length of the conversation and so increases the performance of the bot. Finally, the bot is prompted with the condensed final model to request an analysis of its properties and adherence to specified requirements.

3   Binary Trees

We start with a basic model of binary trees represented by a type graph with a single node type and one attribute for the number of a node's children. We define saveral rules as examples of basic rule features, such as creating and deleting nodes and edges, testing and updating attributes, and negative application conditions.

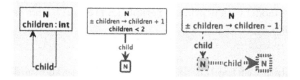

Fig. 1. Binary tree model: type graph, and rules *addChild* and *remChild*

---

[1] https://drive.google.com/file/d/18wq9s7HKsT8B1DZjchUYU8iH0-ZV5HqP/view?usp=sharing.

The first task was to extend this model by adding a *depth* attribute and generating rules for computing the depth of each node:[2]

Create a Groove graph rewrite system extending the type graph and rules shown below [...] Proceed as follows:

1. update the type graph by adding an attribute depth of type integer to the node type N
2. create a rule to set the depth attribute of a node n to 0
3. create a rule to set the depth attribute of a node n to 1 if n has a child, but n does not have a child $n1$ such that $n1$ has another child n2
4. create a rule to set the depth attribute of a node n to $x + 1$ if n has a child $n1$ whose depth attribute is $n1.x$, and such that there is no other child $n2$ with depth attribute $> n2.x$.

Create the type graph and rules in Groove XML notation.

This prompt was followed by GXL representations of the type graph and rules in Fig. 1 with an explanation of the rules and some general instructions on Groove's use of the GXL syntax, especially the notation for types and instructions such as `new:`, `del:`, `not:`, attribute tests and updates.

The bot correctly extended the type graph and created the rules under 2 and 3, but for the rule under 4 did not know Groove's notation using node IDs to refer to an attribute of another node. Instead, it implemented the instruction using `test: depth = x` in one node and the update `let: depth = x+1` in another. After being corrected as follows

To access attribute x of n1, write n1.x in the condition at n2, and add a self loop with label id:n1 to node n1. You can then drop the test at n1. Can you integrate that change into the rule incrementDepth to provide an updated version of the entire GXL code for that rule?

it produced the correct solution visualised using Groove as shown in Fig. 2.

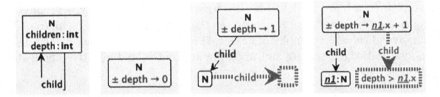

Fig. 2. Binary tree model with depth: type graph, and rules *setDepthZero*, *setDepthOne* and *incrementDepth*

To explore the ability of the bot to analyse and explain the model, we asked

---

[2] Full conversation available at https://chat.openai.com/share/e4579d80-aab3-4e7a-8ee2-9240adde5481. Formatting added for readability.

> What does this model compute as the depth attribute of the root node of a binary tree, assuming that we apply rules in the order 2. - 4. such that each rule is applied for as long as possible before moving on to the next rule.

and received the following reply:

> The model computes the depth attribute of the root node of a binary tree
> by progressively determining the depth of each node based on its position
> in the tree and the depth of its children. The application of rules in the
> specified order (2 to 4) ensures that each node's depth is calculated in a
> bottom-up manner, starting from the leaf nodes up to the root. Here's how
> the depth is computed through these rules:
>
> 1. **Rule 2**: Sets the depth of a node to 0. This rule would initially target
>   leaf nodes, as they do not have any children.
> 2. **Rule 3**: Sets the depth of a node to 1 if the node has a child, but does
>   not have a child with another child. Essentially, this rule is applied
>   to nodes that are parents of leaf nodes, marking them with a depth of 1,
>   indicating they are one level above the leaves.
> 3. **Rule 4** (updated as per your request): Sets the depth of a node n to $x+1$
>   if n has a child $n1$ whose depth attribute is x, and there is no other
>   child $n2$ with a depth attribute greater than x. This rule systemati-
>   cally calculates the depth for each node by looking at the maximum depth
>   of its children and adding one to it. The application of this rule pro-
>   gresses from the leaf nodes upwards towards the root, ensuring that each
>   node's depth is correctly calculated based on its children's depth.
>
> By applying these rules in the given order and ensuring each rule is
> applied as long as possible before moving to the next, the depth of the
> root node is computed as the maximum height of the binary tree. The depth
> attribute at the root node will reflect the longest path from the root to a
> leaf node, effectively calculating the height of the tree. [...]

To summarise, apart from specific features of the Groove syntax such as node IDs and their use in attribute expressions, the bot was able to perform the required tasks, but we had to go through a number of experiments to design the prompts in a sufficiently clear way and provide background materials illustrating and explaining the syntax. In some cases, clear instructions, such as that every node must have a type, were ignored. This is visible in the negative application condition of rule *setDepthOne* in Fig. 2, which is nevertheless correct Groove GXL syntax and has the desired effect for the applicability of the rule.

The discussion of the way the rules compute the depth attribute is convincing and detailed, even if it may be derived from both the informal description of the rules and their formal expression in the Groove GXL syntax. The bot is clearly able to work with both representations (it generated the GXL code for the rules itself) so it may be as much use in creating models from informal requirements as in explaining them.

It is worth noting that the results are not deterministic, i.e., faced with the same prompts at different times GPT produced somewhat different results. This makes a systematic study of its capabilities difficult, and it shows that, while it may be a useful tool in the hands of an expert, it is far from being able to perform these tasks autonomously.

4   A Voting System

As a second case study we developed a simple voting system model. Here we only provided the type graph and instructions for the creation of rules, but no example rules as guidance (Fig. 3).[3]

Can you create rewrite rules in Groove format conforming to the type graph below, with the following preconditions, negative application conditions, and effects:

- *createVoter*: creates a new voter
- *createMotion*: create a new motion and initilise its attribiute to 0 or false
- *issueInvite*: create a new invite for an existing voter and motion and increase the count of eligible voters by 1, under the negative application condition that no invite for this motion and voter exists, and that the motion is not open for voting
- *openVoting*: set the open attribute for an existing motion to true, but only if false
- *voteYes*: voter votes in favour of a motion, if the motion is open for votes and the voter has been invited to vote on it. After the vote is cast, the invitation to vote is deleted.

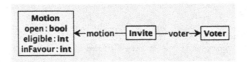

Fig. 3. Type graph for voting system

In the response, all rules were created correctly apart from *issueInvite*, where in the negative application condition the node to ensure there was only one invite for each voter and motion lacked the `type:Invite` label. After requesting this to be added the rule was correctly updated as shown in Fig. 4.

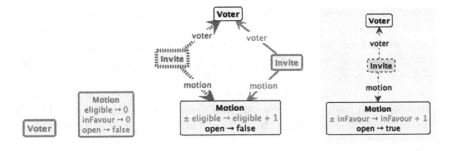

Fig. 4. Rules for voting system: *createVoter*, *createMotion*, *issueInvite*, *voteYes*

We then wanted to explore how to evolve a rewrite system by adding new requirements.

---

[3] https://chat.openai.com/share/dc77d599-2d96-484a-aeb7-df125fe05b33.

> Can you add an attribute 'carried' to Motion, that is false until the motion is carried, when more than half of the eligible voters voted in favour. Recreate the type graph and the 'createMotion' rule, and define a new rule 'carried', such that the motion is closed once it is carried.

This led to a correct extension of the type graph and rules as shown in Fig. 5.[4]

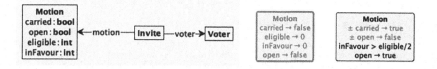

Fig. 5. Updated type graph and rules *createMotion, carried*

To analyse the system, we then asked:

> Assuming that we start our voting model from an empty graph, then proceed to create voters, motions and invites, etc, is it possible for a voter to vote twice on the same motion?

but the answer was inconclusive and somewhat evasive:

```
Based on the rules and definitions provided in the document related to the
voting model, it is not explicitly mentioned that a voter cannot vote twice
on the same motion. [...] the specific rule to prevent double voting on the
same motion would depend on the implementation details of these operations,
particularly how votes are cast and recorded.
In a typical voting system, each voter would have a unique identifier, and
each motion would track the voters who have already cast their votes. This
could be implemented through a relation or attribute in the graph model
that records whether a vote has been cast by a particular voter for a spe-
cific motion. [...]
Without explicit rules or attributes in the provided document to prevent
double voting, one cannot conclusively say that the model prevents a voter
from voting twice on the same motion. [...]
```

We then decided to start a new conversation, providing the model generated thus far as input to ask for the analysis of the possibility of double voting, and received a well-reasoned reply.[5]

---

[4] https://chat.openai.com/share/67a340d9-2a15-4a6c-bf54-341bd680f5fc.
[5] https://chat.openai.com/share/879456a3-1401-40eb-9fba-44b1a0fe1ba9.

Based on the provided type graph, graph transformation rules, and the voting system's structure, it appears designed to prevent a voter from voting twice on the same motion. Here's why:

1. **Invite Mechanism:** The system uses an invite mechanism (*issueInvite* rule) where an *Invite* node is created to link a *Voter* node to a *Motion* node. This *Invite* node acts as a permission or a token that allows a voter to vote on a specific motion. The rule to issue an invite includes a negative application condition to ensure no existing invite for the same voter and motion exists before creating a new one. This setup inherently prevents duplicate votes by requiring a unique invite for each voting action.
2. **Vote Casting and Invite Deletion:** The *voteYes* rule allows a voter to vote in favor of a motion only if an invite exists (i.e., the voter has been invited to vote on the motion). Once the vote is cast, the corresponding *Invite* node is deleted (**del:** label on the *Invite* node within the *voteYes* rule). This deletion means the invite cannot be used again, effectively preventing the voter from casting another vote on the same motion.
3. **Motion Open Check:** Before a vote can be cast, the motion must be open (**test:open=true** in the *voteYes* rule). This check, coupled with the invite mechanism, ensures that votes are cast in a controlled manner.

The system's design around the invite mechanism, coupled with the requirements for a motion to be open and the subsequent deletion of invites upon voting, forms a robust method to prevent double voting on the same motion. A voter cannot vote twice on the same motion because once they use their invite to vote, the invite is deleted, and without an invite, they cannot vote again.

Like in the binary tree example, the bot was ultimately able to create the required rules, adapt them, and analyse the model. Also here, the quality of the results varied over time, and the conversations reported are the more successful ones. It is worth noting that the prompt in this case study did not include any rules, so they were generated entirely by the bot without having seen an example in the prompt itself. This means that it must have learned the rule syntax from either the background resources in the knowledge base or from publicly available sources. The quality of responses was found to degrade with the length of the conversation, an effect we were able to remedy by condensing the final results of a conversation on creating and adapting the model into a new prompt for the analysis task.

5   Related Work

To our knowledge, the use of LLMs in graph rewriting has not been studied so far, but we discuss work on their use in modelling more generally.

In [11], the authors explore the effectiveness of LLMs in generating models that can be passed directly to other frameworks to create a complete web application. LLMs are used to transform natural language description into models of a domain-specific language in the context of model-driven software engineering. Several tests are performed to ensure the semantic and syntactic correctness of the created models.

Arulmohan *et al.* [2] explore the use of LLMs to extract domain models from textual agile product backlogs. Their work demonstrates the potential of

LLMs and tailored Natural language Processing (NLP) in automated requirements engineering. The authors in [4] present an approach for fully automated domain modelling using LLMs. Given a textual description of a domain, the approach aims to derive a complete domain model without any human intervention. Their experimental results show that while LLMs demonstrate excellent domain modelling capabilities, they are still impractical for full automation.

Petrović and Al-Azzoni [12] present an approach for leveraging ChatGPT for the purpose of automated smart contract generation. The approach is a model-driven approach, where a contract specified using a smart contract modelling language is used as input for prompt generation. The evaluation demonstrates the great potential of LLMs in supporting model-driven approaches.

In [6], the authors introduce a new method that uses LLMs, trained on natural language text and code and fine-tuned on proofs, to generate proofs for theorems. This shows the applicability of LLMs in automating formal software verification. More generally, the work by Belzner *et al.* [3] provides an overview of the potential benefits and challenges of adopting LLMs in software engineering, including requirements engineering, system design, code, and test generation.

6   Conclusion

We conducted experiments to assess the potential of GPT-4 to create, extend, analyse and explain complex graph rewriting models. Using the Groove's GXL notation and the Groove tool for syntax checks and visualisation, we were able to create what we consider to be convincing models and analysis results.

Results were not reliably reproducible, sometimes depending on the time of the day, which leads us to believe that performance parameters may be adjusted by the bot's provider depending on demand. We found that the bot tends to make mistakes comparable to a human student, omitting elements requested, claiming that a task is too difficult, or that certain requirements cannot be expressed in Groove syntax, only to provide a good solution at a later run. This syntax, which is not for human consumption, may also add to the challenge. A dedicated human readable textual notation may be more usable also for a bot.

The results show that there is potential for the use of bots in creating and analysing graph rewriting systems, but a more reliable implementation is required for a systematic study. There are already open source models available to download and run locally and they can be customised by techniques such as Retrieval Augmented Generation (RAG) [10] where background documentation can be stored in a special database to augment prompts with additional context.

Less than reliable results are also not a problem if we can verify them automatically. This has already been attempted here by using Groove to verify the syntax, but one could imagine a more semantic verification using a range of other analysis methods, from testing if a rule is or is not applicable to a given graph, via executing certain scenarios and examining the results, to full model checking and the static verification of invariants. This would require a deeper tool integration, considering the bot's analyses as hypotheses to be tested, generating relevant analysis queries, such as start graphs, control expressions, temporal formulas or invariants, evaluating them using graph rewrite tools and translating the results back into feedback and further instructions for the bot.

References

1. GPT-4 Technical Report. arXiv preprint: arXiv:2303.08774 (2024)
2. Arulmohan, S., Meurs, M.J., Mosser, S.: Extracting domain models from textual requirements in the era of large language models. In: 2023 ACM/IEEE International Conference on Model Driven Engineering Languages and Systems Companion (MODELS-C), pp. 580–587. IEEE (2023)
3. Belzner, L., Gabor, T., Wirsing, M.: Large language model assisted software engineering: prospects, challenges, and a case study. In: Steffen, B. (ed.) AISoLA 2023. LNCS, vol. 14380, pp. 355–374. Springer, Cham (2023). https://doi.org/10.1007/978-3-031-46002-9_23
4. Chen, K., Yang, Y., Chen, B., López, J.A.H., Mussbacher, G., Varró, D.: Automated domain modeling with large language models: a comparative study. In: 2023 ACM/IEEE 26th International Conference on Model Driven Engineering Languages and Systems (MODELS), pp. 162–172. IEEE (2023)
5. Fan, A., et al.: Large language models for software engineering: survey and open problems. arXiv preprint arXiv:2310.03533 (2023)
6. First, E., Rabe, M., Ringer, T., Brun, Y.: Baldur: whole-proof generation and repair with large language models. In: ACM Joint European Software Engineering Conference and Symposium on the Foundations of Software Engineering, pp. 1229–1241 (2023)
7. Ghamarian, A.H., de Mol, M., Rensink, A., Zambon, E., Zimakova, M.: Modelling and analysis using GROOVE. Int. J. Softw. Tools Technol. Transf. 14(1), 15–40 (2012). https://doi.org/10.1007/s10009-011-0186-x
8. Heckel, R., Taentzer, G.: Graph Transformation for Software Engineers - With Applications to Model-Based Development and Domain-Specific Language Engineering. Springer, Cham (2020). https://doi.org/10.1007/978-3-030-43916-3
9. Hou, X., et al.: Large language models for software engineering: a systematic literature review. arXiv preprint arXiv:2308.10620 (2023)
10. Lewis, P.S.H., et al.: Retrieval-augmented generation for knowledge-intensive NLP tasks. In: Larochelle, H., Ranzato, M., Hadsell, R., Balcan, M., Lin, H. (eds.) Advances in Neural Information Processing Systems 33: Annual Conference on Neural Information Processing Systems 2020, NeurIPS 2020, 6–12 December 2020, virtual (2020). https://proceedings.neurips.cc/paper/2020/hash/6b493230205f780e1bc26945df7481e5-Abstract.html
11. Netz, L., Michael, J., Rumpe, B.: From natural language to web applications: using large language models for model-driven software engineering. In: Modellierung 2024, pp. 179–195. Gesellschaft für Informatik eV (2024)
12. Petrović, N., Al-Azzoni, I.: Model-driven smart contract generation leveraging ChatGPT. In: Selvaraj, H., Chmaj, G., Zydek, D. (eds.) ICSEng 2023. LNNS, vol. 761, pp. 387–396. Springer, Cham (2023). https://doi.org/10.1007/978-3-031-40579-2_37

A Graph Transformation-Based Engine for the Automated Exploration of Constraint Models

Christopher Stone(✉) , András Z. Salamon , and Ian Miguel

School of Computer Science, University of St Andrews, St Andrews, UK
{cls29,Andras.Salamon,ijm}@st-andrews.ac.uk

Abstract. In this demonstration, we present an engine leveraging graph transformations for the automated reformulation of constraint specifications of combinatorial search problems. These arise in many settings, such as planning, scheduling, routing and design. The engine is situated in the Constraint Modelling Pipeline that, starting from an initial high-level specification, can apply type-specific refinements while targeting solvers from multiple paradigms: SAT, SMT, Mixed Integer Programming, and Constraint Programming. The problem specification is crucial in producing an effective input for the target solver, motivating our work to explore the space of reformulations of an initial specification.

Our system transforms a constraint specification in the ESSENCE language into an Abstract Syntax Tree (AST). These ASTs, considered as directed labelled graphs, serve as inputs to the graph transformation language GP2 (Graph Programs 2) for subsequent reformulation. Our engine currently employs a curated set of handcrafted rewrite rules applied sequentially to the ASTs by the GP2 framework. It is designed to learn the efficacy of various rewrites, prioritising those that yield superior performance outcomes. At this stage, our primary emphasis is ensuring the rewritten specifications' soundness and semantic invariance.

Central to our methodology is constructing a search graph, where nodes represent model specifications and solutions, while edges represent graph transformations and solver performance. Through this search graph our system enables the exploration of constraint specification variants and the evaluation of their effects on lower-level refinement strategies and solvers. Finally, we present a visualisation tool that allows the interactive inspection of the search graph and its content.

Keywords: Model Transformation · Constraint Programming · Graph Search

1 Introduction

Combinatorial search problems arise in many and varied settings. Consider, for example, a parcel delivery problem. Parcels and drivers must be assigned to delivery vehicles respecting capacity constraints, staff resources, shift patterns

© The Author(s), under exclusive license to Springer Nature Switzerland AG 2024
R. Harmer and J. Kosiol (Eds.): ICGT 2024, LNCS 14774, pp. 223–238, 2024.
https://doi.org/10.1007/978-3-031-64285-2_13

and working preferences. At the same time routes must be chosen so as to meet delivery deadlines. The problem is further complicated by the need also to optimise an objective, such as minimising fuel consumption.

Such problems are naturally characterised as a set of decision variables, each representing a choice that must be made to solve the problem at hand (such as which items are assigned to the first van leaving the depot), and a set of constraints describing allowed combinations of variable assignments (for instance, a vehicle cannot be assigned to particular route depending on its specific height or weight). A solution assigns a value to each variable satisfying all constraints. Many decision-making and optimisation formalisms take this general form, including: Constraint Programming (CP), Propositional Satisfiability (SAT), SAT Modulo Theories (SMT), and Mixed Integer Programming (MIP). Metaheuristic methods offer a complementary approach, sacrificing completeness for the ability to explore rapidly a neighbourhood of candidate solutions. These approaches have much in common, but differ in the types of decision variables and constraints they support, and the inference mechanisms used to find solutions.

In all of these formalisms the *model* of the problem is crucial to the efficiency with which it can be solved. A model in this sense is the set of decision variables and constraints chosen to represent a given problem. A model typically has parameters in order to represent a parameterised problem class, with a particular instance of the class derived by giving values for each of its parameters. A *solution* to a problem instance is then an assignment of a value to each decision variable that satisfies all of the constraints. To avoid confusion we note that the notion of a model and solution in our domain correspond broadly to the notions of metamodel and model in the Model-Driven Engineering literature [51]. There are typically many possible models and formulating an effective model is notoriously difficult. Therefore automating modelling is a key challenge.

This work extends our *Constraint Modelling Pipeline*, visible in Fig. 1. A user writes a problem specification in the abstract constraint specification language ESSENCE [21], capturing the structure of the problem in terms of familiar discrete mathematical objects such as sets, relations, functions, and partitions, but *without* committing to detailed modelling decisions. The CONJURE [2] automated modelling system refines an ESSENCE specification into a solver-independent constraint model where the abstract structure is represented by a collection of primitive variables and constraints. The second stage in the pipeline, performed automatically by SAVILE ROW [40], is in preparing the model for a particular constraint solver for one of the formalisms listed above. By following different refinement pathways the pipeline can produce a set of alternative models from a single ESSENCE specification for evaluation or selection via machine learning. We have situated our work in the ESSENCE language and constraint modelling pipeline because ESSENCE allows problems to be expressed in a concise, highly structured manner. We exploit this structure to trigger the transformations we consider.

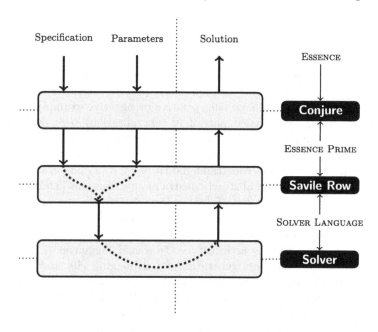

Fig. 1. Diagram of the Constraint Modelling Pipeline, from [2]

The constraint modelling pipeline eliminates the need for specific modelling expertise in constraint programming and SAT solving. However, the pipeline is sensitive to the way a problem is specified in ESSENCE. This is because CONJURE's refinement rules are triggered directly from the abstract variables and expressions in the ESSENCE specification, therefore determining the reachable set of models. This motivates our work herein: an automated reformulation system to explore the space of ESSENCE specifications and therefore increase the size of the set of models from which we can select. Our tool uses graph formalisms throughout taking ESSENCE specifications as inputs and producing alternative ESSENCE specifications as output. An ESSENCE specification is represented as a graph; all reformulations of an ESSENCE specification are graph transformations. Specific properties can be preserved by transformations (such as semantic invariance), and rewrites are applied only when well-defined conditions are present. The scenarios with the highest potential benefits are *class* level transformations, where a large portion of the instances of a given parameterised problem class could be positively affected. The effort expended in rewriting and evaluating the specification of a given class is then amortised over all of the instances of the class that are subsequently solved - similar to the training cost versus prediction/inference cost split in machine learning.

This does not exclude the use of our system on single instances when these are not solvable as is or enough knowledge is available to reliably attempt alternative specifications. Furthermore, our system uses a persistent data structure holding the history of previous reformulations, so supporting the accumulation of data for subsequent learning.

2   Related Work

To solve a constraint model of a problem efficiently, it is crucial to first choose an effective formulation for the model [20]. Various approaches have therefore been proposed to automate the modelling process. Methods learn models from, variously, natural language [30], positive or negative examples [3,10,17], queries (membership, equivalence, partial, or generalisation) [6–8], or from arguments [46]. Other approaches include: automated transformation of medium-level solver-independent constraint models [37–41,45,50]; automatic derivation of implied constraints from a constraint model [9,14,15,23,33]; case-based reasoning [34]; and refinement of abstract constraint specifications [22] in languages such as ESRA [19], ESSENCE [21], $\mathcal{F}$ [25] or Zinc [31,35,44].

Model transformation by means of graph transformations [36] has been applied successfully in a large variety of domains within software engineering, from databases and software architectures to visual languages [29,48]. Their application on automated test generation for debugging [49] and automated software refactoring using local search are particularly relevant [43], making its underlying mechanism an ideal candidate for implementing a system that allows us to explore variants of a constraint model rigorously.

Weber [52] also implements a persistent data structure for a graph rewriting system, a feature at the heart of our system necessary for accumulating data in the prospect of subsequent learning. Automatic search of design spaces via machine learning approaches is highly desirable and has been sought-after in several fields, such as chip design [27] and has been implemented in the context of graph substitution guided by a neural network in [28].

As opposed to [13], we only consider graph rewrites that maintain consistency and therefore do not require repairing the specification after rewrites. Consistency-preserving configuration operators have also been investigated for configurable software [26]. The automated search for rewrite operators using graph transformations has been explored in the context of software engineering [11,18], a highly desirable feature and direction that should naturally be integrated into our current tool.

3   System Description

We now describe our overall system and the main format converters.

3.1   Overview

Reformulating Essence Specifications via Graph Transformation. The toolkit's core objective is rewriting ESSENCE constraint specifications via graph transformations. This is implemented via an ESSENCE parser that constructs an Abstract Syntax Tree (AST) of the specification, which preserves syntactic information by storing it as attributes and is then translated into a GP2 graph. Once translated, any available GP2 rewriting rule can be applied. The question then arises as to how to explore the space of ESSENCE specifications. Since transformations can be chained sequentially, this process can be seen as balancing exploration and exploitation. For example, should a reformulated specification be reformulated (exploited) further or should we explore a new branch of the possible reformulation pathways? This process can be treated as a multi-armed bandit problem [32], informed by evaluating the models produced from the reformulated specification on training instances of the specified problem, which must be selected or generated so as to be *discriminating* [1, 16]. A similar approach has been taken to deciding how to add *streamliner* constraints to constraint models, designed to aggressively focus search on promising areas of the search space [47]. Evaluation results are stored persistently to support the accumulation of knowledge and effective reformulations.

Persistent Knowledge Graph. The exploration produces a persistent knowledge graph where nodes represent ESSENCE specifications and solutions, while edges represent graph transformations, and evaluation results. The graph is implemented using the Python package NetworkX [24] as an attributed nodes and edges list. In practice, starting from a single node representing the initial specification, the knowledge graph is iteratively expanded by successfully applying a GP2 rewrite. Each new specification is evaluated, for a set of parameters if necessary, by refining and solving it, then adding new nodes and edges to the knowledge graph. A simple graph union operation can join knowledge graphs created during separate runs. The knowledge graph can be formatted as JSON or visualised using a web browser thanks to a custom procedurally generated HTML visualisation. The graph is translated into Force-Graph [4], based on D3.js [12], which allows the visual inspection of the knowledge graph, the models and solutions that have been generated. This has already proven useful for examining the behaviour of the system and diagnosing programming errors during development.

Format Translation Suite. Preparatory to the above, we have developed a collection of format converters that form a graph of available translations, pictured in Fig. 2, that allows the conversion of the AST into NetworkX, GP2, Force-Graph, JSON, and can convert back to ESSENCE itself. Some translations are necessary as intermediate steps to reach the GP2 language, while the complete suite allows users to access many tools for visualising, storing and analysing graphs. As the path from the GP2 language to NetworkX and JSON does not check or enforce compliance with ESSENCE grammar the suite brings several further benefits beyond our specific goals. For example, the suite can be used independently to visualise and analyse arbitrary graphs specified in the GP2 language. Likewise, labelled NetworkX graphs can be converted to GP2, giving access to an extensive collection of graphs and graph generating functions.

Python API Hub. We implement our system in Python to facilitate interfacing with heterogeneous software, particularly CONJURE and GP2, acting as a portable, easy-to-access and easy-to-use hub. This enables users to run our system from Jupyter Notebooks and Colab, two popular choices for developing and sharing prototypes. Our wrapper around GP2 provides several quality-of-life features for managing graph transformations. For example, it can detect if a graph transformation rule is already compiled, automated on-the-fly compilation, or using pre-compiled programs generated with GP2.

3.2   Format Converters

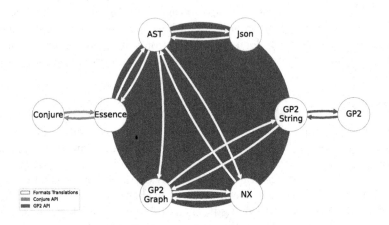

Fig. 2. Format Converters Graph

We now describe some of the main format converters in our system. ESSENCE, NetworkX(NX), GP2, and JSON are well-established formats and languages, while our particular implementation of the AST, GP2Graph and all conversions are novel contributions.

Essence string → Essence AST: parses an ESSENCE specification and returns a tree, as well as performing syntax and grammar checking. The AST is a typed rooted labelled directed tree where each node object stores its adjacency list. More formally it is a tuple $T : (N, A, n, t, l, A_n)$

- N and A are a set of *nodes* and ordered lists of *nodes*.
- $n : N \rightarrow A$ maps each node to an indexable ordered list of *nodes*.
- $t : N \rightarrow OBJECT$ maps each node to a set of types associated with ESSENCE's grammar.
- $l : N \rightarrow STRING$ maps each node to a label holding an atomic unit of the ESSENCE language.
- $A_n : N \rightarrow STRING$ maps each node to a set of attributes associated with ESSENCE's grammar. This is redundant with t, but facilitates format conversion mechanics while t facilitates the engineering of the parser.

Essence AST → NetworkX(NX): Converts a tree of python objects into a NetworkX graph by recursive traversal starting from the root. When translating to NetworkX format it returns an attributed directed graph with vertex and edge labels. The node types are preserved as vertex attributes, and the children ordering is stored as edge attributes. More formally, the tree is turned into a tuple $G : (V, E, A_v, A_e, L_v)$ where

- V is a set of vertices.
- $E \subseteq \{(u, v)|u, v \in V\}$ is a set of edges as ordered tuples with u source and v target.
- $A_v : V \to STRING$ are attributes associated with each vertex, storing grammatical information.
- $A_e : E \to INTEGER$ are attributes associated with each edge, holding the order of the target node.
- $L_v : V \to STRING$ are labels mapped to each vertex holding an atomic unit of the ESSENCE language.

NetworkX or Essence AST → GP2Graph: When transforming a graph into GP2 graph format vertices are tuples of indices and labels while edges are quadruples of indices, source, target and labels. There are no ESSENCE specific checks during this conversion and so it can be used for arbitrary graphs. Formally a GP2 graph is a tuple $G : (V_i, E_i, L_v, s, t, L_e)$

- V_i is a set of indexed vertices.
- E_i is an indexed set of edges with s source, t target vertices from V_i.
- $L_v : V \to STRING$ are labels associated with each vertex, storing the grammatical information and atomic element.
- $L_e : E \to INTEGER$ are attributes associated with each edge, holding the order of the target node.

NetworkX → Essence AST: When converting from a NetworkX graph to an ESSENCE tree, the tree's root must be found. We do this by topological sort, and start from the root to reconstruct the tree recursively.

Essence AST → Essence String: Compiling back to ESSENCE is done by recursively traversing the tree and assembling the string according to the ESSENCE syntax. We call this operation *icing* as it mostly involves adding syntactic sugar.

Essence AST ↔ JSON: We pack and unpack the nested Python objects using a standard JSON library.

In practice, the converters are not used in isolation but via an ESSENCE Format Graph (EFG) module, with formats as vertices and converters as callable edges constructed by scanning and harvesting the functions in the Python module holding the converters via code introspection. Formally, it is a tuple $EFG : (V_f, E_c)$ where

- V_f is a set of vertices corresponding to each available format, which act as indices.
- $E_c : V \to CALLABLE$ is a pair $u, v \subseteq V_f$, a callable function that converts the input from the source format u to the target format v.

The EFG module is used by specifying the source and target formats, then the shortest path is computed, and the input object is iteratively translated. For example, by using the method `EFG.FormToForm(input, ''NX'', ''GP2String'')`, the path NX→GP2Graph→GP2String will be found. Then, the *input* is passed to the source of the callable edge NetworkX → GP2Graph and then to the GP2Graph → GP2String, after these two conversions the outcome is returned.

4   Transforming Models with Graph Transformations

The driving engine of our system is an implementation of GP2 [5], a rule-based graph programming language which uses double pushout graph transformations. For a detailed explanation of the syntax and semantics of the language, see [42]. A Double Pushout Rule $p = (L, K, R)$ is a triplet of graphs known as the left-hand side L that describes the precondition of the rule p, the right-hand side R that describes its post-condition, and an interface K, the intersection of L and R, to apply the rule p. Applying the rule to a graph G means finding a match m of L in the graph. This is used to create the context graph $D = (G - m(L)) \cup m(K)$, in such a way that gluing of L and D via K is equal to G, finally gluing R and D via K leads to the final graph H.

In our system, the source graphs are always graph representations of valid ASTs of the ESSENCE language, and we ensure that the rules we apply produce valid ESSENCE ASTs after transformation. To control exactly which rules are applied, we keep each rule in separate files, trigger them one at a time, record if it is applied to the target graph and all the effects this has to the rest of the system. This includes the effects on CONJURE and SAVILE ROW, where the transformation could cause significant variations in the space or time required to refine the specification before passing it to the solver.

4.1   Concrete Example

We provide an illustrative simple example of an ESSENCE specification and a graph transformation rule written in GP2 that can be applied to it. In the specification 1.1, we have a Boolean parameter a, an integer parameter b bounded between 1 and 10, a constant c equal to 5 and a decision variable d, which is a Boolean that should satisfy the constraint described by the Boolean expression $a =!(d \land (b > c))$. The model's AST in its GP2 form, after all necessary format conversions have been applied, can be seen in Appendix A in Fig. 4.

In Fig. 3, we show the graphical form of a De Morgan rule implemented using GP2. Its raw string format can be found in Appendix A in Listing 1.3. Applying the rule to the initial specification Listing 1.1 leads to the new specification in Listing 1.2. The new specification is semantically equivalent, and the transformation has left intact the subtree of the expression $b > c$ as desired.

Listing 1.1. Original Specification

```
given a : bool
given b : int(1..10)
letting c be 5
find d : bool
    such that
        a = !(d /\ (b>c))
```

Listing 1.2. Rewritten Specification

```
given a : bool
given b : int(1..10)
letting c be 5
find d : bool
    such that
        a = (!d \/ !(b>c))
```

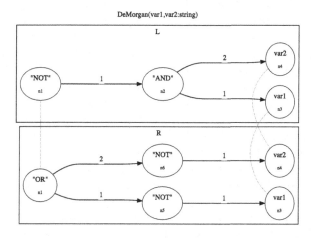

Fig. 3. De Morgan Rule implemented in GP2. Nodes n1,n3,n4 are the interface nodes.

5   Final Remarks and Future Directions

With this demo we presented a new engine for exploring the space of reformulations of a constraint specification of a combinatorial search problem. The
system allows the systematic testing of the Constraint Modelling Pipeline, the
search for improved models of ESSENCE specifications, the evaluation of reformulations that maintain the same level of abstraction, and gathering data useful
for training model selection algorithms.

In the near term, we envision the system evolving towards the automated
production of rewrite rules, starting from a pair of specifications. This allows
the incorporation of good existing examples and expert modellers' knowledge
without handcrafted rules but by rule derivation. As we accumulate data, we
will integrate machine learning algorithms and data driven approaches able to
select models and drive the search for new improved models.

Our system is fundamentally designed around graphs. By introducing the
ability to modify models and components via graph rewriting, we gained adapt-
ability, reliability, and efficiency. As we extend our range of graph transformations, we expect to significantly advance the system's capabilities.

A   Appendix

Listing 1.3. A De Morgan rule implemented in GP2

```
Main = demorgFromNOTconjunction
demorgFromNOTconjunction ( operand1 , operand2 : string )
//LHS
[
    ( n1 , "NOT˜ UnaryExpression " )
    ( n2 , "AND˜ BinaryExpression " )
    ( n3 , operand1 )
    ( n4 , operand2 )
    |
    ( e1 , n1 , n2 , 1)
    ( e2 , n2 , n3 , 1)
    ( e3 , n2 , n4 , 2)
]
=>
//RHS
[
    ( n1 , "OR˜ BinaryExpression " )
    ( n3 , operand1 )
    ( n4 , operand2 )
    ( n5 , "NOT˜ UnaryExpression " )
    ( n6 , "NOT˜ UnaryExpression " )
    |
    ( e4 , n1 , n5 , 1)
    ( e5 , n1 , n6 , 2)
    ( e6 , n5 , n3 , 1)
    ( e7 , n6 , n4 , 1)
]
interface =
{
  n1 , n3 , n4
}
```

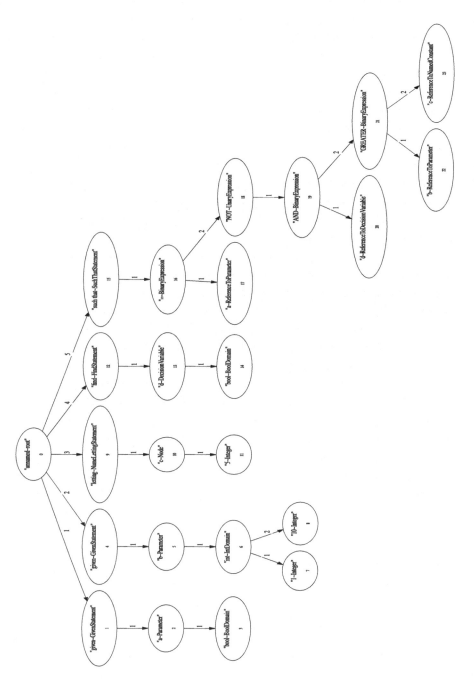

Fig. 4. Example specification in GP2 format before application of the rewrite

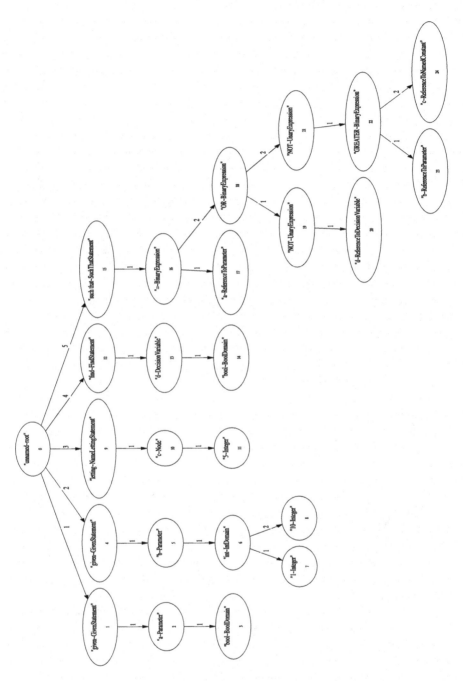

Fig. 5. Example specification in GP2 format after application of the rewrite

References

1. Akgün, Ö., Dang, N., Miguel, I., Salamon, A.Z., Spracklen, P., Stone, C.: Discriminating instance generation from abstract specifications: a case study with CP and MIP. In: Integration of Constraint Programming, Artificial Intelligence, and Operations Research: 17th International Conference, CPAIOR 2020, pp. 41–51 (2020). https://doi.org/10.1007/978-3-030-58942-4_3
2. Akgün, Ö., Frisch, A.M., Gent, I.P., Jefferson, C., Miguel, I., Nightingale, P.: Conjure: automatic generation of constraint models from problem specifications. Artif. Intell. **310**, 103751 (2022). https://doi.org/10.1016/j.artint.2022.103751
3. Arcangioli, R., Bessiere, C., Lazaar, N.: Multiple constraint aquisition. In: IJCAI, pp. 698–704 (2016). https://www.ijcai.org/Proceedings/16/Papers/105.pdf
4. Asturiano, V.: Force-directed graph rendered on HTML5 canvas (2024). https://github.com/vasturiano/force-graph
5. Bak, C., Plump, D.: Compiling graph programs to C. In: Echahed, R., Minas, M. (eds.) ICGT 2016. LNCS, vol. 9761, pp. 102–117. Springer, Cham (2016). https://doi.org/10.1007/978-3-319-40530-8_7
6. Beldiceanu, N., Simonis, H.: A model seeker: extracting global constraint models from positive examples. In: CP, pp. 141–157 (2012). https://doi.org/10.1007/978-3-642-33558-7_13
7. Bessiere, C., Coletta, R., Daoudi, A., Lazaar, N.: Boosting constraint acquisition via generalization queries. In: ECAI, pp. 99–104 (2014). https://doi.org/10.3233/978-1-61499-419-0-99
8. Bessiere, C., et al.: Constraint acquisition via partial queries. In: IJCAI, pp. 475–481 (2013). https://www.ijcai.org/Proceedings/13/Papers/078.pdf
9. Bessiere, C., Coletta, R., Petit, T.: Learning implied global constraints. In: IJCAI, pp. 44–49 (2007). https://www.ijcai.org/Proceedings/07/Papers/005.pdf
10. Bessiere, C., Koriche, F., Lazaar, N., O'Sullivan, B.: Constraint acquisition. Artif. Intell. **244**, 315–342 (2017). https://doi.org/10.1016/j.artint.2015.08.001
11. Bill, R., Fleck, M., Troya, J., Mayerhofer, T., Wimmer, M.: A local and global tour on MOMoT. Softw. Syst. Model. **18**(2), 1017–1046 (2019). https://doi.org/10.1007/s10270-017-0644-3
12. Bostock, M., Ogievetsky, V., Heer, J.: D3: Data-driven documents. IEEE Trans. Visualization & Comp. Graphics (Proc. InfoVis) (2011). http://idl.cs.washington.edu/papers/d3
13. Burdusel, A., Zschaler, S., John, S.: Automatic generation of atomic multiplicity-preserving search operators for search-based model engineering. Softw. Syst. Model. **20**(6), 1857–1887 (2021). https://doi.org/10.1007/s10270-021-00914-w
14. Charnley, J., Colton, S., Miguel, I.: Automatic generation of implied constraints. In: ECAI, pp. 73–77 (2006). https://ebooks.iospress.nl/volumearticle/2653
15. Colton, S., Miguel, I.: Constraint generation via automated theory formation. In: Walsh, T. (ed.) CP 2001. LNCS, vol. 2239, pp. 575–579. Springer, Heidelberg (2001). https://doi.org/10.1007/3-540-45578-7_42
16. Dang, N., Akgün, Ö., Espasa, J., Miguel, I., Nightingale, P.: A framework for generating informative benchmark instances. In: 28th International Conference on Principles and Practice of Constraint Programming, CP. LIPIcs, vol. 235, pp. 18:1–18:18 (2022). https://doi.org/10.4230/LIPIcs.CP.2022.18
17. De Raedt, L., Passerini, A., Teso, S.: Learning constraints from examples. In: AAAI, pp. 7965–7970 (2018). https://doi.org/10.1609/aaai.v32i1.12217

18. Fleck, M., Troya, J., Wimmer, M.: Search-based model transformations with MOMoT. In: Van Van Gorp, P., Engels, G. (eds.) ICMT 2016. LNCS, vol. 9765, pp. 79–87. Springer, Cham (2016). https://doi.org/10.1007/978-3-319-42064-6_6

19. Flener, P., Pearson, J., Ågren, M.: Introducing ESRA, a relational language for modelling combinatorial problems. In: Bruynooghe, M. (ed.) LOPSTR 2003. LNCS, vol. 3018, pp. 214–232. Springer, Heidelberg (2004). https://doi.org/10.1007/978-3-540-25938-1_18

20. Freuder, E.C.: Progress towards the Holy Grail. Constraints **23**(2), 158–171 (2018). https://doi.org/10.1007/s10601-017-9275-0

21. Frisch, A.M., Harvey, W., Jefferson, C., Martínez-Hernández, B., Miguel, I.: Essence: a constraint language for specifying combinatorial problems. Constraints **13**(3), 268–306 (2008). https://doi.org/10.1007/s10601-008-9047-y

22. Frisch, A.M., Jefferson, C., Martínez-Hernández, B., Miguel, I.: The rules of constraint modelling. In: IJCAI, pp. 109–116 (2005). https://www.ijcai.org/Proceedings/05/Papers/1667.pdf

23. Frisch, A.M., Miguel, I., Walsh, T.: CGRASS: a system for transforming constraint satisfaction problems. In: O'Sullivan, B. (ed.) CologNet 2002. LNCS, vol. 2627, pp. 15–30. Springer, Heidelberg (2003). https://doi.org/10.1007/3-540-36607-5_2

24. Hagberg, A.A., Schult, D.A., Swart, P.J.: Exploring network structure, dynamics, and function using NetworkX. In: Varoquaux, G., Vaught, T., Millman, J. (eds.) Proceedings of the 7th Python in Science Conference, Pasadena, CA USA, pp. 11–15 (2008). https://conference.scipy.org/proceedings/SciPy2008/paper_2/

25. Hnich, B.: Function variables for constraint programming. AI Commun. **16**(2), 131–132 (2003). https://content.iospress.com/articles/ai-communications/aic281

26. Horcas, J.M., Strüber, D., Burdusel, A., Martinez, J., Zschaler, S.: We're not gonna break it! consistency-preserving operators for efficient product line configuration. IEEE Transactions on Softw. Eng. **49**(3), 1102–1117 (2023). https://doi.org/10.1109/TSE.2022.3171404

27. Hu, Y., Mettler, M., Mueller-Gritschneder, D., Wild, T., Herkersdorf, A., Schlichtmann, U.: Machine learning approaches for efficient design space exploration of application-specific nocs. ACM Trans. Design Autom. Electr. Syst. (TODAES) **25**(5), 1–27 (2020). https://doi.org/10.1145/3403584

28. Jia, Z., Padon, O., Thomas, J., Warszawski, T., Zaharia, M., Aiken, A.: Taso: optimizing deep learning computation with automatic generation of graph substitutions. In: Proceedings of the 27th ACM Symposium on Operating Systems Principles, pp. 47–62 (2019). https://doi.org/10.1145/3341301.3359630

29. Kahani, N., Bagherzadeh, M., Cordy, J.R., Dingel, J., Varró, D.: Survey and classification of model transformation tools. Softw. Syst. Modeling **18**, 2361–2397 (2019). https://doi.org/10.1007/s10270-018-0665-6

30. Kiziltan, Z., Lippi, M., Torroni, P.: Constraint detection in natural language problem descriptions. In: IJCAI, pp. 744–750 (2016), https://www.ijcai.org/Proceedings/16/Papers/111.pdf

31. Koninck, L.D., Brand, S., Stuckey, P.J.: Data independent type reduction for zinc. In: Proceedings of the 9th International Workshop on Reformulating Constraint Satisfaction Problems (2010)

32. Lattimore, T., Szepesvári, C.: Bandit Algorithms. Cambridge University Press (2020). https://doi.org/10.1017/9781108571401

33. Leo, K., Mears, C., Tack, G., De La Banda, M.G.: Globalizing constraint models. In: CP, pp. 432–447 (2013). https://doi.org/10.1007/978-3-642-40627-0_34

34. Little, J., Gebruers, C., Bridge, D.G., Freuder, E.C.: Using case-based reasoning to write constraint programs. In: CP, p. 983 (2003). https://doi.org/10.1007/978-3-540-45193-8_107

35. Marriott, K., Nethercote, N., Rafeh, R., Stuckey, P.J., de la Banda, M.G., Wallace, M.: The design of the Zinc modelling language. Constraints **13**(3), 229–267 (2008). https://doi.org/10.1007/s10601-008-9041-4

36. Mens, T.: On the use of graph transformations for model refactoring. In: Lämmel, R., Saraiva, J., Visser, J. (eds.) GTTSE 2005. LNCS, vol. 4143, pp. 219–257. Springer, Heidelberg (2006). https://doi.org/10.1007/11877028_7

37. Mills, P., Tsang, E., Williams, R., Ford, J., Borrett, J.: EaCL 1.5: An Easy abstract Constraint optimisation Programming Language. Tech. Rep. CSM-324, Department of Computer Science, University of Essex (1999)

38. Nethercote, N., Stuckey, P.J., Becket, R., Brand, S., Duck, G.J., Tack, G.: MiniZinc: towards a standard CP modelling language. In: Bessière, C. (ed.) CP 2007. LNCS, vol. 4741, pp. 529–543. Springer, Heidelberg (2007). https://doi.org/10.1007/978-3-540-74970-7_38

39. Nightingale, P., Akgün, Ö., Gent, I.P., Jefferson, C., Miguel, I.: Automatically improving constraint models in savile row through associative-commutative common subexpression elimination. In: O'Sullivan, B. (ed.) CP 2014. LNCS, vol. 8656, pp. 590–605. Springer, Cham (2014). https://doi.org/10.1007/978-3-319-10428-7_43

40. Nightingale, P., Akgün, Ö., Gent, I.P., Jefferson, C., Miguel, I., Spracklen, P.: Automatically improving constraint models in Savile Row. Artifi. Intell. **251**, 35–61 (2017). https://doi.org/10.1016/j.artint.2017.07.001

41. Nightingale, P., Spracklen, P., Miguel, I.: Automatically improving SAT encoding of constraint problems through common subexpression elimination in savile row. In: Pesant, G. (ed.) CP 2015. LNCS, vol. 9255, pp. 330–340. Springer, Cham (2015). https://doi.org/10.1007/978-3-319-23219-5_23

42. Plump, D.: From imperative to rule-based graph programs. J. Logical Algebraic Methods Program. **88**, 154–173 (2017). https://doi.org/10.1016/j.jlamp.2016.12.001

43. Qayum, F., Heckel, R.: Local search-based refactoring as graph transformation. In: 1st International Symposium on Search Based Software Engineering, pp. 43–46. IEEE (2009). https://doi.org/10.1109/SSBSE.2009.27

44. Rafeh, R., Jaberi, N.: LinZinc: a library for linearizing Zinc models. Iranian J. Sci. Technol. Trans. Electr. Eng. **40**(1), 63–73 (2016). https://doi.org/10.1007/s40998-016-0005-1

45. Rendl, A.: Effective Compilation of Constraint Models. Ph.D. thesis, University of St Andrews (2010). http://hdl.handle.net/10023/973

46. Shchekotykhin, K., Friedrich, G.: Argumentation based constraint acquisition. In: ICDM. pp. 476–482 (2009). https://doi.org/10.1109/ICDM.2009.62

47. Spracklen, P., Dang, N., Akgün, Ö., Miguel, I.: Automated streamliner portfolios for constraint satisfaction problems. Artifi. Intell. **319**, 103915 (2023). https://doi.org/10.1016/j.artint.2023.103915

48. Taentzer, G., et al.: Model transformation by graph transformation: a comparative study. In: Model Transformation in Practice (MTiP 2005), Montego Bay, Jamaica, p. 17 (2005). https://hdl.handle.net/10486/665862

49. Troya, J., Segura, S., Burgueño, L., Wimmer, M.: Model transformation testing and debugging: A survey. ACM Comput. Surv. **55**(4), 1–39 (2022). https://doi.org/10.1145/3523056

50. Van Hentenryck, P.: The OPL Optimization Programming Language. MIT Press, Cambridge, MA, USA (1999)
51. Wachsmuth, G.: Metamodel adaptation and model co-adaptation. In: Ernst, E. (ed.) ECOOP 2007. LNCS, vol. 4609, pp. 600–624. Springer, Heidelberg (2007). https://doi.org/10.1007/978-3-540-73589-2_28
52. Weber, J.H.: Tool support for functional graph rewriting with persistent data structures – GrapeVine. In: International Conference on Graph Transformation. LNCS, vol. 13349, pp. 195–206. Springer (2022). https://doi.org/10.1007/978-3-031-09843-7_11

Author Index

R. Harmer and J. Kosiol (Eds.): ICGT 2024, LNCS 14774, p. 239, 2024.
https://doi.org/10.1007/978-3-031-64285-2